Ordinary Differential Operators

Mathematical
Surveys
and
Monographs

Volume 245

Ordinary Differential Operators

Aiping Wang
Anton Zettl

American Mathematical Society
Providence, Rhode Island

2010 *Mathematics Subject Classification.* Primary 47E05, 34B05, 47B25, 34B24.

For additional information and updates on this book, visit
www.ams.org/bookpages/surv-245

Library of Congress Cataloging-in-Publication Data

Cataloging-in-Publication Data has been applied for by the AMS.
See http://www.loc.gov/publish/cip/.

∞ The paper used in this book is acid-free and falls within the guidelines
established to ensure permanence and durability.
Visit the AMS home page at `https://www.ams.org/`

10 9 8 7 6 5 4 3 2 1 24 23 22 21 20 19

Contents

Preface ix

Part 1. Differential Equations and Expressions 1

Chapter 1. First Order Systems 3
1. Introduction 3
2. Existence and Uniqueness of Solutions 3
3. Variation of Parameters 8
4. The Gronwall Inequality 8
5. Bounds and Extensions to the Endpoints 10
6. Continuous Dependence of Solutions on the Problem 12
7. Differentiable Dependence of Solutions on the Data 14
8. Adjoint Systems 20
9. Inverse Initial Value Problems 21
10. Comments 22

Chapter 2. Quasi-Differential Expressions and Equations 25
1. Introduction 25
2. Classical Symmetric Expressions 25
3. Quasi-Derivative Formulation of the Classical Expressions 26
4. General Quasi-Differential Expressions 27
5. Quasi-Differential Equations 29
6. Comments 30

Chapter 3. The Lagrange Identity and Maximal and Minimal Operators 31
1. Introduction 31
2. Adjoint and Symmetric Expressions 31
3. The Lagrange Identity 34
4. Maximal and Minimal Operators 35
5. Boundedness Below of the Minimal Operator 38
6. Comments 40

Chapter 4. Deficiency Indices 41
1. Introduction 41
2. The Deficiency Index Continued 45
3. Powers of Differential Expressions and Their Deficiency Index 47
4. Complex Parameter Decompositions of the Maximal Domain 52
5. Comments 57

Part 2. Symmetric, Self-Adjoint, and Dissipative Operators 59

Chapter 5. Regular Symmetric Operators 61
1. Introduction 61
2. Boundary Conditions and Boundary Matrices 61
3. Characterization of Symmetric Domains 62
4. Examples of Symmetric Operators 66
5. Comments 68

Chapter 6. Singular Symmetric Operators 69
1. Introduction 69
2. Singular Boundary Conditions 69
3. Symmetric Domains and Proofs 70
4. Symmetric Domain Characterization with Maximal Domain Functions 75
5. Comments 79

Chapter 7. Self-Adjoint Operators 81
1. Introduction 81
2. LC Solutions and Real Parameter Decompositions of the Maximal Domain 82
3. A Real Parameter Characterization of Self-Adjoint Domains 89
4. The Maximal Deficiency Cases 91
5. Boundary Conditions for the Friedrichs Extension 93
6. Comments 96

Chapter 8. Self-Adjoint and Symmetric Boundary Conditions 97
1. Introduction 97
2. Separated Conditions 98
3. Separated, Coupled, and Mixed Conditions 101
4. Examples and Construction for All Types 105
5. Symmetric Boundary Conditions 112
6. Comments 117

Chapter 9. Solutions and Spectrum 119
1. Introduction 119
2. Only One Singular Endpoint 120
3. Two Singular Endpoints 124
4. Comments 127

Chapter 10. Coefficients, the Deficiency Index, Spectrum 129
1. Introduction 129
2. An Algorithm for the Construction of the Maximal Deficiency Index 130
3. Discreteness Conditions 131
4. Comments 135

Chapter 11. Dissipative Operators 137
1. Introduction 137
2. Concepts for Complex Symplectic Geometry Spaces 137
3. Finite Dimensional Complex Symplectic Spaces and Their Dissipative Subspaces 139

4. Applications of Symplectic Geometry to Ordinary Differential Operators 143
5. LC Representation of Dissipative Operators 149
6. Symplectic Geometry Characterization of Symmetric Operators 155
7. Comments 156

Part 3. Two-Interval Problems 157

Chapter 12. Two-Interval Symmetric Domains 159
1. Introduction 159
2. Two-Interval Minimal and Maximal Operators 161
3. Two-Interval Symmetric Domains 162
4. Discontinuous Symmetric and Self-Adjoint Boundary Conditions 168
5. Examples 172
6. Comments 179

Chapter 13. Two-Interval Symmetric Domain Characterization with Maximal Domain Functions 181
1. Introduction and Main Theorem 181
2. Comments 186

Part 4. Other Topics 187

Chapter 14. Green's Function and Adjoint Problems 189
1. Introduction 189
2. Adjoint Matrices and Green's Functions for Regular Systems 189
3. Green's Functions of Regular Scalar Boundary Value Problems 193
4. Regularization of Singular Problems 196
5. Green's Functions of Singular Boundary Value Problems 198
6. Construction of Adjoint and Self-Adjoint Boundary Conditions 200
7. The Green's Function of the Legendre Equation 206
8. Comments 211

Chapter 15. Notation 213

Chapter 16. Topics Not Covered and Open Problems 215
1. Topics Not Covered 215
2. Open Problems 218

Bibliography 219

Index 249

Preface

In 1836–1837 Sturm and Liouville published a series of papers on second order linear ordinary differential equations and boundary value problems. The influence of their papers was such that this subject became known as Sturm-Liouville theory. Thousands of papers by mathematicians, physicists, engineers, and others have been published on this topic since then. Yet, remarkably, this subject is still a very active field of research today.

In 1910 Herman Weyl published one of the most widely quoted papers of the 20th century in analysis. Just as the 1836–1837 papers of Sturm and Liouville started the study of *regular* Sturm-Liouville problems (SLP) the 1910 paper of Weyl initiated the study of *singular* SLP. The work on the foundations of quantum mechanics in the 1920s and 1930s, the proof of the spectral theorem for unbounded self-adjoint operators in Hilbert space by von Neumann and Stone, provided some of the motivation for the study of differential operators in Hilbert space with particular emphasis on self-adjoint operators and their spectrum.

In his paper, if an endpoint is singular, Weyl constructs a sequence of concentric circles in the complex plane which converge either to a circle or a point. These two cases became known as the limit-circle (LC) and limit-point (LP) cases. At a singular LP endpoint there is no boundary condition required or allowed to determine a self-adjoint operator. In the case of a regular or singular LC endpoint, a boundary condition is required to get a self-adjoint operator. The number of independent endpoint boundary conditions required for an interval $J = (a, b)$, $-\infty \leq a < b \leq \infty$, is 0 if both endpoints are LP, 1 if one endpoint is LP and the other regular or LC, and 2 if neither endpoint is LP. This number is called the deficiency index d on the interval J. The conditions required are known as self-adjoint boundary conditions and the self-adjoint operators they determine are known as self-adjoint extensions of the minimal operator, even though they are actually defined as restrictions of the maximal operator. We will follow this terminology here because it is universally used in mathematics, physics, and other fields.

For differential equations of order $n = 2k$, $k > 1$ with real coefficients, in an attempt to generalize Weyl's LP, LC alternative from the second order case, in 1921 Windau [**582**] for $n = 4$ and Shin [**513**], [**514**], [**515**], [**516**] in 1938–1943 for general even order $n = 2k$ came to the erroneous conclusion that there are only two deficiency indices: $d = k$ (LP) and $d = 2k$ (LC). In a seminal paper in 1950 Glazman [**239**], using Hilbert space methods, showed that all values of d between k and $2k$ are realized. Simpler examples were later found by Orlov [**456**] and Read [**488**]; see also [**326**].

In 1966 McLeod [**417**] showed that for complex coefficients there are two deficiency indices d^+, d^- which, in general, are different. What are their possible values?

This question has received a great deal of attention. In 1975 and 1976 Kogan and Rofe-Beketov [**338**], [**337**] proved that all values of (d^+, d^-) satisfying $|d^+ - d^-| \leq 1$ are realized.

In 1978 and 1979 Gilbert [**237**], [**238**] showed that the difference between d^+ and d^- can be arbitrarily large by proving that $|d^+ - d^-| \geq p$ for any positive integer p provided that the order n of the differential equation is allowed to be $8p$ or larger.

These results gave considerable support to the conjecture that all values of $d^+, d^- \in \{0, \cdots, n-1\}$ which satisfy the well known inequalities (see Section 4.1) are realized.

In their 1999 monograph [**184**] Everitt and Markus state: "...the Deficiency Index Conjecture, ... is essential for a satisfactory completion of the conclusions of Section V." (Section V is Chapter 5 in this monograph.) In Section 4.1 we prove this conjecture.

Also Everitt-Markus [**184**] found a 1-1 correspondence between the Hilbert space and symplectic geometry space characterizations for self-adjoint operators. This was extended by Yao-Sun-Zettl [**591**] to dissipative and strictly dissipative operators. It is exended here to symmetric operators. Thus the tools of symplectic geometry can be used to study self-adjoint, symmetric, and dissipative ordinary differential operators.

For self-adjoint operators this 1-1 correspondence clarifies a comment made by Everitt and Markus in the preface of their monograph, *Boundary Value Problems and Symplectic Algebra for Ordinary and Quasi-Differential Operators* [**184**]:

> We provide an affirmative answer ... to a long-standing open question concerning the existence of real differential expressions of even order ≥ 4 for which there are nonreal self-adjoint differential operators specified by strictly separated boundary conditions. ... This is somewhat surprising because it is well known that for order $n = 2$ strictly separated boundary conditions can produce only real operators (that is, any such given complex boundary condition can always be replaced by real boundary conditions).

Our construction in Chapter 7 of self-adjoint boundary conditions using LC solutions in Hilbert space naturally produces nonreal self-adjoint boundary conditions for regular and singular problems for each order $n = 2k$, $k > 1$. Furthermore, our analysis shows that it is not the order of the equation which is the relevant factor for the existence of nonreal self-adjoint boundary conditions but *the number* of boundary conditions. If there is only one separated complex condition at a given endpoint, as must be the case for $n = 2$, then it can be replaced by an equivalent *real* condition.

In this monograph we discuss self-adjoint, symmetric, and dissipative operators in Hilbert and symplectic geometry spaces. We do not discuss Krein or Pontryagin spaces. Self-adjoint operators and their spectrum are of special interest and will get special attention.

Our Hilbert space approach is motivated by methods used in the well known book by Naimark [**440**] which are based, to a considerable extent, on the work of Glazman. We do not use the method of boundary triplets. Our symplectic geometry space methods were motivated, to a considerable extent, by the Everitt-Markus monograph [**184**]. However, while this monograph focuses on two point boundary conditions in symplectic geometry space for self-adjoint operators, we study self-adjoint, symmetric, and dissipative operators in Hilbert and symplectic geometry spaces. In addition we get some information about the spectrum of self-adjoint operators and find multi-point (two-interval) domain characterizations for self-adjoint and symmetric operators.

Which boundary value problems generate self-adjoint, symmetric, and dissipative differential operators in Hilbert and symplectic geometry spaces? Such operators are generated by two things:

(1) *Symmetric (formally self-adjoint) differential expressions.* (In the older literature the term "formally self-adjoint expression" is commonly used, but the term "symmetric expression" will be used here to avoid confusion between the terms "self-adjoint" and "formally self-adjoint".)
(2) *Boundary conditions.* The number of independent boundary conditions depends on the deficiency index.

Regarding point (1) Naimark [**440**] introduced quasi-differential expressions which are analogues of the classical formally self-adjoint expressions with smooth coefficients as found, e.g., in the books by Coddington and Levinson [**102**] and Dunford and Schwartz [**111**], and showed that by using these quasi-differential expressions the smoothness conditions on the coefficients can be weakened to just local integrability. (There is an error in Naimark's representation of the expressions but not in his definition of the quasi-derivatives; see Chapter 2 for details. This error has been repeated by many authors in the literature.)

Much more general symmetric quasi-differential expressions were found by Shin in 1938–1943 [**513**], [**514**], [**515**], [**516**]. They were rediscovered by Zettl in [**595**], [**608**] in a somewhat different but equivalent form. Special cases of these rediscovered symmetric expressions were previously used by many authors, including Barrett [**42**], Hinton [**289**], Reid [**492**]. The forms given in [**595**], [**608**] for general n were motivated to some extend by the forms used by Barrett for $n = 3$ and $n = 4$. Although the general symmetric forms used by Shin are mentioned in a footnote in the book by Naimark [**440**], they are not used in this book and apparently were not used elsewhere in the literature before the publication of the Zettl paper [**595**] in 1965. After this publication these expressions became widely used and will be used here. For each differential expression of order $n > 2$ they extend the class of these expressions used by Naimark by several dimensions, and we will refer to them here as "general" quasi-differential expressions. In 1974 Zettl [**608**] showed that the Hilbert space techniques used by Naimark [**440**], based largely on the work of Glazman, can be applied to this general class.

Part I consisting of Chapters 1 to 4, is an introduction to quasi-differential expressions and their associated equations, including symmetric expressions, adjoint expressions, and the minimal and maximal operators. Section 3.5 also deals with some issues raised in the book by Weidmann [**574**]. On page 109 Weidmann says:

> Operators of even order with positive coefficient of highest order have turned out to be semibounded from below if the lower order terms are sufficiently small. On the other hand, operators of odd order are usually expected to be not semibounded (from below or above). We shall see this in many examples. ... Anyhow

we do not know of a general result which assures this for every operator of odd order.

For a class of operators larger than that studied by Weidmann, in [**427**] Möller and Zettl proved—*without any smoothness or smallness assumptions on the coefficients*—that:

(1) Odd order operators, regular or singular, and regardless of the sign of the leading coefficient, are unbounded above and below.
(2) Regular operators of even order with positive leading coefficient are bounded below and not bounded above.
(3) Even order regular or singular operators with leading coefficient that changes sign are unbounded above and below.

The results in [**427**] are for matrix coefficients, in Section 3.5 we present these results only for scalar coefficients.

Sections 4.2 and 4.3 discuss the deficiency index of symmetric expressions and sequences of deficiency indices of an expression and its powers. These powers are constructed without any smoothness assumptions on the coefficients.

Part II, consisting of Chapters 5 to 11, discusses symmetric, self-adjoint, and dissipative operators.

Regarding point (2), Theorem 4 on page 75 in [**440**] characterizes self-adjoint domains of singular problems in terms of functions from the maximal domain. The dependence of these functions on the coefficients is implicit and complicated. For regular endpoints Theorem 4 is then used in [**440**] to characterize the self-adjoint domains of regular problems in terms of two point boundary conditions involving only values of the quasi-derivatives at the endpoints. In the singular case such a characterization is not possible because, in general, the quasi-derivatives do not exist at a singular endpoint.

In [**199**] Everitt and Zettl named Theorem 4 the GKN Theorem in honor of the work of Glazman-Krein-Naimark for reasons given in Section 7 of their paper. Today the GKN Theorem and its extensions are widely used by this name to study regular and singular self-adjoint differential operators, difference operators, Hamiltonian systems, multi-interval operators, multi-valued operators, etc.

For singular problems with maximal deficiency index the GKN characterization can be made more explicit by replacing the maximal domain functions with a solution basis for any real or complex value of the spectral parameter λ. In the much more difficult intermediate deficiency cases, Sun [**528**] in 1986 found a self-adjoint characterization in terms of *certain* solutions of the equation

$$My = \lambda wy \quad \text{on } J = (a, b)$$

using nonreal values of the spectral parameter λ for expressions M of Naimark type and equations with only one singular endpoint. His proof is based on a decomposition of the maximal domain in terms of solutions. Shang [**510**] extended this result to the case when both endpoints are singular.

Chapter 5 characterizes *symmetric domains* for regular problems with the self-adjoint domain characterization as a special case. Chapter 6 characterizes *symmetric domains* for singular problems. It is interesting to observe from Theorem 5.3.4 and Theorem 6.3.3 that the characterization of the *symmetric operators* $S = S(U)$ satisfying $S_{\min} \subset S \subset S^* \subset S_{\max}$, where $U = (A : B)$, is completely determined by the rank r of the matrix $C = AEA^* - BEB^*$. With $r = 0$ if and only if S is self-adjoint and $r = 2s$, $0 < s \leq d$ when S is symmetric but not self-adjoint. From this perspective and the long proofs of these theorems, we do not agree with the statement on page 107 of the book, *Spectral Theory and Differential Operators* by Edmund and Evans [**129**] that:

> It is usually a straightforward matter to determine whether or not an operator is symmetric, but self-adjointness is a much more difficult property to establish.

Theorem 6.3.3 in Chapter 6 characterizes the *symmetric* operators for general symmetric expressions and two singular endpoints. It reduces to the case when one or both endpoints are regular. The self-adjoint characterization is a special case. The proof of this theorem is based on a decomposition of the maximal domain

$$D_{\max} = D_{\min} \dotplus \operatorname{span}\{u_1, \dots, u_{m_a}\} \dotplus \operatorname{span}\{v_1, \dots, v_{m_b}\}$$

in terms of certain maximal domain functions u_i and v_i, where u_i are solutions on (a, c) for some λ_a with $\operatorname{Im}(\lambda_a) \neq 0$ which are identically zero in a neighborhood of b, and v_i are solutions on (c, b) for some λ_b with $\operatorname{Im}(\lambda_b) \neq 0$ which are identically zero in a neighborhood of a. There is no restriction on the behavior of these solutions at the singular endpoints a, b. In particular there is no restriction on their asymptotic or oscillatory behavior at either endpoint.

Theorem 6.4.1 characterizes the *symmetric* operators in terms of maximal domain functions. Thus Theorem 6.4.1 can be considered an extension of the above mentioned Theorem 4 in Naimark [**440**] to general symmetric expressions. It is interesting to observe that the proof of Theorem 6.4.1, which characterizes symmetric domains—in terms of maximal domain functions—is based on Theorem 6.3.3, which characterizes the symmetric domains $D(U)$ of $S(U)$ in terms of *symmetric boundary conditions.*

For problems with one regular endpoint, in 2009 Wang et al. [**553**] found a decomposition of the maximal domain in terms of *certain real values of* λ, constructed LC and LP solutions and used these to characterize the self-adjoint domains. In 2012 Hao et al. [**253**] extended this construction to problems with two singular endpoints.

This decomposition of the maximal domain requires an extra hypothesis EH:

Assume there exists a real λ_a and a real λ_b such that there are d_a linearly independent solutions $u_1, \dots, u_{d_a}$ in the Hilbert space $L^2((a, c), w)$ and $v_1, \dots, v_{d_b}$ in the Hilbert space $L^2((c, b), w)$. We use the same notation u_i and v_i for simplicity of presentation but these solutions are for *some real values of* λ_a, λ_b. If EH does not hold, then the essential spectrum of every self-adjoint extension of the minimal operator covers the whole real line. In this case nothing, other than examples, seems to be known about self-adjoint boundary conditions. Thus this is a mild "extra" hypothesis. This decomposition has the form,

$$D_{\max} = D_{\min} \dotplus \operatorname{span}\{u_1, \dots, u_{m_a}\} \dotplus \operatorname{span}\{v_1, \dots, v_{m_b}\},$$

where u_i are solutions for some *real* value λ_a which are identically zero in a neighborhood of b and v_i are solutions for some *real* value λ_b which are identically zero in a neighborhood of a.

Based on this *real* λ decomposition of the maximal domain and the construction of LC and LP solutions a characterization of the self-adjoint domains for two singular endpoints is found in [**253**] and information about the spectrum of self-adjoint operators is obtained. The above mentioned response to the Everitt-Markus comment is based, in part, on this real λ decomposition.

The LC and LP solutions constructed by Wang et al. [**553**] also plays a critical role in the characterization of symmetric domains with the self-adjoint domains as a special case.

For real coefficients and real λ, let $r(\lambda)$ denote the number of linearly independent solutions of $My = \lambda wy$ in $H = L^2(J, w)$. It is well known that if only one endpoint is singular and $r(\lambda) < d$, then λ is in the essential spectrum of every self-adjoint extension of the minimal operator; see Weidmann [**574**]. If both endpoints are singular, Weidmann proves that this result still holds provided an additional hypothesis is satisfied. It follows from Theorem 9.3.2 that Weidmann's extra condition is not needed. This result was established by Hao et al. in [**257**].

The contrasting behavior of $r(\lambda)$ for the two singular endpoint cases from the one singular endpoint case has some interesting consequences. In the case of only one singular endpoint, we have $r(\lambda) \leq d$, whereas in the two singular endpoint case $r(\lambda)$ we may assume values less than d, equal to d, and greater than d. The case $r(\lambda) > d$ leads to the surprising result that this value of λ is an eigenvalue of every self-adjoint extension of $S_{\min}$. In other words, for every given self-adjoint boundary condition there are eigenfunctions of λ which satisfy this condition.

Theorem 9.2.2 extends another result in Weidmann [**574**] for problems with one regular endpoint from the case when $n = 2k$, and $d = k$ to the intermediate deficiency case when $d > k$. When $d = k$ there is no boundary condition required at the singular endpoint to determine a self-adjoint operator. For $d > k$ there are singular boundary conditions required. These pose a significant obstacle which is overcome with an analysis using LC solutions.

The real parameter characterization of self-adjoint domains and the construction of LC solutions are applied in Chapter 8 to classify the self-adjoint boundary conditions for expressions of order $n > 2$ into three types—separated, coupled, and mixed—and we construct examples of each type.

Chapter 10 contains a very brief discussion of the dependence of the deficiency index and the spectrum on the coefficients. There is a vast literature on this topic which involves, among other things, the theory of asymptotics and perturbations. This is beyond the scope of this book. So we briefly illustrate two methods which have not been widely used in the literature.

Chapter 11 describes the symplectic geometry space characterizations for self-adjoint, symmetric, and dissipative operators and gives a 1-1 correspondence between these and their Hilbert space versions. This was motivated to a large extent by the work of Yao et al. [**589**], [**590**], [**591**].

Chapter 12 contains a characterization of the two-interval symmetric domains for intervals which may or may not have a common endpoint. Then apply this theory to intervals which have a common endpoint. This case generates symmetric operators in the Hilbert space $L^2(J, w)$ which are determined by discontinuous boundary conditions. Such boundary conditions are known by various names, including transmission conditions, interface conditions, multi-point conditions, point interactions (in the physics literature), etc. The self-adjoint operators with discontinuous conditions are a special case of this theory.

Chapter 13 characterizes the two-interval symmetric boundary conditions in terms of functions from the maximal domain. The proof uses the maximal domain decomposition from Chapter 4.

Chapter 14 can be considered an appendix. We discuss the construction of regular and singular Green's functions for general one-interval adjoint boundary conditions with the self-adjoint conditions as a special case. The construction of the Green's function for regular problems given here is not the usual one, it has its roots in a construction of Neuberger [**442**] for second order problems which was extended to higher order problems by Coddington-Zettl [**103**]. In contrast to the usual construction of singular such problems as, for example, in the well known book by Coddington and Levinson [**102**], which involves a selection theorem to select a sequence of regular Green's functions on truncated intervals whose limit as the truncated endpoints approach the singular endpoints, our construction is direct, elementary, and explicit in terms of solutions. As an illustration the Green's function of the classical Legendre equation with arbitrary separated or coupled self-adjoint boundary conditions is constructed.

Chapter 15 summarizes our notation; Chapter 16 mentions topics not covered and gives a long, but not up to date, list of references for these; and Chapter 17 discusses some open problems.

ACKNOWLEDGMENT 1. The authors are indebted to Jiong Sun and Manfred Möller for their contributions to the study of ordinary differential operators and boundary value problems. Their results have significantly contributed to the contents of this monograph.

ACKNOWLEDGMENT 2. The second named author was supported by the Ky and Yu-fen Fan US-China Exchange Fund through the American Mathematical Society. This made his two visits to the Department of Mathematical Sciences of Inner Mongolia University in Hohhot possible. Special thanks go to my host Jiong Sun and my translators Xiaoling Hao and Yaping Yuan, who also became co-authors, along with Jijun Ao, Qinglan Bao, Dan Mu, Xiaoxia Lv, Siqin Yao, Chuanfu Yang, Yingchun Zhao, Maozhu Zhang. Many of the projects which led to the contents of this book started in Hohhot with these colleagues.

Last, but certainly not least, the second named author thanks his wife Sandra for her help with the database for the references of this book and, especially, for helping with the hardware and software problems that arose during the typing of this manuscript with Scientific Workplace (SWP). And he thanks her for her tolerance and understanding during this and many other mathematics projects.

Without Mathematics (M), there would be no Science (S), without Science there would be no Engineering (E), without Science and Engineering there would be no modern Technology (T). The world needs more STEM scientists; we hope this book will encourage more people to study these fields. STEM should be spelled MSET.

The world of Mathematics is full of wonders and of mysteries, at least as much so as the physical world.

The Mathematician is an artist whose medium is the mind and whose creations are ideas.

H. S. Wall

Part 1

Differential Equations and Expressions

CHAPTER 1

First Order Systems

1. Introduction

This chapter is devoted to the study of basic properties of first order systems of general dimension $n > 1$.

Notation. An open interval is denoted by (a, b) with $-\infty \leq a < b \leq \infty$; $[a, b]$ denotes the closed interval which includes the left endpoint a and the right endpoint b, regardless of whether these are finite or infinite, $\mathbb{R}$ denotes the reals, $\mathbb{C}$ the complex numbers, and

$$\mathbb{N}_0 = \{0, 1, 2, \ldots\}, \quad \mathbb{N} = \{1, 2, 3, \ldots\}, \quad \mathbb{N}_2 = \{2, 3, 4, \ldots\}.$$

For any interval J of the real line, open, closed, half open, bounded or unbounded, by $L(J, \mathbb{C})$ we denote the linear manifold of complex valued Lebesgue measurable functions y defined on J for which

$$\int_a^b |y(t)| \,\mathrm{dt} \equiv \int_J |y(t)| \,\mathrm{dt} \equiv \int_J |y| < \infty.$$

The notation $L_{loc}(J, \mathbb{C})$ is used to denote the linear manifold of functions y satisfying $y \in L([\alpha, \beta], \mathbb{C})$ for all compact intervals $[\alpha, \beta] \subseteq J$. If $J = [a, b]$ and both of a and b are finite, then $L_{loc}(J, \mathbb{C}) = L(J, \mathbb{C})$. Also, we denote by $AC_{loc}(J)$ the collection of complex-valued functions y which are absolutely continuous on all compact intervals $[\alpha, \beta] \subseteq J$. The symbols $L(J, \mathbb{R})$ and $L_{loc}(J, \mathbb{R})$ are defined similarly.

For a given set S, $M_{n,m}(S)$ denotes the set of $n \times m$ matrices with entries from S. If $n = m$ we write $M_n(S) = M_{n,n}(S)$; also if $m = 1$ we sometimes write S^n for $M_{n,1}(S)$. $M_{n,m}(\mathbb{C})$ is abbreviated to just $M_{n,m}$. Given $U \in M_{n,m+r}(S)$ the notation

$$U = (A : B)$$

is defined to mean that $A \in M_{n,m}(S)$ consists of the first m columns of U in the same order as they appear in U and $B \in M_{n,r}(S)$ consists of the next r columns of U in the same order as they appear in U.

The norm of a constant matrix as well as the norm of a matrix function P is denoted by $|P|$. This may be taken as

$$|P| = \sum |p_{ij}|.$$

2. Existence and Uniqueness of Solutions

DEFINITION 1.2.1 (Solution). Let J be any interval, open, closed, half open, bounded or unbounded; let $n, m \in \mathbb{N}$, let $P : J \to M_n(\mathbb{C})$, $F : J \to M_{n,m}(\mathbb{C})$. By a solution of the equation

$$Y' = PY + F \quad on\ J$$

we mean a function Y from J into $M_{n,m}(\mathbb{C})$ which is absolutely continuous on all compact subintervals of J and satisfies the equation a.e. on J. A matrix function is absolutely continuous if each of its components is absolutely continuous.

THEOREM 1.2.1 (Existence and Uniqueness). *Let J be any interval, open, closed, half open, bounded or unbounded; let $n, m \in \mathbb{N}$. If*

$$P \in M_n(L_{loc}(J, \mathbb{C}))$$

and

$$F \in M_{n,m}(L_{loc}(J, \mathbb{C}))$$

then every initial value problem (IVP)

$$Y' = PY + F \tag{1.2.1}$$

$$Y(u) = C, \quad u \in J, \quad C \in M_{n,m}(\mathbb{C}) \tag{1.2.2}$$

has a unique solution defined on all of J. Furthermore, if C, P, F are all real-valued, then there is a unique real valued solution.

PROOF. We give two proofs of this important theorem; the second one is the standard successive approximations proof. As we will see later the analytic dependence of solutions on the spectral parameter λ follows more readily from the second proof than the first.

For both proofs we note that if Y is a solution of the IVP (1.2.1), (1.2.2) then an integration yields

$$Y(t) = C + \int_u^t (PY + F), \quad t \in J. \tag{1.2.3}$$

Conversely, every solution of the integral equation (1.2.3) is also a solution of the IVP (1.2.1), (1.2.2).

Choose c in J, $c \neq u$. We show that (1.2.3) has a unique solution on $[u, c]$ if $c > u$ and on $[c, u]$ if $c < u$. Assume $c > u$. Let

$$B = \{Y : [u, c] \to M_{n,m}(\mathbb{C}),\ Y \textit{ continuous}\}.$$

Following Bielecki [**59**] we define the norm of any function $Y \in B$ to be

$$\|Y\| = \sup\left\{\exp\left(-K \int_u^t |P(s)|\, \mathrm{ds}\right) |Y(t)|,\ t \in [u, c]\right\},$$

where K is a fixed positive constant $K > 1$. It is easy to see that with this norm B is a Banach space. Let the operator $T\colon B \to B$ be defined by

$$(TY)(t) = C + \int_u^t (PY + F)(s)\, \mathrm{ds}, \quad t \in [u, c], \quad Y \in B.$$

Then for $Y, Z \in B$ we have

$$|(TY)(t) - (TZ)(t)| \leq \int_u^t |P(s)||Y(s) - Z(s)|\mathrm{ds}$$

and hence

$$\begin{aligned}
&\exp\left(-K\int_u^t |P(s)|\,\mathrm{ds}\right)|(TY)(t)-(TZ)(t)| \\
&\leq \|Y-Z\|\int_u^t |P(s)|\exp\left(-K\int_s^t |P(r)|\mathrm{dr}\right)\mathrm{ds} \\
&\leq \frac{1}{K}\,\|Y-Z\|.
\end{aligned}$$

Therefore

$$\|TY-TZ\| \leq \frac{1}{K}\,\|Y-Z\|.$$

From the contraction mapping principle in Banach space it follows that the map T has a unique fixed point and therefore the IVP (1.2.1), (1.2.2) has a unique solution on $[u,c]$. The proof for the case $c<u$ is similar; in this case the norm of B is modified to

$$\|Y\| = \sup\{\exp\left(K\int_u^t |P(s)|\mathrm{ds}\right)|Y(t)|,\ t\in[c,u]\}.$$

Since there is a unique solution on every compact subinterval $[u,c]$ and $[c,u]$ for $c\in J$, $c\neq u$ it follows that there is a unique solution on J. To establish the furthermore part take the Banach space of real-valued functions and proceed similarly. This completes the first proof.

For the second proof we construct a solution of (1.2.3) by successive approximations. Define

$$Y_0(t)=C,\ \ Y_{n+1}(t)=C+\int_u^t (PY_n+F),\quad t\in J,\ \ n=0,1,2,\ldots.$$

Then Y_n is a continuous function on J for each $n\in\mathbb{N}_0$. We show that the sequence $\{Y_n : n\in\mathbb{N}_0\}$ converges to a function Y uniformly on each compact subinterval of J and that the limit function Y is the unique solution of the integral equation (1.2.3) and hence also of the IVP (1.2.1), (1.2.2). Choose $b\in J$, $b>u$ and define

$$p(t)=\int_u^t |P(s)|\mathrm{ds},\ \ t\in J;\quad B_n(t)=\max_{u\leq s\leq t}|Y_{n+1}(s)-Y_n(s)|,\ \ u\leq t\leq b.$$

Then

$$Y_{n+1}(t)-Y_n(t)=\int_u^t P(s)\,[Y_n(s)-Y_{n-1}(s)]\mathrm{ds},\ \ t\in J,\ \ n\in\mathbb{N}.$$

From this we get

$$|Y_2(t)-Y_1(t)|\leq B_0(t)\int_u^t |P(s)|\mathrm{ds}=B_0(t)\,p(t)\leq B_0(b)\,p(b),\ \ u\leq t\leq b.$$

$$\begin{aligned}
|Y_3(t)-Y_2(t)| &\leq \int_u^t |P(s)|\,|Y_2(s)-Y_1(s)|\mathrm{ds}\leq \int_u^t |P(s)|\,B_0(s)\,p(s)\mathrm{ds} \\
&\leq B_0(t)\int_u^t |P(s)|\;p(s)\mathrm{ds}\leq B_0(b)\,\frac{p^2(t)}{2!} \\
&\leq B_0(b)\,\frac{p^2(b)}{2!},\quad u\leq t\leq b.
\end{aligned}$$

From this and mathematical induction we get

$$|Y_{n+1}(t) - Y_n(t)| \le B_0(b)\,\frac{p^n(b)}{n!}, \quad u \le t \le b.$$

Hence for any $k \in \mathbb{N}$

$$\begin{aligned}|Y_{n+k+1}(t) - Y_n(t)| \le & |Y_{n+k+1}(t) - Y_{n+k}(t)| + |Y_{n+k}(t) - Y_{n+k-1}(t)| + \\ & \cdots + |Y_{n+1}(t) - Y_n(t)| \\ \le & B_0(b)\,\frac{p^n(b)}{n!}[1 + \frac{p(b)}{n+1} + \frac{p^2(b)}{(n+2)(n+1)} + \cdots].\end{aligned}$$

Choose m large enough so that $p(b)/(n+1) \le 1/2$, then $p^2(b)/((n+2)(n+1)) \le 1/4$, etc. when $n > m$ and the term in brackets is bounded above by 2. It follows that the sequence $\{Y_n : n \in \mathbb{N}_0\}$ converges uniformly, say to Y, on $[u, b]$. From this it follows that Y satisfies the integral equation (1.2.3) and hence also the IVP (1.2.1), (1.2.2) on $[u, b]$.

To show that Y is the unique solution assume Z is another one; then Z is continuous and therefore $|Y - Z|$ is bounded, say by $M > 0$ on $[u, b]$. Then

$$|Y(t) - Z(t)| = \left|\int_u^t P(s)[Y(s) - Z(s)]\,\mathrm{ds}\right| \le M\int_u^t |P(s)|\,\mathrm{ds} \le M\,p(t), \quad u \le t \le b.$$

Now proceeding as above we get

$$|Y(t) - Z(t)| \le M\,\frac{p^n(t)}{n!} \le M\,\frac{p^n(b)}{n!}, \quad u \le t \le b, \quad n \in \mathbb{N}.$$

Therefore $Y = Z$ on $[u, b]$. There is a similar proof for the case when $b < u$. This completes the second proof. □

It is interesting to observe that the initial approximation $Y_0(t) = C$ can be replaced with $Y_0(t) = G(t)$ for any continuous function G without any essential change in the proof.

To study the dependence of the unique solution on the parameters of the problem we introduce a convenient notation. Let J be an interval. For each $P \in M_n(L_{loc}(J, \mathbb{C}))$, each $F \in M_{n,m}(L_{loc}(J, \mathbb{C}))$, each $u \in J$ and each $C \in M_{n,m}(\mathbb{C})$ there is, according to Theorem 1.2.1, a unique $Y \in M_{n,m}(AC_{loc}(J))$ such that $Y' = PY + F$, $Y(u) = C$. We use the notation

$$Y = Y(\cdot, u, C, P, F) \tag{1.2.4}$$

to indicate the dependence of the unique solution Y on these quantities. Below, if the variation of Y with respect to some of the variables u, C, P, F is studied while the others remain fixed we abbreviate the notation (1.2.4) by dropping those quantities which remain fixed. Thus we may use $Y(t)$ for the value of the solution at $t \in J$ when u, C, P, F are fixed or $Y(\cdot, u)$ to study the variation of the solution function Y with respect to $u, Y(\cdot, P)$ to study Y as a function of P, etc.

THEOREM 1.2.2 (Rank Invariance). *Let $J = (a, b)$, and assume that $P \in M_n(L_{loc}(J, \mathbb{C}))$. If Y is an $n \times m$ matrix solution of*

$$Y' = PY \text{ on } J, \tag{1.2.5}$$

then we have

$$\operatorname{rank} Y(t) = \operatorname{rank} Y(u), \quad t, u \in J. \tag{1.2.6}$$

Moreover, if $m = n$, then for any $u, t \in J$, we have

$$(\det Y)(t) = (\det Y)(u) \exp\left(\int_u^t \operatorname{trace} P(s) ds\right). \tag{1.2.7}$$

PROOF. The formula (1.2.7) follows from the fact that $y = \det Y$ satisfies the first order scalar equation $y' - py = 0$ where $p = \operatorname{trace} P$. To prove the general case, let $Y(u) = C$ and let $\operatorname{rank} C = r$. If $r = 0$, then $Y(t) = 0$ for all t by Theorem 1.2.1. For $r > 0$ let C_i, $i = 1, \ldots, r$ be linearly independent columns of C and construct a nonsingular $n \times n$ matrix D by adding $n - r$ appropriate constant vectors to C_i, $i = 1, \ldots, r$. Denote by Z the solution of (1.2.5) satisfying the initial condition $Z(u) = D$. Then by (1.2.7) $\operatorname{rank} Z(t) = n$, for $t \in J$. Hence the first r columns of $Z(t)$, $Z_1(t), \ldots, Z_r(t)$ are linearly independent. From this and the uniqueness part of Theorem 1.2.1 the (constant) n-vectors $Y_1(t), Y_2(t), \ldots, Y_r(t)$ are linearly independent since $Z_j = Y_j$ on J. Hence $\operatorname{rank} Y(t) \geq r$, for $t \in J$. Now suppose that rank $Y(c) > r$ for some c in J. Then by repeating the above argument with u replaced by c we reach the conclusion that $\operatorname{rank} Y(t) > r$ for all $t \in J$. But this contradicts $\operatorname{rank} Y(u) = r$ and concludes the proof. □

Formula (1.2.7) is sometimes called Abel's formula. It follows from this formula that if a solution is nonsingular at some point $u \in J$ then it is nonsingular at every point of J. The rank invariance of solutions given by (1.2.6) extends this result to the case when the matrix Y is not square but without a formula corresponding to (1.2.7).

THEOREM 1.2.3 (Everitt and Race). *Let $P : J \to M_n(\mathbb{C})$ and $F : J \to M_{n,1}(\mathbb{C})$, $J = (a, b)$, $-\infty \leq a < b \leq \infty$. If, for any $u \in J$ and any linearly independent constant vectors $C_1, \ldots, C_n$, each initial value problem*

$$Y' = PY + F, \quad Y(u) = C_i, \quad i = 1, \ldots, n,$$

has a unique (vector) solution Y_i on J, then

$$P \in M_n(L_{loc}(J, \mathbb{C})) \quad \text{and} \quad F \in M_n(L_{loc}(J, \mathbb{C})).$$

Furthermore, if each Y_i is a C^1 solution, then there exist such P and F which are continuous.

PROOF. We first prove the special case when $F = 0$ on J. Let Y_i be a vector solution satisfying $Y_i(u) = C_i$ and let Y be the matrix whose $i - th$ column is Y_i, $i = 1, \ldots, n$. Then the matrix solution Y is nonsingular in some neighborhood N_u of u. Choose

$$P = Y'Y^{-1} \quad \text{on} \quad N_u.$$

Let K be a compact subinterval of J. Since the open cover $\{N_u : u \in J\}$ of K has a finite subcover, we can conclude that Y is uniquely defined and invertible on K. Since Y is continuous and invertible on K, it follows that Y^{-1} is continuous and hence bounded on K. Also, Y' is integrable on K since Y is absolutely continuous on K by virtue of the fact that it is a solution on J. Therefore $P \in M_n(L_{loc}(J, \mathbb{C}))$.

To establish the case when F is not identically zero on J, let Y be a vector solution of $Y' = PY + F$ satisfying $Y(u) = 0$ and choose a solution Z of this equation such that $Z(u) = C$ and let $V = Z - Y$. Then $V' = PV$ and $V(u) = C$. Since this holds for arbitrary C we may conclude from the special case established above that $P \in M_n(L_{loc}(J, \mathbb{C}))$. Hence $F = V' - PV \in M_n(L_{loc}(J, \mathbb{C}))$. The furthermore statement is clear from the proof. □

3. Variation of Parameters

Let $P \in M_n(L_{loc}(J))$. From Theorem 1.2.1 we know that for each point u of J there is exactly one matrix solution X of (1.2.5) satisfying $X(u) = I_n$ where I_n denotes the $n \times n$ identity matrix.

DEFINITION 1.3.1 (Primary Fundamental Matrix). For each fixed $u \in J$, let $\Phi(\cdot, u)$ be the fundamental matrix of (1.2.5) satisfying

$$\Phi(u, u) = I_n \, .$$

Note that for each fixed u in J, $\Phi(\cdot, u)$ belongs to $M_n(AC_{loc}(J))$. Furthermore, if J is compact and $P \in M_n(L(J, \mathbb{C}))$, then u can be an endpoint of J and $\Phi(\cdot, u)$ belongs to $M_n(AC(J))$. (This is clear from the proof of Theorem 1.2.1, or see Theorem 1.5.2 below.) By Theorem 1.2.2, $\Phi(t, u)$ is invertible for each $t, u \in J$ and we note that

$$\Phi(t, u) = Y(t)Y^{-1}(u) \tag{1.3.1}$$

for any fundamental matrix Y of (1.2.5).

We call Φ the primary fundamental matrix of the system $Y' = PY$, or just the primary fundamental matrix of P and also write

$$\Phi = \Phi(P) = (\Phi_{rs})_{r,s=1}^{n} \, , \quad \Phi(P)(t, u) = \Phi(t, u, P).$$

Observe that for any constant $n \times m$ matrix C, ΦC is also a solution of (1.2.5). If C is a constant nonsingular $n \times n$ matrix then ΦC is a fundamental matrix solution and every fundamental matrix solution has this form.

The next result is called the variation of parameters formula and is fundamental in the theory of linear differential equations.

THEOREM 1.3.1 (Variation of Parameters Formula). *Let J be any interval, let $P \in M_n(L_{loc}(J, \mathbb{C}))$ and let $\Phi = \Phi(\cdot, \cdot, P)$ be the primary fundamental matrix of* (1.2.5) *defined above. Let $F \in M_{n,m}(L_{loc}(J, \mathbb{C}))$, $u \in J$ and $C \in M_{n,m}(\mathbb{C})$. Then*

$$Y(t) = \Phi(t, u, P)\, C + \int_u^t \Phi(t, s, P)\, F(s) \mathrm{ds}, \quad t \in J \tag{1.3.2}$$

is the solution of (1.2.1), (1.2.2). *Note that if J is compact and $P \in M_n(L(J))$, $F \in M_{n,m}(L(J))$, then $Y \in M_{n,m}(AC(J))$, and u can be an endpoint or an interior point of J.*

PROOF. Clearly $Y(u) = C$. Differentiate (1.3.2) and substitute into the equation (1.2.1). □

4. The Gronwall Inequality

Since we need a Gronwall inequality which is slightly more general than the one usually found in the literature we state and proof it here.

THEOREM 1.4.1 (The Gronwall Inequality). *(i) (The "right" Gronwall inequality) Let $J = [a, b]$. Assume g in $L(J, \mathbb{R})$ with $g \geq 0$ a.e., f real valued and continuous on J. If y is continuous, real valued, and satisfies*

$$y(t) \leq f(t) + \int_a^t g(s)\, y(s) \, \mathrm{ds} \, , \quad a \leq t \leq b, \tag{1.4.1}$$

then

$$y(t) \le f(t) + \left(\int_a^t f(s)\, g(s) \exp\left(\int_s^t g(u)\, \mathrm{du} \right) \mathrm{ds} \right), \quad a \le t \le b. \tag{1.4.2}$$

For the special case when $f(t) = c$, *a constant, we get*

$$y(t) \le c \exp\left(\int_a^t g(s)\, \mathrm{ds} \right), \quad t \in J.$$

For the special case when f *is nondecreasing on* $[a, b]$ *we get*

$$y(t) \le f(t) \exp\left(\int_a^t g(s)\, \mathrm{ds} \right), \quad a \le t \le b.$$

(ii) (The "left" Gronwall inequality) Let $J = [a, b]$. *Assume* g *is in* $L(J, \mathbb{R})$, $g \ge 0$ *a.e. and* f *real valued and continuous on* J. *If* y *is continuous, real valued, and satisfies*

$$y(t) \le f(t) + \int_t^b g(s)\, y(s)\, \mathrm{ds}, \quad a \le t \le b,$$

then

$$y(t) \le f(t) + \left(\int_t^b f(s)\, g(s) \exp\left(\int_t^s g(u)\, \mathrm{du} \right) \mathrm{ds} \right), \quad a \le t \le b.$$

For the special case when $f(t) = c$, *a constant, we get*

$$y(t) \le c \exp\left(\int_t^b g(s)\, \mathrm{ds} \right), \quad a \le t \le b.$$

For the special case when f *is nondecreasing on* $[a, b]$ *we have*

$$y(t) \le f(t) \exp\left(\int_t^b g(s)\, \mathrm{ds} \right), \quad a \le t \le b.$$

PROOF. For part (i) let $z(t) = \int_a^t gy$, $t \in J$ and note that

$$z' = gy \le gf + gz; \quad z' - gz \le gf \quad \text{a.e.}$$

Hence we have

$$\exp(-\int_a^s g(u)\mathrm{du})[z'(s) - g(s)z(s)] = [\exp(-\int_a^s g(u)\mathrm{du})z(s)]'$$
$$\le g(s)\ f(s) \exp(-\int_a^s g(u)\mathrm{du}), \quad a \le s \le b.$$

Integrating from a to t we get

$$\exp(-\int_a^t g(u)\mathrm{du})z(t) \le \int_a^t g(s)\ f(s) \exp(-\int_a^s g(u)\mathrm{du})\mathrm{ds}, \quad a \le t \le b.$$

From (1.4.1) and the above line we obtain

$$\begin{aligned} y(t) &\le f(t) + z(t) \\ &\le f(t) + \exp(\int_a^t g(u)\mathrm{du}) \int_a^t g(s)\ f(s) \exp(-\int_a^s g(u)\mathrm{du})\mathrm{ds} \\ &= f(t) + \int_a^t g(s)\ f(s) \exp(\int_s^t g(u)\mathrm{du})\mathrm{ds}, \quad a \le t \le b. \end{aligned}$$

This concludes the proof of (1.4.2). The two special cases follow from (1.4.2). For part (ii) let $z(t) = \int_t^b g\, y$ and note that

$$z' + gz \geq -g\, f,$$

then proceed as in part (i). □

5. Bounds and Extensions to the Endpoints

In this section we investigate bounds for solutions and the continuous extension of solutions to the endpoints of the underlying interval.

THEOREM 1.5.1. *Let $J = (a,b)$, $-\infty \leq a < b \leq \infty$, let $n, m \in \mathbb{N}$. Suppose that $P \in M_n(L(J,\mathbb{C}))$; $F \in M_{n,m}(L(J,\mathbb{C}))$. Assume that for some $u \in J$, $C \in M_{n,m}(\mathbb{C})$, we have*

$$Y' = PY + F \quad \text{on } J, \quad Y(u) = C. \tag{1.5.1}$$

Then

$$|Y(t)| \leq \left(|C| + \int_a^b |F|\right) \exp\left(\int_a^b |P|\right), \quad a < t < b. \tag{1.5.2}$$

PROOF. Note that (1.5.1) is equivalent to

$$Y(t) = C + \int_u^t (P(s)\,Y(s) + F(s)\,)\mathrm{ds}, \quad a < t < b. \tag{1.5.3}$$

Case 1. $u \leq t < b$. From (1.5.3) we get

$$\begin{aligned}|Y(t)| &\leq |C| + |\int_u^t (PY + F)\,| \\ &\leq |C| + \int_u^t (|P|\,|Y| + |F|\,) \\ &\leq \left(|C| + \int_u^b |F|\right) + \int_u^t (|P||Y|)\,, \quad u \leq t < b.\end{aligned}$$

From this and Gronwall's inequality we obtain

$$|Y(t)| \leq \left(|C| + \int_u^b |F|\right) \exp\left(\int_u^t |P|\right) \leq \left(|C| + \int_u^b |F|\right)\left(\int_u^b |P|\right), \quad u \leq t < b.$$

Case 2. $a < t \leq u$. From (1.5.3)

$$\begin{aligned}|Y(t)| &\leq |C| + \left|\int_u^t (PY + F)\right| \leq |C| + \int_t^u (|P|\,|Y| + |F|) \\ &\leq \left(|C| + \int_a^u |F|\right) + \int_t^u (|P||Y|)\,, \quad a < t \leq u.\end{aligned}$$

From this and the "left" Gronwall inequality we get

$$\begin{aligned}|Y(t)| &\leq \left(|C| + \int_a^u |F|\right) \exp\left(\int_t^u |P|\right) \\ &\leq \left(|C| + \int_a^u |F|\right) \exp\left(\int_a^u |P|\right), \quad a < t \leq u.\end{aligned}$$

Combining the two cases we conclude that (1.5.2) holds. □

Below we will show that, under the conditions of Theorem 1.5.1, the inequality $a < t < b$ can be replaced with $a \leq t \leq b$ in (1.5.2). For this $Y(a)$ and $Y(b)$ are defined as limits. This holds for both finite and infinite endpoints a, b.

THEOREM 1.5.2 (Continuous extensions to endpoints). *Let $J = (a,b)$, $-\infty \leq a < b \leq \infty$. Assume that*

$$P \in M_n(L_{loc}(a,b),\mathbb{C}); \quad F \in M_{n,m}(L_{loc}(a,b),\mathbb{C}). \tag{1.5.4}$$

i) Suppose, in addition to (1.5.4), *that*

$$P \in M_n(L(a,c),\mathbb{C}); \quad F \in M_{n,m}(L(a,c),\mathbb{C})$$

for some $c \in (a,b)$. For some $u \in J$ and $C \in M_{n,m}(\mathbb{C})$, let Y be the solution of the IVP (1.2.1), (1.2.2) *on J. Then*

$$Y(a) = \lim_{t\to a^+} Y(t) \tag{1.5.5}$$

exists and is finite.

ii) Suppose that, in addition to (1.5.4), *P, F satisfy*

$$P \in M_n(L(c,b),\mathbb{C}); \quad F \in M_{n,m}(L(c,b),\mathbb{C})$$

for some $c \in (a,b)$. For some $u \in J$ and $C \in M_{n,m}(\mathbb{C})$, let Y be the solution of the IVP (1.2.1), (1.2.2) *on J. Then*

$$Y(b) = \lim_{t\to b^-} Y(t) \tag{1.5.6}$$

exists and is finite.

PROOF. We establish Theorem 1.5.2 for b; the proof for the endpoint a is similar and hence omitted. It follows from (1.5.2) that $|Y|$ is bounded on $[c,b)$ for $c \in J$, say by B. Let $\{b_i\}$ be any strictly increasing sequence converging to b. Then for $j > i$ we have

$$|Y(b_j) - Y(b_i)| = \left| \int_{b_i}^{b_j} PY \right| \leq B \int_{b_i}^{b_j} |P|.$$

From this and the absolute continuity of the Lebesgue integral it follows that $\{Y(b_i) : i \in \mathbb{N}\}$ is a Cauchy sequence and hence converges to a finite limit. □

The next result establishes the rank invariance of solutions of homogeneous systems at the endpoints of the underlying interval and establishes the existence and uniqueness of solutions of initial value problems when the initial condition is specified at an endpoint.

THEOREM 1.5.3 (Rank invariance at endpoints). *Let $J = (a,b)$, $-\infty \leq a < b \leq \infty$. Assume that*

$$P \in M_n(L_{loc}(a,b),\mathbb{C}). \tag{1.5.7}$$

i) Suppose, in addition to (1.5.7), *that*

$$P \in M_n(L(a,c),\mathbb{C})$$

for some $c \in (a,b)$. Let, for some $u \in J$ and $C \in M_{n,m}(\mathbb{C})$, Y be the solution of the IVP (1.2.1), (1.2.2) *with $F = 0$ on J. Then*

$$\operatorname{rank} Y(a) = \operatorname{rank} Y(u), \tag{1.5.8}$$

where $Y(a)$ is given by (1.5.5). Moreover, given any $C \in M_{n,m}(\mathbb{C})$ there exists a unique solution Y of the "endpoint" value problem:

$$Y' = PY, \quad Y(a) = C. \tag{1.5.9}$$

ii) Suppose, in addition to (1.5.7), that

$$P \in M_n(L(c,b), \mathbb{C})$$

for some $c \in (a,b)$. Let, for some $u \in J$ and $C \in M_{n,m}(\mathbb{C})$, Y be the solution of the IVP (1.2.1), (1.2.2) with $F = 0$ on J. Then

$$\operatorname{rank} Y(b) = \operatorname{rank} Y(u) \tag{1.5.10}$$

where $Y(b)$ is given by (1.5.6). Moreover, given any $C \in M_{n,m}(\mathbb{C})$ there exists a unique solution Y of the "endpoint" value problem:

$$Y' = PY, \quad Y(b) = C.$$

Note that (1.5.8) and (1.5.10) do not follow directly from (1.2.6) and (1.5.5) or (1.5.6) since the rank of a matrix is not a continuous function of the matrix. We argue as follows: Let $Y(u) = C$, $\operatorname{rank} C = r$. If $r = 0$, then $Y(t) = 0$ for all $t \in J$ and $Y(b) = 0$ by (1.5.6). If $r > 0$, let $C_1, \ldots, C_r$ be linearly independent columns of $Y(u)$ and construct a nonsingular $n \times n$ matrix D by adding $n - r$ appropriate columns to C_j, $j = 1, \ldots, r$. Let Z denote the solution of (1.2.5) determined by the initial condition $Z(u) = D$. It follows from (1.2.7) that $Z(t)$ is nonsingular for each $t \in J$ and hence $Z(b)$ is nonsingular by (1.5.6) and (1.2.7). Therefore $Z_1(b), ..., Z_r(b)$ are linearly independent. By Theorem 1.2.1 $Y_r(t) = Z_r(t)$ for $t \in J$ and hence also for $t = b$ by (1.5.6); thus $\operatorname{rank} Y(b) \geq r$. If $\operatorname{rank} Y(b) = k > r$ then $\sum_1^k c_j Y_j(t) = 0$ for $t \in J$ and hence also for $t = b$, contradicting $k > r$. The proof for the endpoint a is similar. This establishes (1.5.8) and (1.5.10). To prove the moreover parts of the Theorem consider the primary fundamental matrix Φ, choose $u \in J$ and determine the solution Y of (1.2.5) by the initial condition $Y(u) = \Phi(b,u)\, C$. Then $Y(b) = C$. Note that $\Phi(b,u)$ exists by (1.5.6). The proof of (1.5.9) is similar.

6. Continuous Dependence of Solutions on the Problem

The next result establishes bounds for solutions of initial value problems; these are then used in Theorem 1.6.2 to show that solutions of initial value problems depend continuously on all parameters of the problem.

THEOREM 1.6.1. *Let $u, v \in J = (a,b)$, $-\infty \leq a < b \leq \infty$, $C, D \in M_{n,m}(\mathbb{C})$, $P, Q \in M_n(L(J,\mathbb{C}))$, $F, G \in M_{n,m}(L(J,\mathbb{C}))$. Assume*

$$Y' = PY + F \text{ on } J, \quad Y(u) = C; \quad Z' = QZ + G \text{ on } J, \quad Z(v) = D.$$

Then

$$|Y(t) - Z(t)| \leq K \exp\left(\int_a^b |Q|\right), \quad a \leq t \leq b, \tag{1.6.1}$$

where

$$K = |C - D| + \left|\int_u^v |F|\right| + M\left|\int_u^v |P|\right| + \int_a^b |F - G| + M\int_a^b |P - Q|, \tag{1.6.2}$$

and

$$M = \left(|C| + \int_a^b |F|\right) \exp\left(\int_a^b |P|\right).$$

PROOF. For $a < t < b$ this follows from the Gronwall inequality as in the proof of Theorem 1.5.1. The case $t = a$ and $t = b$ then follows from Theorem 1.5.2. □

THEOREM 1.6.2 (Continuous dependence). *Let $J = (a, b), -\infty \leq a < b \leq \infty$, $u \in J$, $C \in M_{n,m}(\mathbb{C})$, $P \in M_n(L(J, \mathbb{C}))$, and $F \in M_{n,m}(L(J, \mathbb{C}))$. Let $Y = Y(\cdot, u, C, P, F)$ be the solution of (1.2.1), (1.2.2) on J. Then Y is a continuous function of all its variables u, C, P, F uniformly on the closure of J ; more precisely, for fixed P, F, u, C; given any $\epsilon > 0$ there is a $\delta > 0$ such that if $v \in J$, $D \in M_{n,m}(\mathbb{C})$, $Q \in M_n(L(J, \mathbb{C}))$, and $G \in M_{n,m}(L(J, \mathbb{C}))$ satisfy*

$$|u - v| + |C - D| + \int_a^b |P - Q| + \int_a^b |F - G| < \delta, \tag{1.6.3}$$

then

$$|Y(t, u, C, P, F) - Y(t, v, D, Q, G)| < \epsilon\,, \quad a \leq t \leq b.$$

Note that $Y(t, u, C, P, F)$ is jointly continuous in u, C, P, F, uniformly for t in the closure of J.

PROOF. The absolute continuity of the Lebesgue integral and (1.6.3) imply that the constant K in (1.6.2) can be made arbitrarily small. The conclusion then follows from (1.6.1). □

THEOREM 1.6.3. *Let $J = (a, b)$, $-\infty \leq a < b \leq \infty$, let $P_k \in M_{n,n}(L_{loc}(J, \mathbb{C}))$, $F_k \in M_{n,m}(L_{loc}(J, \mathbb{C}))$, $C_k \in M_{n,m}$, $u_k \in J$, $k \in \mathbb{N}_0 = \{0, 1, 2, \ldots\}$. Assume*

(i) $P_k \to P_0$ as $k \to \infty$ locally in $L_{loc}(J, \mathbb{C})$ in the sense that for each compact subinterval K of J we have

$$\int_K |P_k - P_0| \to 0 \quad as \quad k \to \infty;$$

(ii) $F_k \to F_0$ as $k \to \infty$ locally in $L_{loc}(J)$ in the sense that for each compact subinterval K of J we have

$$\int_K |F_k - F_0| \to 0 \quad as \quad k \to \infty;$$

(iii) $C_k \to C_0$ as $k \to \infty$;
(iv) $u_k \to u_0 \in J$ as $k \to \infty$.
Then

$$Y(t, u_k, C_k, P_k, F_k) \to Y(t, u_0, C_0, P_0, F_0) \quad as \quad k \to \infty$$

locally uniformly on J, i.e., uniformly in t on each compact subinterval of J.

Moreover, if $P_k \in M_n(L(J), \mathbb{C})$, $F_k \in M_{n,m}(L(J), \mathbb{C})$, and (i), (ii) hold in $L(J, \mathbb{C})$, i.e. with K replaced by J and (iii), (iv) hold, then

$$Y(t, u_k, C_k, P_k, F_k) \to Y(t, u_0, C_0, P_0, F_0) \quad as \quad k \to \infty$$

uniformly on the closure of J.

PROOF. This follows from Theorem 1.6.2. □

7. Differentiable Dependence of Solutions on the Data

Theorem 1.6.2 shows that the solution of the initial value problem (1.2.1), (1.2.2) with $P \in M_{n,n}(L_{loc}(J,\mathbb{C}))$, $F \in M_{n,m}(L_{loc}(J,\mathbb{C}))$, $C \in M_{n,m}(\mathbb{C})$ depends continuously on the problem. In this section we show that this dependence is differentiable with respect to each parameter of the problem.

DEFINITION 1.7.1. A map T from a Banach space X into a Banach space Z, $T : X \to Z$, is differentiable at a point $x \in X$ if there exists a bounded linear map $T'(x) : X \to Z$ such that

$$|T(x+h) - T(x) - T'(x)\,h| = o\,(|h|), \quad as\ h \to 0\ \ in\ X.$$

That is, for each $\varepsilon > 0$ there is a $\delta > 0$ such that

$$|T(x+h) - T(x) - T'(x)\,h| \leq \varepsilon\ (|h|)\ for\ all\ h \in X\ with\ |h| < \delta.$$

If such a map $T'(x)$ exists, it is unique and is called the Frechet derivative of T at x. A map T is differentiable on a set $S \subset X$ if it is differentiable at each point of S. In this case the derivative is a map : $x \to T'(x)$ from S into the Banach space $L(X,Z)$ of all bounded linear operators from X into Z denoted by T'. To say that T' is continuously differentiable on S or T is C^1on S means that the map T' is continuous in the operator topology of the Banach space $L(X,Z)$.

The differentiability of the solution

$$Y \;=\; Y(t,u,C,P,F)$$

with respect to t follows from the definition of solution. The differentiability of Y with respect to u is established in the next lemma.

LEMMA 1.7.1. *Let the hypotheses and notation of Theorem* 1.2.1 *hold. Fix* t, C, P, F *and consider* Y *as a function of* u. *Then* $Y \in AC_{loc}(J)$.

PROOF. It follows from the representation (1.3.1) that the primary fundamental matrix $\Phi(t,u)$ is differentiable with respect to u, since the inverse of a differentiable matrix is differentiable. The differentiability of Y with respect to u then follows from the variation of parameters representation

$$Y(t,u) = \Phi(t,u)\,C + \int_u^t \Phi(t,s)\,F(s)\,\mathrm{ds}.$$

This concludes the proof. □

For fixed t, u, Y is a function of C, P, F mapping

$$M_{n,m}(\mathbb{C}) \times M_n(L(J,\mathbb{C})) \times M_{n,m}(L(J,\mathbb{C}))$$

into $M_{n,m}(L(J,\mathbb{C}))$. By Theorem 1.6.2, Y is continuous in C, P, F. Is it differentiable in C? in P? in F?

THEOREM 1.7.1. *For fixed* $t, u \in J$, $P \in M_n(L(J,\mathbb{C}))$ *and* $F \in M_{n,m}(L(J,\mathbb{C}))$, *the solution* $Y = Y(t,u,C,P,F)$ *of* (1.2.1), (1.2.2) *is differentiable in* C*; its derivative is given by*

$$Y'(C) \;=\; \frac{\partial Y}{\partial C}(t,u,C,P,F) = \Phi(t,u,P).$$

Thus we have

$$Y'(C)\,H = \Phi(t,u,P)\,H, \quad H \in M_{n,m}(\mathbb{C}).$$

The derivative $Y'(C)$ *is constant in* C *and in* F.

PROOF. This follows directly from the variation of parameters formula and the definition of derivative. □

THEOREM 1.7.2. *Let $J = [a, b]$. Fix $t, u \in J$, $C \in M_{n,m}(\mathbb{C})$, $P \in M_n(L(J, \mathbb{C}))$, and $F \in M_{n,m}(L(J, \mathbb{C}))$; let $Y = Y(t, u, C, P, F)$. We have*

(1.7.1)
$$Y'(F)(H) = \frac{\partial Y}{\partial F}(t, u, C, P, F)(H) = \int_u^t \Phi(t, s)H(s)\mathrm{ds}, \quad H \in M_{n,m}(L(J)).$$

Here the right side of equation (1.7.1) *defines a bounded linear operator on the space $M_{n,m}(L(J, \mathbb{C}))$. The derivative $Y'(F)$ is constant in F.*

PROOF. From the variation of parameters formula (1.3.2) we get

$$Y(t, u, C, P, F + H) - Y(t, u, C, P, F) = \int_u^t \Phi(t, s, P)H(s)\mathrm{ds}.$$

The conclusion follows from this equation and the definition of derivative. □

Before stating the next Theorem we give two lemmas. These may be of independent interest.

LEMMA 1.7.2. *Let $J = (a, b)$, $-\infty \le a < b \le \infty$, let $P \in M_n(L_{loc}(J, \mathbb{C}))$, let $u \in J$. Then for any $t \in J$ we have*

$$\begin{aligned}\Phi(t, u, P) &= I + \int_u^t P + \int_u^t P(r) \int_u^r P(s)\,\mathrm{dsdr} \\ &+ \int_u^t P(r) \int_u^r P(s) \int_u^s P(x)\,\mathrm{dxdsdr} + \cdots .\end{aligned}$$

PROOF. This follows directly from the successive approximations proof of the existence-uniqueness Theorem : Start with the first approximation $\Phi_0 = I$; then $\Phi_1 = I + \int_u^t P$, $\Phi_2 = I + \int_u^t P + \int_u^t P(r) \int_u^r P(s)\mathrm{dsdr}$, etc. □

The next Lemma establishes a product formula for fundamental solutions. This can be viewed as an extension of the exponential law; see Lemma 1.7.3, Remark 1.7.1 and Corollary 1.7.1 below.

LEMMA 1.7.3 (Product Formula). *Let $P, H \in M_n(L_{loc}(J, \mathbb{C}))$. Then for any $t, u \in J$ we have*

$$\Phi(t, u, P + H) = \Phi(t, u, P)\ \Phi(t, u, S),$$

where

$$S = \Phi^{-1}(\cdot, u, P)\ H\ \Phi(\cdot, u, P). \tag{1.7.2}$$

PROOF. The proof consists in showing that both sides satisfy the same initial value problem and then using the existence-uniqueness Theorem. □

LEMMA 1.7.4 (Exponential Law). *Let $P, H \in M_n(L_{loc}(J, \mathbb{C}))$. If P commutes with the integral of H in the sense that*

$$P(t) \left(\int_u^s H \right) = \left(\int_u^s H \right) P(t), \quad s, t, u \in J, \tag{1.7.3}$$

then the exponential law holds:

$$\Phi(t, u, P + H) = \Phi(t, u, P)\ \Phi(t, u, H).$$

PROOF. It follows from Lemmas 1.7.2, 1.7.3 and hypothesis (1.7.3) that

$$\Phi(\cdot, u, P)\ H = H\ \Phi(\cdot, u, P)$$

and hence $S = H$ in (1.7.2). □

THEOREM 1.7.3. *Let J be a compact interval $[a, b]$. Fix $t, u \in J$, $C \in M_{n,m}(\mathbb{C})$, $F \in M_{n,m}(L(J, \mathbb{C}))$. For $P \in M_n(L(J, \mathbb{C}))$ let $Y = Y(t, u, C, P, F)$ be the unique solution of the initial value problem* (1.2.1), (1.2.2). *Then the map*

$$P \to Y(t, u, C, P, F)$$

from the Banach space $M_n(L(J, \mathbb{C}))$ to the (finite dimensional) Banach space $M_{n,m}(\mathbb{C})$ is differentiable and its derivative

$$Y'(P) = \frac{\partial Y}{\partial P}(t, u, C, P, F)$$

is the bounded linear transformation from the Banach space $M_n(L(J, \mathbb{C}))$ to the Banach space $M_{n,m}(\mathbb{C})$ given, for any $H \in M_n(L(J, \mathbb{C}))$ by

$$\begin{aligned} Y'(P)\, H \ &= \Phi(t, u, P) \left(\int_u^t \Phi^{-1}(r, u, P) H(r) \Phi(r, u, P) \mathrm{d}r \right) C \\ &+ \int_u^t \Phi(t, r, P) \left(\int_r^t \Phi^{-1}(s, u, P) H(s) \Phi(s, u, P) \mathrm{d}s \right) F(r) \mathrm{d}r. \end{aligned} \tag{1.7.4}$$

PROOF. Fix t, u, C, F and let $Y(t, P) = Y(t, u, C, P, F)$. From the variation of parameters formula it follows that for $H \in M_{n,n}(L(J, \mathbb{C}))$ and S defined by (1.7.2) we have

$$\begin{aligned} &Y(t, P + H) - Y(t, P) \\ &= \Phi(t, P + H)\, C + \int_u^t \Phi(t, s, P + H)\, F(s) \mathrm{d}s \\ &\quad - \Phi(t, P)\, C - \int_u^t \Phi(t, s, P)\, F(s) \mathrm{d}s \\ &= \Phi(t, u, P)\, [\Phi(t, u, S) - I]\, C + \int_u^t \Phi(t, s, P) [\Phi(t, r, S) - I] F(r) \mathrm{d}r \\ &= \Phi(t, u, P) \left[\int_u^t S + \int_u^t S(x) \int_u^x S(y) \mathrm{d}y \mathrm{d}x \right. \\ &\qquad \left. + \int_u^t S(x) \int_u^x S(y) \int_u^y S(z) \mathrm{d}z \mathrm{d}y \mathrm{d}x + \cdots \right] C \\ &\qquad + \int_u^t \Phi(t, r, P) \left[\int_r^t S + \int_r^t S(x) \int_r^x S(y) \mathrm{d}y \mathrm{d}x \right. \\ &\qquad \left. + \int_r^t S(x) \int_r^x S(y) \int_r^y S(z) \mathrm{d}z \mathrm{d}y \mathrm{d}x + \cdots \right] F(r) \mathrm{d}r. \end{aligned}$$

Hence

$$
\begin{aligned}
Y(t,u,P+H) &- Y(t,u,P) - \Phi(t,u,P)\left(\int_u^t S(r)\mathrm{dr}\right)C \\
&- \int_u^t \Phi(t,r,P)\left(\int_r^t S(x)\mathrm{dx}\right)F(r)\mathrm{dr} \\
= {}& \Phi(t,u,P)\left[\int_u^t S(x)\int_u^x S(y)\mathrm{dydx}\right. \\
&\qquad \left. + \int_u^t S(x)\int_u^x S(y)\int_u^y S(z)\mathrm{dzdydx} + \cdots\right]C \\
&\qquad + \int_u^t \Phi(t,r,P)\left[\int_r^t S(x)\int_r^x S(y)\mathrm{dydx}\right. \\
&\qquad \left. + \int_r^t S(x)\int_r^x S(y)\int_r^y S(z)\mathrm{dzdydx} + \cdots\right]F(r)\mathrm{dr} \\
= {}& E(H).
\end{aligned}
$$

Noting that $|S|(b-a)| \leq |k\,H|$ for some $k \in \mathbb{R}$, that

$$|\Phi(t,u,P)| \quad \text{and} \quad |\Phi^{-1}(t,u,P)|$$

are bounded on J, there exists an $M > 0$ such that

$$
\begin{aligned}
|E(H)| &\leq M|C|\,[|S(b-a)|^2 + |S(b-a)|^3 + \cdots] \\
&\quad + M|F|\,[|S(b-a)|^2 + |S(b-a)|^3 + \ldots] \\
&\leq M|C|\,|kH|\,[|kH| + |kH|^2 + \ldots] \\
&\quad + M|F|\,|kH|\,[|kH| + |kH|^2 + \ldots].
\end{aligned}
$$

From this it follows that

$$\frac{|E(H)|}{|H|} \to 0 \textit{ as } |H| \to 0 \textit{ in } M_n(L(J)).$$

This completes the proof. □

THEOREM 1.7.4. *Let the hypotheses and notations of Theorem* 1.7.3 *hold and assume, in addition, that the commutativity hypothesis* (1.7.3) *is satisfied. Then*

(1)
$$H(t)\;\Phi(t,u,P) = \Phi(t,u,P)\;H(t), \quad t,u \in J.$$

(2) *The exponential law holds, i.e.*
$$\Phi(t,u,P+H) = \Phi(t,u,P)\,\Phi(t,u,H), \quad t,u \in J.$$

(3) *Formula* (1.7.4) *reduces to*

$$
\begin{aligned}
Y'(P)(H) = {}&\Phi(t,u,P)\left(\int_u^t H(s)\mathrm{ds}\right)C \\
&+ \int_u^t \Phi(t,r,P)\left(\int_r^t H(s)\mathrm{ds}\right)F(r)\mathrm{dr}, \quad t,u \in J.
\end{aligned}
\tag{1.7.5}
$$

Note however that $Y'(P)$ is not the operator defined by the right hand side of (1.7.5) *since H cannot be restricted to satisfy the commutativity hypothesis* (1.7.3) *due to the definition of the derivative $Y'(P)$.*

PROOF. This follows from Theorem 1.7.3 and Lemma 1.7.4. □

REMARK 1.7.1. In the special case when P and H are constant matrices we have

$$(1.1) \qquad Y'(P)(H) = e^{(t-u)P}\left(\int_u^t e^{(u-r)P} H e^{(r-u)P} \mathrm{dr}\right) C$$

$$(1.7.6) \qquad + \int_u^t e^{(t-r)P}\left(\int_r^t e^{(u-s)P} H e^{(s-u)P} \mathrm{ds}\right) F(r)\mathrm{dr}.$$

Note that if P and H are constant and commute, then (1.7.6) reduces to

$$Y'(P)(H) = (t-u)e^{(t-u)P} HC + \int_u^t (t-r)e^{(t-r)P} HF(r)\mathrm{dr}.$$

But this reduction does not hold, in general, for constant matrices which do not commute.

COROLLARY 1.7.1. *Consider the exponential map of matrices :*

$$E(A) = e^A, \quad A \in M_n(\mathbb{C}).$$

The Frechet derivative of E is the bounded linear operator from $M_n(\mathbb{C})$ into $M_n(\mathbb{C})$ given by

$$(1.7.7) \qquad E'(A)H = e^A \int_0^1 e^{-rA} H e^{rA} \mathrm{dr}, \quad H \in M_n(\mathbb{C}).$$

PROOF. This is the special case of Theorem 1.7.4 when $a = 0 = u,\ b = 1,\ P(t) = A$ for all $t \in [0,1],\ F \equiv 0,\ C = I.$ □

REMARK 1.7.2. Note that (1.7.7) reduces to the more familiar formula $E'(A) = E(A)$ for all $A \in M_n(\mathbb{C})$ only in the one dimensional case $n = 1$. When $n > 1$ (1.7.7) reduces to $E'(A) = E(A)$ only for constant multiples $A = cI_n$, $c \in \mathbb{C}$, of the identity matrix I_n, since only multiples of the identity satisfy the commutativity condition with respect to all matrices in $M_n(\mathbb{C})$.

REMARK 1.7.3. In Corollary 1.7.1 $M_n(\mathbb{C})$ can be replaced by $M_n(\mathbb{R})$; in fact $M_n(\mathbb{C})$ can be replaced by an arbitrary Banach algebra. See [**351**].

For fixed t, u, C, F replace P by $P + zW$ and fix P and W. It is well known that the solution

$$Y = Y(t, u, C, P + zW, F)$$

of the intial value problem (1.2.1), (1.2.2) with P replaced by $P + zW$ is an entire function of z. What is its derivative:

$$Y'(z) = \frac{\partial Y}{\partial z}\,?$$

This question is answered by

THEOREM 1.7.5. *Let $J = (a,b)$, $-\infty \le a < b \le \infty$, $t, u \in J$, $C \in M_{n,m}$, $P, W \in M_n(L(J,\mathbb{C}))$, $F \in M_{n,m}(L(J,\mathbb{C}))$; let $Y = Y(t,u,C,P+zW,F)$ denote the* *unique solution of* (1.2.1), (1.2.2) *for each* $z \in \mathbb{C}$. *Then* Y *is an entire function of* z *and*

$$Y'(z) = \Phi(t,u,P+zW)\left(\int_u^t \Phi^{-1}(r,u,P+zW)\,W(r)\,\Phi(r,u,P+zW)\mathrm{dr}\right)C$$
$$+\int_u^t \Phi(t,r,P+zW)$$
$$\cdot\left(\int_r^t \Phi^{-1}(s,u,P+zW)\,W(s)\,\Phi(s,u,P+zW)\mathrm{ds}\right)F(r)\mathrm{dr}.$$

PROOF.

$$\begin{aligned}&[Y(t,u,C,P+(z+h)W,F)-Y(t,u,C,P+zW,F)]\\ =&[\Phi(t,u,P+(z+h)W)-\Phi(t,u,P+zW)]\,C\\ &+\int_u^t [\Phi(t,r,P+(z+h)W)-\Phi(t,r,P+zW)]\,F(r)\mathrm{dr}.\end{aligned}$$

Let

$$S(z) = \Phi^{-1}(\cdot,u,P+zW)W(\cdot)\,\Phi(\cdot,u,P+zW).$$

Proceeding similarly to the proof of Theorem 1.7.3 we get

$$\begin{aligned}&\Phi(t,u,P+(z+h)W)-\Phi(t,u,P+zW)\\ &=\Phi(t,u,P+(z+h)W)[\Phi(t,u,hS(z))-I]\\ &=\Phi(t,u,P+zW)[h\int_u^t S(z)+o(h)]\\ &=h\Phi(t,u,P+zW)\int_u^t S(z)+o(h).\end{aligned}$$

Combining these two identities we get

$$\begin{aligned}&[Y(t,u,C,P+(z+h)W,F)-Y(t,u,C,P+zW,F)]\\ &= h\left[\Phi(t,u,P+zW)\left(\int_u^t S(z)\right)C\right.\\ &\qquad\left.+\int_u^t \Phi(t,r,P+zW)\left(\int_r^t S(z)\right)F(r)\mathrm{dr}\right]+o(h).\end{aligned}$$

And the result follows. □

THEOREM 1.7.6. *Let* $J=(a,b)$, $-\infty \le a < b \le \infty$, $t,u\in J$, $C\in M_{n,m}(\mathbb{C})$, $P\in M_n(L(J,\mathbb{C}))$, $F\in M_{n,m}(L(J,\mathbb{C}))$; *and for each* $z\in\mathbb{C}$ *let* $Y=Y(t,u,C,P+zW,F)$ *denote the unique solution of* (1.2.1), (1.2.2) *for each* $W\in L(J)$. *Then* Y *is a differentiable function of* W *and*

$$Y'(W)\,H = z\,\Phi(t,u,P+zW)\left(\int_u^t \Phi^{-1}(r,u,P+zW)H(r)\,\Phi(r,u,P+zW)\mathrm{dr}\right)C$$
$$+\ z\int_u^t \Phi(t,r,P+zW)$$
$$\cdot\left(\int_r^t \Phi^{-1}(s,u,P+zW)H(s)\,\Phi(s,u,P+zW)\mathrm{ds}\right)F(r)\mathrm{dr},$$

for $H\in L(J,\mathbb{C})$.

PROOF. The proof is similar to that of Theorem 1.7.3 and hence omitted. □

8. Adjoint Systems

In this section we discuss adjoint systems.

DEFINITION 1.8.1. For $n > 1$, let

$$\begin{aligned} Z_n(J) := \{Q = (q_{rs})_{r,s=1}^n \in M_n(L_{loc}(J)) \\ q_{r,r+1} \neq 0 \text{ a.e. on } J,\ q_{r,r+1}^{-1} \in L_{loc}(J),\ 1 \le r \le n-1, \\ q_{rs} = 0 \text{ a.e. on } J,\ 2 \le r+1 < s \le n; \\ q_{rs} \in L_{loc}(J),\ s \neq r+1,\ 1 \le r \le n-1\}. \end{aligned}$$

For $Q \in Z_n(J)$ we define

$$V_0 := \{y : J \to \mathbb{C},\ y \text{ is measurable}\}$$

and

$$y^{[0]} = y \qquad (y \in V_0).$$

Inductively, for $r = 1, \dots, n$, we define

$$V_r = \{y \in V_{r-1} : y^{[r-1]} \in (AC_{loc}(J))\},$$

$$y^{[r]} = q_{r,r+1}^{-1}\{y^{[r-1]'} - \sum_{s=1}^{r} q_{rs} y^{[s-1]}\} \qquad (y \in V_r),$$

where $q_{n,n+1} := 1$. Finally we set

$$M\,y = M_Q\,y = i^n\,y^{[n]} \qquad (y \in V_n).$$

The expression $M = M_Q$ is called the quasi-differential expression associated with or generated by Q. For V_n we also use the notations $D(Q)$ and $D(M)$. The function $y^{[r]}$ $(0 \le r \le n)$ is called the r-th quasi-derivative of y. Since the quasi-derivative depends on Q, we sometimes write $y_Q^{[r]}$ instead of $y^{[r]}$.

LEMMA 1.8.1. *Let $Q, P \in Z_n(J)$. Let F, G be n by m matrix functions on J. Assume the constant matrix $C \in M_n(\mathbb{C})$. If*

$$Y' = QY + F,\quad Z' = PZ + G \quad \text{on } J, \tag{1.8.1}$$

Then

$$(Z^*CY)' = Z^*(P^*C + CQ)Y + G^*CY + Z^*CF.$$

PROOF. From (1.8.1) we obtain

$$\begin{aligned} (Z^*CY)' = Z^{*\prime}CY + Z^*CY' &= (Z^*P^* + G^*)CY + Z^*C(QY + F) \\ &= Z^*(P^*C + CQ)Y + G^*CY + Z^*CF. \end{aligned}$$

□

LEMMA 1.8.2. *Assume $Q, P \in Z_n(J)$, $C \in M_n(\mathbb{C})$ is invertible and $Q = -C^{-1}P^*C$. If $Y' = QY + F$ and $Z' = PZ + G$ on J, where F, G are n by m matrices on J, then*

$$(Z^*CY)' = G^*CY + Z^*CF.$$

PROOF. Since $Q = -C^{-1}P^*C$ implies that $P^*C + CQ = 0$ the conclusion follows from Lemma 1.8.1. □

The fundamental matrices of adjoint systems are closely related to each other. The next theorem gives this relationship. It plays an important role in the theory of adjoint and, in particular, self-adjoint boundary value problems.

THEOREM 1.8.1. *Let $P \in M_n(L_{loc}(J,\mathbb{C}))$, $C \in M_n(\mathbb{C})$, and let $\Phi(t,s,P)$ be the primary fundamental matrix of $Y' = PY$. Assume*

$$C^* = C^{-1} = (-1)^{n+1}C \tag{1.8.2}$$

holds, and define

$$P^+ = -C^{-1}\,P^*\,C. \tag{1.8.3}$$

Then

$$\Phi(t,s,P) = C^{-1}\,\Phi^*(s,t,P^+)\,C, \quad s,t \in J. \tag{1.8.4}$$

PROOF. Fix $s \in J$ and let

$$Z(t) = (C^{-1})^*\Phi^*(t,s,P)\,C^*\Phi(t,s,P^+), \quad t \in J.$$

Then we have

$$\begin{aligned}
Z'(t) &= (C^{-1})^*[P(t)\Phi(t,s,P)]^*\,C^*\,\Phi(t,s,P^+) \\
&\quad + (C^{-1})^*\Phi^*(t,s,P)\,C^*\,P^+(t)\,\Phi(t,s,P^+) \\
&= (C^{-1})^*\Phi^*(t,s,P)\,CC^{-1}P^*(t)CC^{-1}\,C^*\,\Phi(t,s,P^+) \\
&\quad + (C^{-1})^*\Phi^*(t,s,P)\,C^*\,P^+(t)\,\Phi(t,s,P^+) \\
&= \,-\,(C^{-1})^*\Phi^*(t,s,P)\,C^*\,P^+(t)\,\Phi(t,s,P^+) \\
&\quad + (C^{-1})^*\Phi^*(t,s,P)\,C^*\,P^+(t)\,\Phi(t,s,P^+) \\
&= 0, \quad t \in J,
\end{aligned}$$

by (1.8.2) and (1.8.3), i.e. using the fact that

$$CC^{-1}P^*(t)CC^{-1}\,C^* = -CP^+(t)C^{-1}\,C^* = -(-1)^{n+1}CP^+(t) = -C^*P^+(t).$$

Note that $Z(s) = I$. Hence $Z(t) = I$, for $t \in J$. That this is equivalent to (1.8.4) follows from the representation $\Phi(t,s,P^+) = Y(t)\,Y^{-1}(s)$, $s,t \in J$, for any fundamental matrix Y of $Y' = P^+\,Y$. □

9. Inverse Initial Value Problems

Notation. Given d n-dimensional vectors $Y_1, Y_2, \ldots, Y_d$ we denote the $n \times d$ matrix whose $i-th$ column is Y_i, $i = 1, \ldots, d$, by

$$Y = [Y_1, Y_2, \ldots, Y_d].$$

Above we started with a coefficient matrix P and, possibly, a nonhomogeneous term F and then studied the existence of solutions and their properties. Here we reverse this. Given a number of functions, under what conditions are they solutions of a first order linear system? For the sake of completeness we state the theorem for both the direct and the inverse problems.

THEOREM 1.9.1. *Let $1 \leq d \leq n$, $P \in M_n(L_{loc}(J,\mathbb{C}))$. Assume that Y_i, $i = 1, \ldots, d$ are vector solutions of*

$$Y' = PY. \tag{1.9.1}$$

If

$$\operatorname{rank}[Y_1, Y_2, \ldots, Y_d](t) = d$$

for some t in J, then this is true for every t in J.

Conversely, let $Y_i \in M_{n,1}(AC_{loc}(J))$, $i = 1, \ldots, d$, $1 \le d \le n$ and assume that

$$\operatorname{rank}[Y_1, \ldots, Y_d](t) = d, \quad for\ t \in J.$$

Then there exists an $n \times n$ matrix $P \in M_n(L_{loc}(J, \mathbb{C}))$ such that Y_i, $i = 1, \ldots, d$, are solutions of (1.9.1).

Furthermore, if $Y_i \in M_{n,1}(C^1(J), \mathbb{C})$, $i = 1, \ldots, d$ then there exists a continuous such P.

Proof. The first part is contained in Theorem 1.2.2 so we only prove the second part. If $d = n$ take $P = Y'Y^{-1}$. If $d < n$ we construct an $n \times n$ matrix

$$M = [Y_1, Y_2, \ldots, Y_d, Y_{d+1}, \ldots, Y_n]$$

as follows. For each $t_1 \in J$ there is a $d \times d$ nonsingular submatrix of the $n \times d$ matrix $[Y_1, \ldots, Y_d](t_1)$. Let its rows be numbered by $r_1, \ldots, r_d$. To the right of the first row which in not one of these place the first row of the $(n-d) \times (n-d)$ identity matrix; to the right of the second row which is not one of these place the second row of the $(n-d) \times (n-d)$ identity matrix, and so on. Thus each of Y_i for $i > d$ is a constant matrix with all components zero except one which is the number 1. For each $t_1 \in J$ the matrix M so constructed is nonsingular at t_1 and by continuity $\det M(t) \neq 0$ for all t in some neighborhood N_{t_1} of t_1. Take

$$P(t) = M'(t)\, M^{-1}(t)\,, \ for\ t \in N_{t_1}.$$

Any compact subinterval of J can be covered by a finite number of such neighborhoods N_{t_1} and hence P can be defined on J. On points which are covered by more than one such neighborhood, P is multiply defined, we just choose one definition, say the one determined by the lowest numbered neighborhood. Clearly $P \in M_n(L_{loc}(J, \mathbb{C}))$ and Y_i, $i = 1, \ldots, d$ are solutions. This completes the proof of the first part.

To prove the furthermore part we note that the constructed matrix P is piecewise continuous by construction. Thus to get a continuous P we remove the multiply defined aspect of the above construction as follows: On a subinterval which is covered by two or more of the neighborhoods N_t discard all definitions of M used above - just on this subinterval - then connect the two remaining pieces together in such a way as to keep M nonsingular on J. Then construct a new P from the new M as above for all $t \in J$. This results in a continuous P and completes the proof. □

10. Comments

Much of Chapter 1 is based on the book [**620**] and the paper [**354**] by Kong and Zettl. We comment on each section separately.

(1) The notation for matrix functions such as $M_n(L(J, \mathbb{C}))$ is taken from [**427**].
(2) The sufficiency of the local integrability conditions of Theorem 1.2.1 are well known - see [**440**] or [**574**]; the necessity given by Theorem 1.2.3 is due to Everitt and Race - see [**192**]. Except for the use of the Bielecki norm the first proof of Theorem 1.2.1 is the standard successive approximations argument, although it is dressed in the clothes of the Contraction Mapping

Theorem in Banach space here. The advantage of the Bielecki norm is that it yields a global proof; the sup norm would only give a local proof and then one has to patch together the intervals of existence. The second proof is a minor variant of the usual successive approximations argument.

The constancy of the rank of solutions given by Theorem 1.2.2 is known - see [**283**] but we haven't seen it stated under these general conditions. It is surprising how many authors, including the two just mentioned, assume continuity of the coefficients when local Lebesgue integrability suffices. This is of some consequence both theoretically and numerically when coefficients are approximated by piece-wise constants, piece-wise linear functions, etc.

(3) The variation of parameters formula given by Theorem 1.3.1 is standard, but our notation is not. We use a notation which shows the dependence of the primary fundamental matrix on the coefficient matrix P. This is useful for the differentiation results that follow.
(4) A detailed discussion of the Gronwall inequality is given here because it is a very useful tool and no additional assumptions on f and g are needed.
(5) Theorem 1.5.1 is elementary but we have not seen it stated in this generality. The continuous extensions of solutions given by Theorem 1.5.2 are a special case of much more powerful results e.g. Coddington and Levinson's asymptotic theorem -[**102**]. Often the existence of limits of solutions are stated only for infinite endpoints. *We want to emphasize here that the relevant consideration is not whether the endpoint is finite or infinite but whether the coefficient matrix P and the inhomogeneous term F are integrable or not all the way to the endpoint.* Theorem 1.5.3 may be new in [**354**].
(6) Theorems 1.6.1, 1.6.2 and 1.6.3 illustrate clearly that the natural space in which to study solutions of linear ordinary differential equations is $L^1_{loc}(J,\mathbb{C})$ in the singular case and $L^1(J,\mathbb{C})$ in the regular case.
(7) Sections 7 and 9 were influenced to some extend by the treatment of the inverse spectral theory for regular Sturm-Liouville problems by Pöschel and Trubowitz in [**471**]. Theorems 1.7.1 and 1.7.2 are trivial consequences of the variation of parameters formula; Theorems 1.7.3, 1.7.5 and 1.7.6 are not so trivial consequences of the Variation of Parameters Formula and may be new in [**354**].
(8) Adjoint systems of this type were used by Atkinson [**23**]. Theorem 1.8.1, the Adjointness Identity, is due to Zettl, see [**594**], [**596**], [**607**] for the case when $C = E = ((-1)^r \delta_{r,n+1-s})^n_{r,s=1}$.
(9) These kinds of inverse problems are discussed by Hartman [**283**] but not in this generality.

CHAPTER 2

Quasi-Differential Expressions and Equations

1. Introduction

In this chapter we study quasi-differential expressions defined by Definition 1.8.1 and the associated equations. For a more comprehensive discussion of quasi-differential expressions, the reader is referred to [**607**], [**200**] in the scalar coefficient case and to [**427**], [**210**], [**574**] for matrix coefficients.

Just as the second order quasi-differential expression $-(py')' + qy$ enjoys many advantages over the more classical $py'' + ry' + qy$ so one can formulate quasi-differential expressions of higher order to replace the classical expression:

$$My = p_n y^{(n)} + p_{n-1} y^{(n-1)} + \cdots + p_1 y' + p_0 y \quad \textit{on } J.$$

Among the advantages of these quasi-differential expressions over the classical ones are the following:

(1) They are more general.
(2) An adjoint expression can be defined which has the same form as the original - in contrast to the classical case. (See below for details.)
(3) The Lagrange Identity is much simpler. It involves a sesquilinear form with constant coefficients in contrast to the classical form which depends on the coefficients in a complicated way.
(4) The fact that the adjoint of the adjoint is the original is immediately clear.
(5) Powers of expressions can be formed in a natural way without any smoothness or other additional conditions on the coefficients, see Chapter 4.

2. Classical Symmetric Expressions

Given a classical differential expression M of the form

$$My = p_n y^{(n)} + p_{n-1} y^{(n-1)} + \cdots + p_1 y' + p_0 y \quad \textit{on } J, \tag{2.2.1}$$

the expression M^+ given by

$$M^+ y = (-1)^n (\overline{p}_n y)^{(n)} + (-1)^{n-1} (\overline{p}_{n-1} y)^{(n-1)} + \cdots + (-1) \overline{p}_1 y' + \overline{p}_0 y \quad \textit{on } J \tag{2.2.2}$$

is called the adjoint expression of M. And M is called symmetric (formally self-adjoint) if $M^+ = M$.

Thus to check an expression (2.2.1) for symmetry one must write (2.2.1) in the same form as (2.2.2) and compare coefficients. To do this one must assume that the coefficients p_j are sufficiently smooth, i.e. $p_j \in C^j(J)$.

It is well known [**111**, pp.1285-1289], that if $n > 1$, $M = M^+$ and $p_j \in C^j(J)$, then M given by (2.2.1) can be expressed in the form

(2.2.3)
$$My = \sum_{j=0}^{[n/2]} (-1)^j (a_j y^{(j)})^{(j)} + i \sum_{j=0}^{[(n-1)/2]} (-1)^j [(b_j y^{(j)})^{(j+1)} + (b_j y^{(j+1)})^{(j)}] \quad on\ J,$$

where a_j, b_j are real, i is the complex number $\sqrt{-1}$ and $[x]$ denotes the greatest integer $\leq x$; and that M given by (2.2.3) is equal to its adjoint expression M^+. Thus (2.2.3) is a closed form for all classical symmetric expressions (2.2.1) with sufficiently smooth coefficients. (In [**111**] it is assumed that the p_j are in $C^\infty(J)$ but the proof given there clearly is valid for $p_j \in C^j(J)$.)

Note that if the coefficients p_j in (2.2.2) are all real, then the complex second term in (2.2.3) vanishes. Hence a real classical symmetric expression M given by (2.2.1) with sufficiently smooth coefficients p_j must be of even order $n = 2k$ and have the form

$$My = \sum_{j=0}^{k} (-1)^j (p_j y^{(j)})^{(j)} \tag{2.2.4}$$

with p_j real, $j = 0, 1, 2, \cdots, k$.

For real coefficients (2.2.4) is the familiar Sturm-Liouville form

$$My = -(p_1 y')' + p_0 y \tag{2.2.5}$$

when $n = 2$ and for $n = 4$ we have

$$My = (a_2 y'')'' - (a_1 y')' + a_0 y.$$

In the odd order case $n = 3$ (2.2.3) takes the form

$$My = -(a_1 y')' + a_0 y + i\{-[(b_1 y')'' + (b_1 y'')'] + [(b_0 y)' + (b_0 y')]\}$$

with a_j, b_j real, $j = 0, 1$.

If the coefficients a_j, b_j are not sufficiently smooth then the form (2.2.3) does not reduce to the form (2.2.1). Nevertheless, as we shall see below, analogues of (2.2.3) are "symmetric" without any smoothness assumption on the coefficients at all. Thus if one wishes to study general symmetric differential expressions with nonsmooth coefficients one is forced to consider so-called quasi-differential expressions. In (2.2.5), $(p_1 y')$ is called the quasi-derivative of y and the reason for the parenthesis in $(p_1 y')$ is that it follows from the general theory of linear differential equations [**620**] that the product $(p_1 y')$ is continuous at all points of the underlying open interval J but the separated terms $p_1(t) y'(t)$ may not exist for all t in J.

There exist much more general quasi-differential symmetric expressions than those analogous to (2.2.3) or (2.2.4). These will be identified in Section 2.4 below.

3. Quasi-Derivative Formulation of the Classical Expressions

Before introducing general quasi-derivatives we discuss the quasi-derivative formulation of the classical real expression (2.2.4) and make a comment about its misuse in the literature.

It is well known [**620**] that the classical Sturm-Liouville theory, including the operators it generates in the Hilbert space $L^2(J, w)$, applies to the equation

$$My = -(py')' + qy = \lambda wy \quad on \quad J$$

with coefficients satisfying

$$\frac{1}{p}, q, w \in L_{loc}(J, \mathbb{R}), \;\; w > 0, \;\; a.e. \; on \;\; J. \tag{2.3.1}$$

The local integrability conditions of (2.3.1) (without the positivity condition on w) are necessary and sufficient for every initial value problem to have a unique solution defined on the whole interval J [**192**]. In this sense the local integrability conditions (2.3.1) are minimal conditions for the classical modern Sturm-Liouville theory (with Caratheodory solutions), including the operators it generates in $L^2(J, w)$, to hold. We note for later reference that there is no positivity condition on p in (2.3.1).

In the fourth order case

$$My = (py'')'' - (ry')' + qy = \lambda wy \;\; on \;\; J \tag{2.3.2}$$

the conditions $p \in C^2(J)$, $r \in C^1(J)$, $q, w \in C(J)$, $p > 0$, $w > 0$ can be weakened to

$$\frac{1}{p}, r, \; q, w \in L_{loc}(J, \mathbb{R}), \;\; w > 0, \;\; a.e. \; on \;\; J, \tag{2.3.3}$$

provided the expression M in (2.3.2) is modified to

$$My = [(py'')' - (ry')]' + qy = \lambda wy \;\; on \;\; J. \tag{2.3.4}$$

This modification i.e. the use of the extra bracket [ ] and the quasi-derivative $[(py'')' - (ry')]$ seems to be a small price to pay for weakening the smoothness conditions to just local integrability. In particular this allows the coefficients to be piece-wise constant which is important for both the theoretical and numerical approximations of the equation.

Similarly, in the higher order cases $n = 2k$, $k > 2$, the smoothness conditions on the coefficients of the classical equation (2.2.1) can be replaced by the corresponding local integrability conditions of the type (2.3.3) provided the classical equation is replaced by its quasi-differential analogue containing an appropriate number of parentheses. For $k = 3$ this requires the introduction of two additional parentheses (brackets) rather than just one as in case $k = 2$. See the next section for details.

REMARK 2.3.1. In his well known book [**440**] Naimark uses the classical form (2.2.4) with just local integrability assumptions on the coefficients but neglects to use the required additional parentheses (brackets) as in (2.3.4). This error has been repeated in the literature by many authors. See Everitt and Zettl [**200**] for more details.

4. General Quasi-Differential Expressions

In this section we discuss some basic properties of quasi-differential expressions defined by Definition 1.8.1.

For the benefit of the reader we repeat the definition of quasi-differential expressions given in Section 1.8.

DEFINITION 2.4.1. For $n > 1$, let

$$\begin{aligned} Z_n(J) := \{Q = (q_{rs})_{r,s=1}^n \in M_n(L_{loc}(J)) \\ & q_{r,r+1} \neq 0 \;\text{ a.e. on } J, \; q_{r,r+1}^{-1} \in L_{loc}(J), \;\; 1 \leq r \leq n-1, \\ & q_{rs} = 0 \quad \text{a.e. on } J, \;\; 2 \leq r+1 < s \leq n; \\ & q_{rs} \in L_{loc}(J), \; s \neq r+1, \, 1 \leq r \leq n-1\}. \end{aligned}$$

For $Q \in Z_n(J)$ we define

$$V_0 := \{y : J \to \mathbb{C},\ y \text{ is measurable}\}$$

and

$$y^{[0]} = y \qquad (y \in V_0).$$

Inductively, for $r = 1, \ldots, n$, we define

$$V_r = \{y \in V_{r-1} : y^{[r-1]} \in (AC_{loc}(J))\},$$

$$y^{[r]} = q_{r,r+1}^{-1}\{y^{[r-1]'} - \sum_{s=1}^{r} q_{rs} y^{[s-1]}\} \qquad (y \in V_r),$$

where $q_{n,n+1} := 1$. Finally we set

$$M\,y = M_Q\,y = i^n\,y^{[n]} \qquad (y \in V_n). \tag{2.4.1}$$

The expression $M = M_Q$ is called the quasi-differential expression associated with or generated by Q. For V_n we also use the notations $D(Q)$ and $D(M)$. The function $y^{[r]}$ $(0 \le r \le n)$ is called the r-th quasi-derivative of y. Since the quasi-derivative depends on Q, we sometimes write $y_Q^{[r]}$ instead of $y^{[r]}$.

Definition 2.4.2. In this definition if q_{rs} are real valued functions, $1 \le r, s \le n$, we use the notation $Q \in Z_n(J, \mathbb{R})$.

Definition 2.4.3. When $n = 2k$, the coefficient $q_{k,k+1}$ is called the leading coefficient of M_Q. We say that the leading coefficient $q_{k,k+1}$ changes sign on J if it assumes positive and negative values, each on a subset of J which has positive Lebesgue measure. The sign of the leading coefficient $q_{k,k+1}$ plays an important role in the semiboundedness of even order operators generated by M_Q as we will see below in Chapter 3.

Definition 2.4.4. Let $Q \in Z_n(J)$, $J = (a, b)$. The expression $M = M_Q$ is said to be regular (R) at a or we say a is a regular endpoint, if for some c, $a < c < b$, we have

$$q_{r,r+1}^{-1} \in L(a,c), \quad r = 1, \cdots, n-1;$$
$$q_{rs} \in L(a,c), \quad 1 \le r, s \le n, \quad s \ne r+1.$$

Similarly the endpoint b is regular if for some c, $a < c < b$, we have

$$q_{r,r+1}^{-1} \in L(c,b), \quad r = 1, \cdots, n-1;$$
$$q_{rs} \in L(c,b), \quad 1 \le r,\ s \le n, \quad s \ne r+1.$$

Note that from the definition of $Q \in Z_n(J)$ it follows that if the above hold for some $c \in J$, then they hold for any $c \in J$. We say that M is regular on J, or just M is regular, if M is regular at both endpoints. An endpoint is singular if it is not regular.

Remark 2.4.1. For a given $Q \in Z_n(J)$, $M = M_Q$ is determined by (2.4.1). However, M_Q does not determine Q uniquely, in general, i.e. there may be a $P \in Z_n(J)$ such that $M_Q = M_P$. See [**621**] for an example. In this example $M_P = M_Q$ and M_P is regular at both endpoints while M_Q is singular at one endpoint.

REMARK 2.4.2. In much of the literature when an endpoint of J is infinite this endpoint and the associated problem is automatically classified as singular. Note that in the above definition $a = -\infty$ is allowed; similarly for $b = \infty$. For any interval $J = (a, b)$ observe that M is regular on any compact subinterval of J.

To illustrate the use of quasi-derivatives we give a simple example.

EXAMPLE 2.4.1. Let $Q = (q_{rs}) \in Z_2(J)$. Then $y^{[0]} = y$, $y^{[1]} = q_{12}^{-1}(y' - q_{11}y)$,

$$y^{[2]} = [q_{12}^{-1}(y' - q_{11}y)]' - q_{21}y - q_{12}^{-1}q_{22}(y' - q_{11}y)$$

and $M = M_Q$ is given by

$$My = i^2 y^{[2]} = -[q_{12}^{-1}(y' - q_{11}y)]' + q_{21}y + q_{12}^{-1}q_{22}(y' - q_{11}y).$$

Note that if $q_{11} = 0 = q_{22}$ then this reduces to the familiar Sturm-Liouville form $My = -(py')' + qy$ with the notation $p = q_{12}^{-1}$, $q = q_{21}$ but without the assumption that p and q are real valued and, if p is real valued, without the hypothesis that p is of one sign as long as $\frac{1}{p} \in L_{loc}(J)$. In particular $\frac{1}{p}$ may be identically zero on one or more subintervals of J; see [**620**] for more details.

5. Quasi-Differential Equations

The next proposition establishes the 'connection' between the first order systems studied in Chapter 1 and the scalar equations of order n studied in this chapter.

The first-order vector system $Y' = QY + F$ and the quasi-differential equation $y_Q^{[n]} = f$ are equivalent in the following sense:

PROPOSITION 2.5.1. *Let $Q \in Z_n(J)$ and $f \in L_{loc}(J)$. Let $M = M_Q$,*

$$F = \begin{pmatrix} 0 \\ \vdots \\ 0 \\ f \end{pmatrix}.$$

(i) If $Y \in (AC_{loc}(J))^n$ is a solution of

$$Y' = QY + F,$$

then there is a unique $y \in D(M)$ such that

$$Y = \begin{pmatrix} y^{[0]} \\ y^{[1]} \\ \vdots \\ y^{[n-1]} \end{pmatrix}$$

and

$$y^{[n]} = f; \tag{2.5.1}$$

(ii) If $y \in D(M)$ is a solution of (2.5.1), *then*

$$Y = \begin{pmatrix} y^{[0]} \\ y^{[1]} \\ \vdots \\ y^{[n-1]} \end{pmatrix} \in (AC_{loc}(J))^n$$

and

$$Y' = QY + F.$$

PROOF. This follows from a straightforward computation, for details see Section 2 in [**427**]. □

From the existence and uniqueness theorem it follows that the initial value problems associated with $Y' = QY + F$ have a unique solution:

COROLLARY 2.5.1. *For each $F \in (L_{loc}(J))^n$, each α in J and each $C \in \mathbb{C}^n$ there is a unique $Y \in (AC_{loc}(J))^n$ such that*

$$Y' = QY + F \quad \text{on } J \quad \text{and } Y(\alpha) = C.$$

For each $f \in L_{loc}(J)$, each $\alpha \in J$ and $c_0, \ldots, c_{n-1} \in \mathbb{C}$ there is a unique $y \in D(Q)$ such that

$$y^{[n]} = f \quad \text{and} \quad y^{[r]}(\alpha) = c_r \quad (r = 0, \ldots, n-1).$$

If $f \in L(J)$, J is bounded and all components of Q are in $L(J)$, then $y \in AC(J)$.

6. Comments

These are made separately for each section.

1. In the classic books by Coddington and Levinson [**102**] and Dunford and Schwartz [**111**], adjoint ordinary differential expressions are defined for expressions of the form

$$p_n y^{(n)} + p_{n-1}\, y^{(n-1)} + \cdots + p_1\, y' + p_0\, y$$

with smooth coefficients.

2. The closed form formulas for the symmetric (formally self-adjoint) real and complex expressions are derived in [**111**].

3. The quasi-derivative formulation of the classical expressions is generally attributed to Naimark [**440**] but, as mentioned in Remark 2.3.1, he uses the classical form (2.2.4) with just local integrability assumptions on the coefficients but neglects to use the required additional parentheses (brackets) as in (2.3.4). This error has been repeated in the literature by many authors. See Everitt and Zettl [**200**] for more details.

4. Very general quasi-differential expressions, in particular symmetric ones, were discovered by Shin [**513**], [**514**], [**515**], [**516**]. They were rediscoverd by Zettl [**607**], [**200**] in a slightly different but equivalent form. Special cases of these have been used by many authors, including Barrett [**42**], Glazman [**239**], Hinton [**289**], Kogan and Rofe-Beketov [**337**], Naimark [**440**], Reid [**492**], Stone [**524**], Walker [**544**], Weyl [**580**]. The forms given in [**607**], [**200**] for general $n \in \mathbb{N}_2$ were motivated to some extent by the forms used by Barrett [**42**] for the cases when $n = 3$ and $n = 4$.

5. The development of the theory of symmetric differential operators in the books by Naimark [**440**] and Akhiezer and Glazman [**5**] is based on the real symmetric form analogous to (2.2.4). Although these authors refer to Shin's more general symmetric expressions they do not use them. In [**607**] Zettl showed that the Hilbert space techniques in these books, based largely on the work of Glazman, can be applied to the much larger class of symmetric operators generated by the very general symmetric expressions studied here.

CHAPTER 3

The Lagrange Identity and Maximal and Minimal Operators

1. Introduction

In this chapter we construct the maximal and minimal operators $S_{\max}$ and $S_{\min}$ in the weighted Hilbert space $L^2(J, w)$ for general quasi-differential expressions and discuss their basic properties. These operators depend only on the coefficients and the weight function w. In Chapters 5 and 6 we characterize the regular and singular *symmetric* operators S satisfying

$$S_{\min} \subset S \subset S^* \subset S_{\max}$$

in terms of two-point boundary conditions. The *self-adjoint* characterization i.e. when $S = S^*$ is a special case.

2. Adjoint and Symmetric Expressions

In this section we briefly review and illustrate Lagrange symmetric expressions $M = M_Q$ where Q is a Lagrange symmetric matrix. Recall that these expressions are called Lagrange symmetric, or L-symmetric for short, because they generate symmetric operators in Hilbert space.

The following symplectic matrix E plays an important role in both, the study of general symmetric differential expressions and in the characterization of domains of symmetric and self-adjoint operators.

DEFINITION 3.2.1. For $k \in \mathbb{N}_2$ let E_k be defined by

$$E_k = ((-1)^r \delta_{r,k+1-s})_{r,s=1}^{k}, \tag{3.2.1}$$

where $\delta_{i,j}$ is the Kronecker δ. Note that

$$E_k^* = E_k^{-1} = (-1)^{k+1} E_k.$$

DEFINITION 3.2.2. Let $Q \in Z_n(J)$ and suppose that Q satisfies

$$Q = -E^{-1} Q^* E, \quad \textit{where} \quad E = E_n = ((-1)^r \delta_{r,n+1-s})_{r,s=1}^{n}, \quad \textit{i.e.}$$

$$q_{rs} = (-1)^{r+s-1} \overline{q}_{n+1-s,n+1-r}, \qquad 1 \leq r, s \leq n. \tag{3.2.2}$$

Then Q is called a Lagrange symmetric or L-symmetric matrix and the expression $M = M_Q$ is called a Lagrange symmetric, or just a symmetric, differential expression.

The next proposition gives a 'visual' interpretation of the L-symmetry condition (3.2.2). It follows immediately from (3.2.2) by inspection.

PROPOSITION 3.2.1. *Suppose $Q = (q_{rs}) \in Z_n(J)$ is Lagrange symmetric. Then Q is invariant under the composition of the following three operations:*

(1) *Flipping the elements* q_{rs} *about the secondary diagonal.*
(2) *Replacing* q_{rs} *by its conjugate* $\overline{q}_{rs}$.
(3) *Changing the sign of* q_{rs} *when* $r+s$ *is even.*

Note that these three operations commute with each other and each one is idempotent.

To illustrate Definition 3.2.2 we give some examples.

EXAMPLE 3.2.1. Assume $Q = (q_{rs}) \in Z_2(J)$ is Lagrange symmetric. Then q_{12}, q_{21} are both real and $q_{11} = -\overline{q}_{22}$ implies that $iq_{11} = q_{22}$ with q_{11} real. Thus $M = M_Q$ is given by

$$\begin{aligned} My &= -[q_{12}^{-1}(y' - q_{11}y)]' + q_{21}y + q_{12}^{-1}q_{22}(y' - q_{11}y) \\ &= -[q_{12}^{-1}(y' + iq_{22}y)]' + q_{21}y + q_{12}^{-1}q_{22}(y' - q_{11}y). \end{aligned}$$

If $q_{11} = 0$ then this reduces to

$$My = -(q_{12}^{-1}\, y')' + q_{21}\, y$$

which is the classical Sturm-Liouville form

$$My = -(p\, y')' + q\, y$$

with the notation $p = q_{12}^{-1}$ and $q = q_{21}$.

REMARK 3.2.1. Note that for this example our assumptions on q_{rs} are that $q_{12} \neq 0$ a.e. on J, q_{12}, q_{21} are both real, $iq_{11} = q_{22}$ with q_{11} real, and

$$q_{12}^{-1}, q_{21}, q_{11} \in L_{loc}(J).$$

REMARK 3.2.2. We make the following observations:

(1) When $q_{11} \neq 0$ this is a second order symmetric expression with nonreal coefficients in contrast with the classical case ($q_{11} = 0$) where the coefficients of second order symmetric expressions must be real.
(2) Although $q_{12} \neq 0$ a.e. on J, $p = q_{12}^{-1}$ may change sign, may be 0 at any number of points in J, may even be identically zero on one or more subintervals of J. (For an elaboration of the latter statement see the paper by Kong, Volkmer and Zettl [**341**] where a class of self-adjoint Sturm-Liouville problems is identified each of which is equivalent to a finite dimensional matrix problems. But we don't discuss the degenerate case when q_{12}^{-1} is zero on the entire interval J.

Since the general pattern for even order Lagrange symmetric matrices is not evident from the second order case $n = 2$, the next example is for $n = 4$. It is followed by $n = 3$ to illustrate the odd order case which has some features significantly different from the even order cases.

EXAMPLE 3.2.2. Assume $Q = (q_{rs}) \in Z_4(J)$ is Lagrange symmetric. Then Q has the form

$$Q = \begin{bmatrix} iq_{11} & q_{12} & 0 & 0 \\ q_{21} & iq_{22} & q_{23} & 0 \\ iq_{31} & q_{32} & iq_{22} & q_{12} \\ q_{41} & iq_{31} & q_{21} & iq_{11} \end{bmatrix},$$

where the q_{rs} are real.

Here

$$\begin{aligned}
y^{[0]} &= y, \\
y^{[1]} &= q_{12}^{-1}(y' - iq_{11}y), \quad y \in V_1, \\
y^{[2]} &= q_{23}^{-1}\{y^{[1]\prime} - q_{21}y - iq_{22}y^{[1]}\}, \quad y \in V_2, \\
y^{[3]} &= q_{12}^{-1}\{y^{[2]\prime} - iq_{31}y - q_{32}y^{[1]} - iq_{22}y^{[2]}\}, \quad y \in V_3, \\
y^{[4]} &= y^{[3]\prime} - q_{41}y - iq_{31}y^{[1]} - q_{21}y^{[2]} - iq_{11}y^{[3]}, \quad y \in V_4.
\end{aligned}$$

and $M = M_Q$ is given by

$$My = i^4 y^{[4]} = y^{[4]}, \quad y \in V_4 = D(Q).$$

If $q_{11} = 0 = q_{22} = q_{31} = q_{21}$ and $q_{12} = 1$, then this reduces to the modified Naimark form

$$My = [(q_{23}^{-1}y'')' - q_{32}y']' - q_{41}y, \quad y \in V_4 = D(Q).$$

REMARK 3.2.3. Without the extra assumptions that $q_{21} = 0$ and $q_{12} = 1$ this a fourth order symmetric expression *with real coefficients* which has two more independent components than the Naimark form, namely q_{21} and q_{12}. Removing the restriction that $q_{11} = 0 = q_{22} = q_{31}$ produces symmetric quasi-differential expressions M_Q with three additional and complex coefficients.

EXAMPLE 3.2.3. Assume $Q = (q_{rs}) \in Z_3(J)$ is Lagrange symmetric. Then Q has the form

$$Q = \begin{bmatrix} iq_{11} & q_{12} & 0 \\ q_{21} & iq_{22} & q_{12} \\ iq_{31} & q_{21} & iq_{11} \end{bmatrix}$$

with q_{rs} real.

Here

$$\begin{aligned}
y^{[0]} &= y, \\
y^{[1]} &= q_{12}^{-1}(y' - iq_{11}y), \quad y \in V_1, \\
y^{[2]} &= q_{12}^{-1}\{y^{[1]\prime} - q_{21}y - iq_{22}y^{[1]}\}, \quad y \in V_2, \\
y^{[3]} &= \{y^{[2]\prime} - iq_{31}y - q_{21}y^{[1]} - iq_{11}y^{[2]}\}, \quad y \in V_3.
\end{aligned}$$

Then $M = M_Q$ is given by

$$My = i^3 y^{[3]} = -iy^{[3]}.$$

The special case when $q_{11} = 0$, $q_{12} = -1/p$, $q_{21} = q_0/p$, $q_{22} = -iq_1/p^2$, $q_{31} = -iq_0$, and $p^2 = 2q_1$ reduces to the classical form for $n = 3$ when q_0, q_1 and y are sufficiently differentiable. This follows from a direct computation and the observation that

$$(p(py')')' = (b_1 y')'' + (b_1 y'')'.$$

Note that in this special case our coefficient hypotheses are that

$$q_{12}, q_{21}, q_{22}, q_{31} \in L_{loc}(J, \mathbb{R})$$

and the relationship $p^2 = 2b_1$ imposes the sign restriction $b_1 > 0$ a.e. on J in contrast with the even order case where there is no sign restriction on the leading coefficient. See [**200**] for additional comments about this point which illustrates an important difference between the even and odd order cases.

3. The Lagrange Identity

The Lagrange Identity is fundamental to the study of boundary value problems. The proofs of this identity given in Weidmann [**574**] and Zettl [**607**] are long and cumbersome using integration by parts and mathematical induction. Here we use an elegant method of Everitt and Neuman [**191**] which makes direct use of the construction of the differential expression $M = M_Q$ and its domain $D(M)$.

THEOREM 3.3.1 (Lagrange Identity). *Let $Q \in Z_n(J)$, $P = -E^{-1}Q^*E$ where $E = E_n = ((-1)^r\delta_{r,n+1-s})_{r,s=1}^n$. Then $P \in Z_n(J)$ and for any $y \in D(M_Q)$, $z \in D(M_P)$ we have*

$$\overline{z}\, M_Q\, y - y\, \overline{M_P\, z} = [y, z]', \tag{3.3.1}$$

where

$$[y, z] = i^n \sum_{r=0}^{n-1} (-1)^{n+1-r} \overline{z}_P^{[n-r-1]} y_Q^{[r]} = -i^n Z^* EY.$$

PROOF. That $P \in Z_n(J)$ follows directly from the definition of $Z_n(J)$. Note that E satisfies (1.8.2) and recall that $M_Q y = i^n y^{[n]}$ and $M_P z = i^n z^{[n]}$. Let $f = y_Q^{[n]}$, $g = z_P^{[n]}$ and let

$$Y = \begin{pmatrix} y^{[0]} \\ y^{[1]} \\ \vdots \\ y^{[n-1]} \end{pmatrix}, \quad Z = \begin{pmatrix} z^{[0]} \\ z^{[1]} \\ \vdots \\ z^{[n-1]} \end{pmatrix}, \quad F = \begin{pmatrix} 0 \\ \vdots \\ 0 \\ f \end{pmatrix}, \quad G = \begin{pmatrix} 0 \\ \vdots \\ 0 \\ g \end{pmatrix}.$$

Then

$$Y' = QY + F, \quad Z' = PZ + G.$$

By computation, we have

$$G^*EY + Z^*EF = -\overline{z}f + (-1)^n \overline{g} y = -(-i)^n(\overline{z} M_Q y - y\overline{M_P z}).$$

Moreover,

$$Z^*EY = -\sum_{r=0}^{n-1} (-1)^{n+1-r} \overline{z}_P^{[n-r-1]} y_Q^{[r]} = -(-i)^n [y, z].$$

From Lemma 1.8.2, we have

$$(Z^*EY)' = G^*EY + Z^*EF.$$

Therefore (3.3.1) holds. □

COROLLARY 3.3.1. *If $My = \lambda wy$ and $Mz = \overline{\lambda} wz$ on some interval $(\alpha, \beta) \in (a, b)$, then $[y, z]$ is constant on (α, β). In particular, if λ is real and $My = \lambda wy$, $Mz = \lambda wz$ on some interval $(\alpha, \beta) \in (a, b)$, then $[y, z]$ is constant on (α, β).*

PROOF. This follows directly from the Lagrange Identity (3.3.1). Since the coefficients are real, the last statement follows from the observation that if $M\, y = \lambda w\, y$ then $M\overline{y} = \lambda w \overline{y}$. □

REMARK 3.3.1. We comment on the difference between the Lagrange Identity of Theorem 3.3.1 and the classical Lagrange Identity one as found, for example, in the well known books by Coddington and Levinson [**102**] and Dunford and Schwartz [**111**]. The fundamental differences are: (i) the matrix E is a simple constant matrix where in the classical case it is a complicated nonconstant function depending on the coefficients, (ii) we assume only that the coefficients are locally Lebesgue integrable in contrast to [**102**], [**111**] where strong smoothness conditions are required. The price we pay for this generalization and simplification is the use of quasi-derivatives $y^{[r]}$ in place of the classical derivatives $y^{(r)}$. These quasi-derivatives $y^{[r]}$ depend on the coefficients.

For the rest of this section we assume that $Q = Q^{+} = -E^{-1}Q^{*}E$. Let $M = M_Q$, then $M = M^{+}$.

LEMMA 3.3.1. *For any y, z in $D(M)$ we have*

$$\int_{c_1}^{c} \{\overline{z}My - y\overline{Mz}\} = [y,z](c) - [y,z](c_1), \tag{3.3.2}$$

for any $c, c_1 \in J = (a,b)$.

PROOF. This follows from the Lagrange Identity and integration. □

LEMMA 3.3.2. *For any y, z in $D(M)$ the limits*

$$\lim_{t\to b^-}[y,z](t), \quad \lim_{t\to a^+}[y,z](t) \tag{3.3.3}$$

exist and are finite, and

$$\int_{a}^{b} \{\overline{z}My - y\overline{Mz}\} = [y,z](b) - [y,z](a).$$

PROOF. This follows from (3.3.2) by taking limits as $c_1 \to a$, $c \to b$. That the limits exist and are finite can be seen from the definition of $D(M)$. □

REMARK 3.3.2. The finite limits (3.3.3) play a critical role in the characterization of the singular symmetric and self-adjoint operators studied below; they 'play the role' of the quasi-derivatives at regular endpoints.

LEMMA 3.3.3. *Suppose M is regular at $c \in (a,b)$. Then for any $y \in D(M)$ the limits*

$$y^{[r]}(c) = \lim_{t\to c} y^{[r]}(t)$$

exist and are finite, $r = 0,\cdots,n-1$. In particular this holds at any regular endpoint and at each interior point of J. At each endpoint the limit is the appropriate one sided limit.

PROOF. See [**440**] or [**573**]. Although our result is more general than those stated in these references the same method of proof can be used here. □

4. Maximal and Minimal Operators

In this section we define the maximal and minimal operators and establish their basic properties. In a sense the maximal operator is the largest operator on which the differential expressions can 'operate' and map the result into $L^2(J,w)$.

DEFINITION 3.4.1. Let $Q \in Z_n(J)$, let w be a weight function, and let $H = L^2(J,w)$. The maximal operator $S_{\max}=S_{\max}(Q,J)$ with domain $D_{\max}=D_{\max}(Q,J)$ is defined by:

$$D_{\max}(Q,J) = \{y \in H : y \in D(Q),\ w^{-1}M_Q\,y \in H\},$$
$$S_{\max}(Q,J)\,y = w^{-1}M_Q\,y, \quad y \in D_{\max}(Q,J).$$

We use the next theorem to define the minimal operator.

THEOREM 3.4.1. *Let $Q \in Z_n(J)$, let w be a weight function, and let*

$$Q^+ = -E^{-1}Q^*E, \quad \text{where } E = E_n = ((-1)^r \delta_{r,n+1-s})_{r,s=1}^n. \tag{3.4.1}$$

Then

(1) $Q^+ = (q_{ij}^+) \in Z_n(J)$.
(2) $D_{\max}(Q,J)$ *is dense in H. Let* $S_{\min}(Q,J) = S_{\max}^*(Q,J)$ *and let* $D_{\min}(Q,J)$ *denote the domain of* $S_{\min}(Q,J)$.
(3) $S_{\min}(Q,J)$ *is a closed operator in H with dense domain and we have*

$$S_{\min}^*(Q,J) = S_{\max}(Q^+,J), \quad S_{\min}(Q,J) = S_{\max}^*(Q^+,J).$$

(4) *If $Q^+ = Q$, then $S_{\min}(Q,J)$ is a closed symmetric operator in H with dense domain and*

$$S_{\min}^*(Q,J) = S_{\max}(Q,J), \quad S_{\min}(Q,J) = S_{\max}^*(Q,J).$$

PROOF. Part (1) follows directly from the definition. The method of Naimark [**440**, Chapter V] can be adapted to prove this theorem with minor modifications. See also [**200**], [**427**]. □

DEFINITION 3.4.2 (Lagrange Adjoint Matrix). We call Q^+ defined by (3.4.1) the Lagrange adjoint matrix of Q and note that $Q^+ \in Z_n(J)$; $M^+ = M_{Q^+}$ is called the adjoint expression of $M = M_Q$. Note that for $P, Q \in Z_n(J)$ we have

$$(P^+)^+ = P, \quad (P+Q)^+ = P^+ + Q^+, \quad (PQ)^+ = -Q^+P^+, \quad (cP)^+ = \overline{c}P^+, \quad c \in \mathbb{C}.$$

This terminology 'Lagrange adjoint' is introduced due to its relationship to boundary value problems which are adjoint in the sense of Lagrange as we will see below. An important special case arises when $Q^+ = Q$, this is the Lagrange Hermitian case which generates, as we will see below, symmetric differential operators.

REMARK 3.4.1. The relationship (3.4.1) can be described in terms of the components of the matrices as follows: Q^+ is obtained from Q by performing the following three commutative operations:

(1) 'flip' the components $q_{i,j}$ across the secondary diagonal:

$$q_{i,j} \to q_{n+1-j,n+1-i};$$

(2) take conjugates:

$$q_{i,j} \to \overline{q}_{i,j};$$

(3) change the sign for all the even positions:

$$q_{i,j} \to (-1)^{i+j+1} q_{i,j}.$$

We note that for the special case considered by Naimark [**440**] there is a change of sign: in (3) the sign is changed for the odd positions.

NOTATION 3.4.1. Below we will also use minimal and maximal domain functions and their restrictions on subintervals (α,β) of $J = (a,b)$, particularly for $(\alpha,\beta) = (a,c)$ and $(\alpha,\beta) = (c,b)$ with $c \in (a,b)$. Since $Q \in Z_n(J)$ implies that $Q \in Z_n((\alpha,\beta))$ the above Definitions and Theorems can be applied in the Hilbert space $L^2((\alpha,\beta),w)$ with J replaced by (α,β). Below when we use the notation $D_{\max}(\alpha,\beta)$, $D_{\min}(\alpha,\beta)$ it is understood that we use the above definitions and theorems with J replaced by (α,β), Q and w replaced by their restrictions to (α,β)

and $L^2(J,w)$ replaced by $L^2((\alpha,\beta),w)$. Below, Q, J, (α,β), as well as the Hilbert space, may be omitted when these are clear from the context.

Below for $Q \in Z_n(J)$, $M = M_Q$, if we consider the equation $My = \lambda wy$ and say that w is a 'weight' function, we mean that $w > 0$ on (a,c) and in $L^1(a,c)$ if a is regular and in $L_{loc}(a,c)$ if a is singular. Similarly for the interval (c,b).

The next lemma is known as the Naimark Patching Lemma or just the Patching Lemma.

LEMMA 3.4.1 (Naimark Patching Lemma). *Let $Q \in Z_n(J)$ and assume that both endpoints are regular. Let $\alpha_0,\dots,\alpha_{n-1}$, $\beta_0,\dots,\beta_{n-1} \in \mathbb{C}$. Then there is a function $y \in D_{\max}$ such that*

$$y^{[r]}(a) = \alpha_r, \quad y^{[r]}(b) = \beta_r \quad (r = 0,\dots,n-1).$$

PROOF. The proof in Naimark [**440**] can readily be adapted to prove this more general Patching Lemma. □

COROLLARY 3.4.1. *Let $a < c < d < b$ and $\alpha_0,\dots,\alpha_{n-1},\beta_0,\dots,\beta_{n-1} \in \mathbb{C}$. Then there is a $y \in D_{\max}$ such that y has compact support in J and satisfies :*

$$y^{[r]}(c) = \alpha_r, \quad y^{[r]}(d) = \beta_r \quad (r = 0,\dots,n-1).$$

PROOF. The Patching Lemma gives a function y_1 on $[c,d]$ with the desired properties. Let c_1, d_1 with $a < c_1 < c < d < d_1 < b$. Then use the Patching Lemma again to find y_2 on (c_1,c) and y_3 on (d,d_1) such that

$$y_2^{[r]}(c_1) = 0, \quad y_2^{[r]}(c) = \alpha_r, \quad y_3^{[r]}(d) = \beta_r, \quad y_3^{[r]}(d_1) = 0 \qquad (r = 0,\dots,n-1).$$

Now set

$$y(x) := \begin{cases} y_1(x) & for \quad x \in [c,d] \\ y_2(x) & for \quad x \in (c_1,c) \\ y_3(x) & for \quad x \in (d,d_1) \\ 0 & for \quad x \in J \setminus (c_1,d_1). \end{cases}$$

Clearly y has compact support in J. Since the quasi-derivatives at c_1, c, d, d_1 coincide on both sides, $y \in D_{\max}$ follows. □

COROLLARY 3.4.2. *Let $a_1 < \cdots < a_k \in J$, where a_1 and a_k can also be regular endpoints. Let $\alpha_{jr} \in \mathbb{C}(j = 1,\dots,k;\ r = 0,\dots,n-1)$. Then there is a $y \in D_{\max}$ such that*

$$y^{[r]}(a_j) = \alpha_{jr} \quad (j = 1,\dots,k;\ r = 0,\dots,n-1).$$

PROOF. This follows from repeated applications of the previous Corollary. □

The next theorem gives a characterization of $D(S_{\min})$.

THEOREM 3.4.2. *Let $Q = Q^+ \in Z_n(J)$. Then*

$$D(S_{\min}) = \{y \in D_{\max} : [y,z](a) = 0 = [y,z](b),\ for\ all\ z \in D_{\max}\}. \tag{3.4.2}$$

PROOF. This follows from the Lagrange Identity and the Naimark Patching Lemma. □

5. Boundedness Below of the Minimal Operator

On page 109 of his well known book "Spectral Theory of Ordinary Differential Operators" [**574**] Weidmann states:

"Operators of even order with positive coefficient of highest order have turned out to be semibounded from below if the lower order terms are sufficiently small. On the other hand, operators of odd order are usually expected to be not semibounded (from below or above). We shall see this in many examples ... Anyhow we do not know of a general result which assures this for every operator of odd order."

For a class of operators larger than that studied by Weidmann, Möller and Zettl [**428**] proved - *without any smoothness or smallness assumptions on the coefficients* - that:

(1) Odd order operators, regular or singular, and regardless of the sign of the leading coefficient, are unbounded above and below.
(2) Regular operators of even order with positive leading coefficient are bounded below and not bounded above.
(3) Even order regular or singular operators with leading coefficient that changes sign are unbounded above and below.

The results in [**428**] are for matrix coeffcients, here we present these results only for scalar coefficients.

Recall that if Q is L-symmetric, $M = M_Q$ and w is a weight function, then M is a symmetric expression and $S_{\min}(M)$ is a closed symmetric operator in $L^2(J, w)$ with dense domain.

THEOREM 3.5.1 (Möller-Zettl). *Assume $Q \in Z_n(J)$, $Q = Q^+$ and let $M = M_Q$ be the symmetric expression generated by Q as above. Then*

(1) *If $n = 2k + 1$, $k > 0$, then $S_{\min}(Q)$ is unbounded above and below in $H = L^2(J, w)$ for any weight function w.*
(2) *If $n = 2k$, $k > 0$, and M is regular with positive leading coefficient, then $S_{\min}(Q)$ is bounded below and unbounded above in $H = L^2(J, w)$ for any weight function w.*
(3) *If $n = 2k$, $k > 0$, and the leading coefficient of M changes sign on J, then $S_{\min}(Q)$ is unbounded above and unbounded below in $H = L^2(J, w)$ for any weight function w. Note that M_Q may be regular or singular.*

PROOF. See [**428**]. □

The hypothesis that M is regular with positive leading coefficient cannot be removed in part (2) of Theorem 3.5.1. For $k = 1$, the hypothesis 'M is regular with positive leading coefficient' can be extended to 'M has a positive leading coefficient and is regular or LCNO at each endpoint'. (See definition below for LCNO or see [**620**]). What is the corresponding extension for $k > 1$? In [**416**] Marletta and Zettl prove the following result:

THEOREM 3.5.2 (Marletta-Zettl). *Suppose $Q = (q_{rs}) \in Z_n(J, \mathbb{R})$ with $n = 2k$, $k > 1$ is Lagrange symmetric and assume that Q is of GN type i.e. $q_{r,r+1} = 1$, for $1 \leq r < k$ and $k < r < n$; $q_{k,k+1} > 0$ on J and all other $q_{rs} = 0$ except $q_{r,n+1-r}$ for $r = 1, \cdots, k$. If the left endpoint a is regular and the right endpoint b is disconjugate in the sense of Reid, then $S_{\min}(Q)$ is bounded below in $H = L^2(J, w)$ for any weight function w.*

PROOF. See [**416**]. This proof uses methods from the theory of Hamiltonian systems and the theory of disconjugacy as developed by Reid [**492**]. Disconjugacy theory is closely related to nonoscillation theory. □

Next we discuss the semi-boundedness results for $S_{\min}$ in the second order case and then comment on the comparison with the higher order cases.

Consider the equation

$$(3.5.1)\quad My = -(py')' + qy = \lambda wy \ \ on\ J = (a,b);\ \ \frac{1}{p}, q, w \in L_{loc}(J,\mathbb{R}),\ \ w > 0.$$

Note that $M = M_Q$ where Q is a Lagrange symmetric matrix of the form

$$Q = \begin{bmatrix} 0 & \frac{1}{p} \\ -q & 0 \end{bmatrix}.$$

We briefly recall the needed definitions and basic facts for equation (3.5.1).

DEFINITION 3.5.1. Let (3.5.1) hold and assume, in addition, that $p > 0$ on J. Recall that the endpoint a of equation (3.5.1) is said to be in the limit-circle (LC) case if all solutions of (3.5.1) are in $L^2((a,c),w)$ for some c in J and some $\lambda \in \mathbb{C}$. If this holds for one $\lambda \in \mathbb{C}$ then it holds for all $\lambda \in \mathbb{C}$ real or complex. Otherwise a is said to be in the limit-point (LP) case. The endpoint a is classified as oscillatory (O) if there is a nontrivial solution with an infinite number of zeros in (a,c) for some $c \in J$ and nonoscillatory (NO) otherwise. This classification is independent of λ if a is LC. (But not if a is LP.) Similar definitions are made at b. Each regular endpoint (when $p > 0$ on J) is limit-circle and nonoscillatory (LCNO). A singular endpoint may be LCNO or LCO - limit-circle and oscillatory. See the book [**620**] for details.

The next theorem summarizes well known results for the second order case [**620**]:

THEOREM 3.5.3. *Let* (3.5.1) *hold, and let* $H = L^2(J,w)$.

(1) *If p changes sign on J, then $S_{\min}(Q)$ is unbounded below and unbounded above in H.*
(2) *Suppose $p > 0$ a.e. on J. If each endpoint is either regular or LCNO, then $S_{\min}(Q)$ is bounded below in H, but not above.*
(3) *Suppose $p > 0$ a.e. on J. If one endpoint is LCO, then $S_{\min}(Q)$ is unbounded below and unbounded above in H.*

PROOF. Part (1) was proven by Möller in [**425**]. For parts (2) and (3) see Niessen and Zettl [**454**] or [**425**]; also see [**620**]. □

In the next remark we comment on the comparison between the second order results and the corresponding higher order results of Theorem 3.5.1 and Theorem 3.5.2.

REMARK 3.5.1. Comparison between $n = 2$ and $n > 2$.

(1) Part (1) and part (3) of Theorem 3.5.1 extend part (1) of Theorem 3.5.3 to the higher order case n for any $n > 2$ even or odd.
(2) Part (2) of Theorem 3.5.1 extends part (2) of Theorem 3.5.3 to the even higher order case $n = 2k$, $k > 1$, *but only for regular problems. (Regular endpoints can be considered LCNO.)*

(3) Regarding the extension of part (2) of Theorem 3.5.1 from $n = 2$ to $n = 2k$ with $k > 1$ given by *Theorem* 3.5.2 *there are a number of natural questions:* (i) Is there a condition more transparent i.e. easier to check than disconjugacy in the sense of Reid? What if both endpoints are singular?
(4) How does part (3) of Theorem 3.5.3 extend from $n = 2$ to $n = 2k$ with $k > 1$? What is the 'appropriate' definition for LCO when the deficiency index $d = n$? When $d < n$?

6. Comments

These are made separately for each section.

1. The symmetric operators studied here are 'between' the minimal and maximal operators as indicated in the Introduction.

2. The adjoint and symmetric expressions are defined using the constant matrix E. This matrix also plays an important role in the characterization of the symmetric and self-adjoint domains.

3. Theorem 3.3.1 could be called the E Lagrange Identity. The proof given here is due to Everitt and Neuman [**191**]. It is much simpler than the earlier proofs of Weidmann [**574**] and Zettl [**607**].

4. We comment on the denseness of the minimal domain $D_{\min}(Q, J)$ and its proof. It was shown by Zettl [**607**] that Naimark's method can be applied to prove that for any $Q \in Z_n(J)$, $D_{\min}(Q)$ is dense in H, see also Everitt-Zettl [**200**], Möller-Zettl [**427**]. In general, under the local integrability assumptions used here, $C_0^\infty(J)$ may not be contained in $D_{\min}(Q, J)$. So the standard argument, widely used in the literature, that $D_{\min}(Q, J)$ is dense because it contains $C_0^\infty(J)$ does not work here. Moreover, it may not be easy to find explicit functions in $D_{\min}(Q, J)$ other than the zero function. Nevertheless, the proof given in [**440**, p. 68], with minor modifications, can be used to prove that $D_{\min}(Q, J)$ is dense under the local integrability assumptions used here in the definition of $Q \in Z_n(J)$.

In much of the current literature in Mathematics and Mathematical Physics, even for the expression $My = -y'' + qy$, many authors assume that $q \in L^2_{loc}(J)$ because this implies that $C_0^\infty(J) \subset D_{\min}$ and therefore $D_{\min}$ is dense. For $q \in L_{loc}(J)$, $C_0^\infty(J)$ may not be contained in $D_{\min}$ so this proof is not valid, but Naimark's proof does 'work', see [**200**], [**427**].

5. This section is based on the papers Marletta-Zettl [**416**], Möller [**425**], Möller-Zettl [**428**], [**427**], and Niessen-Zettl [**454**], [**453**].

CHAPTER 4

Deficiency Indices

1. Introduction

For a symmetric expression $M = M_Q$, $Q \in Z_n(J)$, $Q = Q^+$, a weight function w, an interval $J = (a, b)$, $-\infty \le a < b \le \infty$, the deficiency indices d^+, d^-, are defined as the number of linear independent solutions of the equation

$$My = \lambda wy \quad on\ J, \quad \lambda \in \mathbb{C} \tag{4.1.1}$$

which are in the Hilbert space $H = L^2(J, w)$. They depend on M, w, J and λ. For M, w, J fixed we denote them by $d^+(\lambda)$, $d^-(\lambda)$. Their dependence on λ is very different for real and complex λ as we will see below. The next theorem shows that $d^+(\lambda)$ and $d^-(\lambda)$ are constant in the upper and lower half planes in $\mathbb{C}$.

THEOREM 4.1.1. *For fixed (M, w, J), $d(\lambda_1) = d(\lambda_2)$ if $\mathrm{Im}(\lambda_j) > 0$ or if $\mathrm{Im}(\lambda_j) < 0$ for $j = 1, 2$.*

PROOF. This follows from a well known abstract result. The proof given on p.33 of Naimark's book [**440**, Theorem 5] can readily be adapted to the much more general class of symmetric expressions M considered here. □

DEFINITION 4.1.1. Let $d^+ = d^+(M, w, J) = d^+(i)$ and $d^- = d^-(M, w, J) = d^-(-i)$.

What are the possible values of d^+, d^-? This is the question we answer in this section.

We start with two lemmas.

LEMMA 4.1.1. *If d^+, d^- can be achieved on any one open interval, then they can be achieved on any other open interval. These intervals may have finite or infinite endpoints.*

PROOF. This follows from a change of variable, see pages 85-86 in the Everitt-Markus monograph [**184**] for detailed computations of this change. □

REMARK 4.1.1. In the rest of this section we will apply this lemma repeatedly without further comment. In other words, having established that certain values of d^+, d^- have been achieved on some interval, these same valued can be achieved on any other interval.

LEMMA 4.1.2. *If $Q_a \in Z_n(a, c)$ and $Q_b \in Z_n(c, b)$, then*

$$Q_{ab}(t) = \begin{cases} Q_a(t), & a < t < c \\ Q_b(t), & c < t < b \end{cases}$$

is in $Z_n(a, b)$. Also the sign of the leading coefficient does not change.

PROOF. This follows from the definition of Q in Section 2.4 and the observation that $Q_{ab}(t)$ does not need to be defined for $t = c$ since the integrals in the definition of $Q = (q_{ij})$ are Lebesgue integrals and c is a regular point. $\square$

It is clear from the definition that

$$0 \leq d^+,\ d^- \leq n \tag{4.1.2}$$

where n is the order of M since equation (4.1.1) cannot have more than n linearly independent solutions.

If all the coefficients of M are real valued, i.e. $M = M_Q$, $Q \in Z_n(J, \mathbb{R})$, $Q = Q^+$ then a solution y is in H if and only if its conjugate $\overline{y}$ is in H. Hence, for real valued coefficients, we have $d^+ = d^-$ and (4.1.2) simplifies to

$$0 \leq d^+ = d^- \leq n.$$

In a seminal paper in 1950 Glazman [**239**] proved that all values of $d = d^+ = d^-$ are realized. This result has an interesting history as mentioned in the Preface. In 1966 McLeod [**417**] showed that $d^+ \neq d^-$ is possible if the coefficients are complex valued.

It is well known that if all solutions of equation (4.1.1) are in H for some $\lambda \in \mathbb{C}$ then this is true for all $\lambda \in \mathbb{C}$, including the real values of λ. From this it follows that $d^+ = n$ if and only if $d^- = n$. Therefore the search for all possible values of d^+, d^- can be reduced to

$$0 \leq d^+,\ d^- \leq n - 1. \tag{4.1.3}$$

The claim that all possible values of d^+, d^- satisfying (4.1.3) are realized also has long and interesting history. It is known as the Deficiency Index Conjecture which we now state formally.

CONJECTURE 1 (Deficiency Index Conjecture). *Consider the equation*

$$My = \lambda wy \quad on\ J, \quad \lambda \in \mathbb{C}$$

where $M = M_Q$, $Q \in Z_n(J)$, $Q = Q^+$, w is a weight function, and $J = (a, b)$, $-\infty \leq a < b \leq \infty$, is an open interval on the real line with finite or infinite endpoints. Let the deficiency indices d^+, d^-, be defined as above. For any

$$s, t \in \{0, 1, 2, \cdots, n - 1\}$$

there exists an M and w such that

$$s = d^+(M, w, J) \ \ and \ \ t = d^-(M, w, J).$$

PROOF. The proof will be given at the end of this section. $\square$

The next theorem considers the case when one endpoint of J is regular.

THEOREM 4.1.2. *Let $Q \in Z_n(J)$ satisfy $Q = Q^+$, let $M = M_Q$ be the symmetric expression generated by Q and let w be a weight function. Fix (M, w, J) and assume one endpoint of J is regular. Then:*

(1) *If $n = 2k$, then*

$$k \leq d^+,\ d^- \leq n. \tag{4.1.4}$$

(2) *If $n = 2k + 1$, $k \geq 1$, and the leading coefficient of M is positive on J, then*

$$k \leq d^+ \leq n; \quad k - 1 \leq d^- \leq n. \tag{4.1.5}$$

(3) *If* $n = 2k+1$, $k \geq 1$, *and the leading coefficient of* M *is negative on* J, *then*

$$k - 1 \leq d^+ \leq n; \quad k \leq d^- \leq n. \tag{4.1.6}$$

These inequalities are best possible.

PROOF. The right inequalities are clear since there cannot be more than n linearly independent solutions in H. For the inequalities on the left see the operator-theoretic proof of Kogan and Rofe-Beketov [**338**] or the more classical proof of Everitt [**146**], [**160**], [**161**]. For the inequalities on the left of (2) and (3) the methods given in [**146**] and [**161**] can be adapted; see also [**338**] and [**326**]. □

The next result is sometimes called the Kodaira formula. We will apply it to the interval $J = (a,b)$ using an interior point c, $a < c < b$, the subintervals $J_a = (a,c)$ and $J_b = (c,b)$ and the restrictions of M, w and equation (4.1.1) to these subintervals. When both endpoints of $J = (a,b)$ are singular the Kodaira formula reduces the problem of finding $d^+(a,b)$ to that of finding $d^+(a,c)$ and $d^+(c,b)$ where $a < c < b$. Similarly for $d^-(a,b)$.

THEOREM 4.1.3. *Let* $Q \in Z_n(J)$ *satisfy* $Q = Q^+$, *let* $M = M_Q$, w *be a weight function, and let* $J_a = (a,c)$, $J_b = (c,b)$ *for some* $c \in (a,b)$. *Then*

$$\begin{aligned} d^+(M,w,J) &= d_a^+(M,w,J_a) + d_b^+(M,w,J_b) - n, \\ d^-(M,w,J) &= d_a^-(M,w,J_a) + d_b^-(M,w,J_b) - n. \end{aligned} \tag{4.1.7}$$

PROOF. The proof in [**440**, p.72] for a special case can readily be adapted to this more general result. □

Let $c \in (a,b)$. Note that from the definition of M in Section 2.4 it follows that c is a regular endpoint for both intervals $J_a = (a,c)$ and $J_b = (c,b)$. In particular the inequalities (4.1.4), (4.1.5), (4.1.6) hold for the intervals J_a and J_b. In (4.1.7) $d_a^+(M,w,J_a)$, $d_a^-(M,w,J_a)$ and $d_b^+(M,w,J_b)$, $d_b^-(M,w,J_b)$ denote the deficiency indices for the intervals J_a and J_b, respectively.

Formula (4.1.7) reduces the problem of finding d^+ on an interval $J = (a,b)$ with both endpoints singular to the problem of finding d^+ on intervals with just one singular endpoint. Similarly for d^-. Formula (4.1.7) is independent of the choice of $c \in (a,b)$.

Since the fundamental work of Glazman and the example of McLeod many people have investigated the possible values of d^+ and d^-. In 1975 [**338**] and 1976 [**337**] Kogan and Rofe-Beketov achieved major results for the one singular endpoint case.

THEOREM 4.1.4 (Kogan and Rofe-Beketov). *Suppose one endpoint is regular and the other singular. For all integers* d^+, d^- *which satisfy conditions* (4.1.4), (4.1.5), (4.1.6) *and the inequality*

$$|d^+ - d^-| \leq 1 \tag{4.1.8}$$

there exists a symmetric expression M *of order* n *and a weight function* w *such that*

$$d^+ = d^+(M,w,J), \quad d^- = d^-(M,w,J).$$

Thus all possible values of d^+, d^- which satisfy conditions (4.1.4), (4.1.5), (4.1.6) can be achieved provided they don't differ by more than 1.

In 1978 and 1979 Gilbert [**237**], [**238**] showed that the difference between d^+ and d^- can be arbitrarily large by proving that

$$|d^+ - d^-| \geq p$$

for any positive integer p provided that the order n of M is allowed to be $8p$ or larger. This result gave considerable support to the conjecture that all possible values of d^+, d^- which satisfy (4.1.4), (4.1.5), (4.1.6) are realized when one endpoint is regular and the other singular.

Indeed this clearly follows from these inequalities and (4.1.8) for $n = 2, 3$, and 4. But not for

$$n = 5, \quad d^+ = 2, \quad d^- = 4$$

or

$$n = 6, \quad d^+ = 3, \quad d^- = 5$$

since the difference between d^+ and d^- is greater than 1. And so on for $n > 6$.

1.1. Proof of the Deficiency Index Conjecture. Recall Lemmas 4.1.1 and 4.1.2 and Remark 4.1.1; these will be used here without further comment.

Next we prove the even order case $n = 2k$, $k > 2$.

I. Suppose $k = 3$. Then by Theorem 4.1.2 we have

$$3 \leq d_a^+,\ d_a^-, d_b^+, d_b^- \leq 6.$$

Assume that at least one of $d_a^+,\ d_a^-, d_b^+, d_b^-$ is not 6. From the Kogan and Rofe-Beketov Theorem, there exists $Q_a \in Z_6(J_a)$ and $Q_b \in Z_6(J_b)$ such that $|d_a^+ - d_a^-| \leq 1$ and $|d_b^+ - d_b^-| \leq 1$. Now define Q_{ab} as above and let d^+, d^- be the deficiency indices of $M = M_{Q_{ab}}$. Then from the Kodaira formula we have that

$$|d^+ - d^-| = |d_a^+ + d_b^+ - 6 - (\,d_a^- + d_b^- - 6)| \leq 2.$$

Therefore all possibilities satisfying $|d^+ - d^-| \leq 2$ are realized for the intervals (a, c) and (c, b). From this and the Kodaira formula it follows that all

$$d^+,\ d^- \in \{0, 1, 2, 3, 4, 5\}$$

are realized.

For $k > 3$ we proceed by using an induction type of argument based on (4.1.8) together with the Kodaira formula, e.g. For $k = 4$ by (4.1.4), (4.1.5), (4.1.6) we have that

$$4 \leq d_a^+,\ d_a^-, d_b^+, d_b^- \leq 8$$

and all the differences $0, 1, 2$ can be achieved. So assume $|d_a^+ - d_a^-| \leq 2$ and $|d_b^+ - d_b^-| \leq 1$ and proceed as in case $k = 3$ using the Kodaira formula to obtain

$$|d^+ - d^-| = |d_a^+ + d_b^+ - 8 - (d_a^- + d_b^- - 8)| \leq 3$$

Therefore all possibilities satisfying $|d^+ - d^-| \leq 3$ are realized for the intervals (a, c) and (c, b).

From this and the Kodaira formula it follows that all values

$$d^+,\ d^- \in \{0, 1, 2, 3, 4, 5, 6, 7\}$$

are realized.

Proceeding by mathematical induction we obtain that all possibilities

$$d^+, d^- \in \{0, 1, ..., n-1\}$$

are realized for M of order $n = 2k$. This completes the proof for the even order case.

Next we show that the argument for the odd order case can be reduced to the even order case.

II. Assume $n = 2k + 1$, $k \geq 1$ and the leading coefficient of M is positive. The case $k = 1$ follows from the inequality (4.1.5) and the Kodaira formula.

For $n = 2k + 1$, we have

$$k \leq d^+ \leq n; \quad k - 1 \leq d^- \leq n.$$

From the Kogan and Rofe-Beketov Theorem we can get that $d^- = k$ on some interval. From this and Lemma 4.1.1 we can then proceed as in the even order case.

2. The Deficiency Index Continued

In this section we make some general remarks about deficiency indices.

As mentioned above if the deficiency indices are realized for one interval, say $(0, \infty)$, they are realized for any interval

$$J = (a, b), \quad -\infty \leq a < b \leq \infty.$$

There is a vast literature on the determination of these indices for the interval $(0, \infty)$. We make some remarks.

REMARK 4.2.1. $n = 2$. This case has a voluminous literature dating back at least to the seminal 1910 paper of Weyl [**580**] discussing the dependence of d on p and q when $w = 1$. In this case, with one regular and one singular endpoint, we have either $d = 1$ (LP) or $d = 2$ (LC). Many sufficient and some necessary conditions are known for LP. There is also a necessary and sufficient condition known [**326**] but it is such that it cannot be checked in every case. So in general - despite the vast literature on this problem (partly because of its interest in Quantum Mechanics where $p = 1$ and q is the potential function for the one dimensional Schrödinger equation) - the problem is still open. See Kauffman, Read and Zettl [**326**] for an extensive but not comprehensive (and not up to date) discussion of the case $n = 2$. Also see [**620**], [**102**], [**111**].

REMARK 4.2.2. $n = 2k$, $k > 1$. This case has an interesting history as mentioned in the Preface and also an extensive literature. The classification for real coefficients when $d^+ = d^- = d : k \leq d \leq n = 2k$ was established by Weyl [**580**] in 1910 for the case $n = 2$ and by Glazman in 1950 for general n. Glazman showed that all values of d are realized. Simpler examples were later found by Orlov [**456**] and by Read [**488**], also see [**326**]. In this 40 year interval Windau [**582**] in 1921 and Shin [**513**], [**514**], [**515**], [**516**] in 1936-1943 claimed to have established that only the two cases $d(M) = k$ (LP) and $d(M) = 2k$ (LC) are possible. In these papers Shin also discovered general Lagrange symmetric quasi-differential expressions. These were rediscovered, in a somewhat different but equivalent form, by Zettl [**595**], [**607**]. These rediscovered forms are used here; they are generated in Chapters 2 and 3.

REMARK 4.2.3 (Delicate Nature of the Deficiency Index). Assume one endpoint is regular and $Q \in Z_n(J, \mathbb{R})$, $Q = Q^+$, $M = M_Q$. Then $d(M) = d^+(M) = d^-(M)$.

For $n = 2$ and $My = -(py')' + qy$ it is known that when $d(M) = 2$, given any positive number ε it is possible to change the coefficients p, q on a set of Lebesgue measure less than ε such that for the changed M, we have $d(M) = 1$. It is possible to find sufficient conditions for $d(M) = 1$ which are given only on an infinite sequence of intervals. See [**326**] Chapter III.

For $n = 2k$, $k > 1$, and one regular endpoint, $k \leq d(M) \leq 2k$. For any $d(M) > k$ it is possible to change the coefficients of M only on a sequence of intervals such that for the changed M, we have $d(M) = k$ [**326**].

Thus for $n = 2$ and $n = 2k$, $k > 1$, there are strong 'interval-type' sufficient conditions for the LP case $d(M) = k$ but not for any of the cases $d(M) > k$. Thus the LP case is 'special' and for all the other cases $d(M) > k$, including the LC case, the dependence of $d(M)$ on the coefficients is very delicate. There are no 'interval-type' sufficient conditions known for the cases $d(M) > k$. See Kauffman-Read-Zettl [**326**], especially page 45 for some details.

Next we recall the definition of matrices $Q = (q_{ij}) \in Z_n(J)$ which are of GN type and give some illustrations and remarks. These have been studied most extensively.

DEFINITION 4.2.1. A Lagrange symmetric matrix $Q = (q_{ij}) \in Z_n(J, \mathbb{R})$ is of GN type if $n = 2k$, $k = 1, 2, 3, \ldots$; and all $q_{ij} = 0$ except for the following:

(1) $q_{i,i+1} = 1$ for all $i = 1, \cdots, n-1$ except when $i = k$, then $q_{k,k+1}^{-1} \in L_{loc}(J, \mathbb{R})$; and
(2) $q_{n,1}, q_{n-1,2}, \cdots, q_{k,k+1} \in L_{loc}(J, \mathbb{R})$.

Thus for $n = 2, 4, 6$ a GN type matrix Q has the following form:

$$Q = \begin{bmatrix} 0 & q_{12} \\ q_{21} & 0 \end{bmatrix}, \qquad Q = \begin{bmatrix} 0 & 1 & 0 & 0 \\ 0 & 0 & q_{23} & 0 \\ 0 & q_{32} & 0 & 1 \\ q_{41} & 0 & 0 & 0 \end{bmatrix},$$

$$Q = \begin{bmatrix} & 1 & 0 & 0 & 0 & 0 \\ & & 1 & 0 & 0 & 0 \\ & & & q_{34} & 0 & 0 \\ & & q_{43} & & 1 & 0 \\ & q_{52} & & & & 1 \\ q_{61} & & & & & \end{bmatrix}.$$

Recall that the matrices of GN type generate the quasi-derivative forms of the real classical symmetric expressions as shown in Chapter 2. In Naimark's book [**440**] $q_{34} = -1$ in the fourth order case and $q_{45} = -1 = q_{56}$ in the $n = 6$ case. Nevertheless, we will continue to call these matrices of GN type.

For symmetric expressions M generated by matrices Q of GN type there is a vast literature studying the dependence of the deficiency index

$$d(M) = d(q_{n,1}, q_{n-1,2}, \cdots, q_{k,k+1}; w) = d(q_0, q_1, \cdots, q_k; w)$$

on these coefficients when $w = 1$. Not much is known for general w but see [**544**] and [**373**] for exceptions. (Here the notation $q_{n,1}, q_{n-1,2}, \cdots, q_{k,k+1}$ corresponds to $q_0, q_1, \cdots, q_k$.)

REMARK 4.2.4. Assume that one endpoint is regular. Most of the known conditions on the coefficients in the even order case are for $d = k$, see [**144**], also Kauffman, Read and Zettl [**326**]. The few results known for the even order intermediate cases $k < d < 2k$ and the odd order case are based on the asymptotic form of solutions, see [**111**], [**440**], Gilbert [**237**], [**238**], the book by Rofe-Beketov and Kholkin [**502**] and its 941 references. There are some exceptions. For the fourth order case Eastham [**117**] found some conditions for $d = 3$ by proving that $d \neq 2$ and $d \neq 4$.

3. Powers of Differential Expressions and Their Deficiency Index

If M is a classical symmetric expression with real C^∞coefficients then each power M^s, $s = 1, 2, 3, \cdots$, is also a classical symmetic expression. Thus the coefficients of M uniquely determine a sequence $d(M)$, $d(M^2)$, $d(M^3), \cdots$ of deficiency indices. In [**326**] Kauffman, Read, and Zettl give a complete description of these possible sequences $d(M)$, $d(M^2)$, $d(M^3), \cdots$.

Given a Lagrange symmetic matrix $Q \in Z_n(J)$ and its associated symmetric expression $M = M_Q$ the powers M^s, $s \in \mathbb{N}_2$, of M can be defined naturally by induction: $M^2(y) = M(My)$, $M^3(y) = M^2(My)$, $\cdots$ without any additional hypothesis on the coefficients and these powers are symmetric expressions generated by Lagrange symmetric matrices $Q^{[s]} \in Z_{ns}(J)$. These powers can be constructed using a methof of Zettl [**607**]. Many proofs given in [**326**] make no use whatsoever of the strong smoothness assumptions on the coefficients other than for the construction of the powers of the classical expressions M^s so that these also have the form (2.2.3), (2.2.4). Thus the proofs in [**326**], when combined with the extension of the Glazman-Naimark theory given above, will yield the results in [**326**] for the sequences $\{M^s,\ s \in \mathbb{N}\}$ where $M = M_Q$ with $Q = (q_{ij}) \in Z_n(J)$, $n > 1$ without any additional assumptions on the coefficients of Q. In particular, no smoothness assumptions. Thus we will state our results in this section for general symmetric expressions and their powers and refer to [**326**] for the proofs not given here.

We start with the construction of powers of quasi-differential expressions given in [**615**].

THEOREM 4.3.1. *Assume that $Q = (q_{ij}) \in Z_n(J)$ is a Lagrange symmetric matrix. Let $M = M_Q$ and define M^2 by $M^2y = M(My)$, $\cdots$, $M^sy = M(M^{s-1}y)$. Let $Q^{[1]} = Q$ and for $s \in \mathbb{N}_2$, let $Q^{[s]}$ denote the block diagonal matrix*

$$Q^{[s]} = \begin{bmatrix} Q & & \\ & \ddots & \\ & & Q \end{bmatrix},$$

where there are s matrices Q on the diagonal and all other entries in this $sn \times sn$ matrix are zero except for the entries in positions $(n, n+1)$, $(2n, 2n+1)$, $\cdots$, $((s-1)n, (s-1)n+1)$, these are all equal to 1. Then the matrices $Q^{[s]}$ are in $Z_{sn}(J)$, and are Lagrange symmetric and the symmetric differential expression M^s is given by

$$M^s = M_{Q^{[s]}}, \quad s \in \mathbb{N}.$$

PROOF. The Lagrange symmetry follows from the characterization

$$q_{ij} = (-1)^{i+j-1}\overline{q}_{n+1-j,n+1-i}.$$

The construction follows from a direct computation using the construction of M in Section 2.4. Note that for the construction of M^2, y is replaced by My due to the 1 in the $(n, n+1)$ position to get M^2, then My is replaced by M^2y and the construction is again repeated using a 1 in the $(2n, 2n+1)$ position to obtain M^3, etc. □

REMARK 4.3.1. It is interesting to observe that if, in the above construction, Q is of GN type, the matrices $Q^{[s]}$ for $s > 1$ *are not of GN type.* Thus the theory developed by Naimark [**440**] and by Glazman [**239**] for matrices Q of GN type does not directly apply to $M^s = M_{Q^{[s]}}$ for $s > 1$ but the extension of this theory developed by Zettl [**607**] and by Everitt and Zettl [**200**] does apply. The extended theory in [**607**], [**200**] uses some of the abstract Hilbert space results of Glazman [**239**], [**240**] which were developed by Glazman in extending the Weyl limit-point, limit-circle theory from $n = 2$ to general even order $n = 2k$ using Hilbert space methods rather than extending the Weyl concentric circles approach.

REMARK 4.3.2. We also note that in the classical theory the minimal domain contains the C_0^∞ functions and is therefore dense. With just local integrability assumptions on the coefficients the minimal domain may not contain the C_0^∞ functions. With just local integrability assumptions on the coefficients it is, in general, not easy to find non zero functions in the minimal domain. Nevertheless, Naimark [**440**] contains a proof that the domain of the minimal operator is dense under only local integrability conditions on the coefficients for expressions of GN type and Zettl [**607**] has shown that Naimark's proof extends to the expressions used here.

REMARK 4.3.3. Many authors who have studied products and powers of differential expressions have placed strong differentiability assumptions on the coefficients, in many cases it was assumed that the coefficients are C^∞ e.g. Dunford and Schwarz [**111**]. These assumptions were then used only to construct these powers and products in the classical way. It follows from our development based on $Z_n(J)$ and the associated quasi-derivatives that (i) the results of these authors are valid with just the local integrability assumptions used here and (ii) these results hold for the much more general differential expressions developed here. In particular, these conclusions apply to the following papers: [**99**], [**126**], [**324**], [**325**], [**326**], [**144**], [**140**], [**139**], [**167**], [**168**].

There are not many sufficient conditions on the coefficients known for the maximal deficiency case. It is well known that if all solutions of equation (4.1.1) are in $L^2(J, w)$ for some $\lambda \in \mathbb{C}$, then this is true for all λ real or complex. Next we strengthen this well known result thereby constructing higher order equations with maximal deficiency index. The elementary proof is based on two lemmas which we establish first. Note that the weight function $w = 1$ in this section.

LEMMA 4.3.1. *Let $M = M_Q$ with Q a Lagrange symmetric matrix in $Z_n(J)$, λ a complex number and p a real polynomial. If $My = \lambda y$, then $p(M)y = p(\lambda)y$.*

PROOF. Let $p(x) = a_k x^k + a_{k-1}x^{k-1} + \cdots + a_1 x + a_0$, a_j real. Then

$$\begin{aligned} p(M)y &= (a_k M^k + a_{k-1}M^{k-1} + \cdots + a_1 M + a_0)y \\ &= (a_k \lambda^k + a_{k-1}\lambda^{k-1} + \cdots + a_1\lambda + a_0)y \\ &= p(\lambda)y. \end{aligned}$$

□

LEMMA 4.3.2. *Let the hypotheses and notation of Lemma 4.3.1 hold. If* $\lambda_1, \cdots, \lambda_k$ *are distinct complex numbers and, for each* j, z_q^j *with* $q = 1, \cdots, m$ *are linearly independent solutions of* $My = \lambda_j y$, *then the set of functions* $\{z_q^j,\ j = 1, \cdots, k;\ q = 1, \cdots, m\}$ *is linearly independent.*

PROOF. Suppose there exist complex numbers such that c_q^j, $j = 1, \cdots, k$; $q = 1, \cdots, m$ such that $y_1 + \cdots + y_k = 0$ where $y_j = c_1^j z_1^j + c_2^j z_2^j + \cdots + c_m^j z_m^j$ for $j = 1, \cdots, k$. Then $M(y_1 + \cdots + y_k) = \lambda_1 y_1 + \cdots + \lambda_k y_k = 0$. Repeated applications of M yield

$$(\lambda_1)^r y_1 + \cdots + (\lambda_k)^r y_k = 0, \quad r = 0, \cdots, k-1.$$

The $k \times k$ coefficient of this homogeneous system is a Vandermonde matrix whose determinant is not zero. Hence $y_i = 0$ for $i = 1, \cdots, k$.

Therefore by the linear independence of $\{z_1^j, z_2^j, \cdots, z_m^j\}$ it follows that $c_q^j = 0$, $j = 1, \cdots, k$; $q = 1, \cdots, m$. □

THEOREM 4.3.2. *Let* Q *in* $Z_n(J)$ *be Lagrange symmetric and let* $M = M_Q$. *If for some* $\lambda \in \mathbb{C}$ *all solutions of*

$$My = \lambda y \quad on\ J$$

are in $L^2(J, w)$, *then this is true for all solutions of*

$$M^s y = \lambda y \quad on\ J$$

for every $\lambda \in \mathbb{C}$ *and every* $s \in \mathbb{N}$.

PROOF. The case $s = 1$ is well known. Choose a complex number C such that the roots, say $\lambda_1, \cdots, \lambda_k$, of the real polynomial equation $p(x) = C$ are distinct. For each $j \in \{1, \cdots, k\}$ let $z_1^j, z_2^j, \cdots, z_m^j$ be linearly independent solutions of $My = \lambda_j y$. By Lemma 4.3.1 each z_q^j is a solution of

$$p(M)y = Cy$$

and by Lemma 4.3.2 the z_q^j, $j = 1, \cdots, k$; $q = 1, \cdots, m$ form a fundamental set of solutions of $p(M)y = Cy$. Hence $p(M)$ is LC since each z_q^j is in $L^2(J, 1)$.

On the other hand if $p(M)$ is LC, choose λ_j as above and conclude that all solutions of $My = \lambda_j y$ are in $L^2(J, 1)$ i.e. M is LC in $L^2(J, 1)$. □

Next we extend the general classification result given in Section 4.2 for M to powers M^s. Recall that by the Kodaira formula the case when both endpoints are singular reduces to the case when one endpoint is regular. So we state the results for the case when a is regular; the case when b is regular is entirely similar.

THEOREM 4.3.3. *Let* $Q \in Z_n(J)$ *be L-symmetric,* $M = M_Q$, *and assume that the endpoint* a *is regular. Let* $M^s = M_{Q^{[s]}}$, $s \in \mathbb{N}_2$ *be constructed as above. Then for any polynomial* $p(x) = a_s x^s + a_{s-1} x^{s-1} + \cdots + a_1 x + a_0$ *with* $a_s \neq 0$ *and real coefficients* a_j *we have*

a: *For* $s = 2r$, $r > 0$,

$$d^+(p(M)), \quad d^-(p(M)) \geq r\,(d^+(M) + d^-(M)).$$

b: *For* $s = 2r + 1$, $r > 0$,

$$d^+(p(M)) \geq (r+1)d^+(M) + r\,d^-(M),$$

$$d^-(p(M)) \geq r d^+(M) + (r+1)d^-(M).$$

Strict inequality can occur in each of these inequalities.

PROOF. See [**607**], [**608**]; see also [**326**]. □

COROLLARY 4.3.1. *If for some $s > 1$, one of $d^+(M^s)$ or $d^-(M^s)$ take on the minimum value possible according to the general classification Theorem 4.1.2, then both $d^+(M)$ and $d^-(M)$ are minimal, i.e. M is LP. In particular, if some power M^s is LP then M is LP.*

The converse of this Corollary is false in general. Next we take up the question of when does the converse hold?

DEFINITION 4.3.1. Let the hypotheses and notation of Theorem 4.3.3 hold and let $H = L^2(J,1)$, $s \in \mathbb{N}_2$. Let $p(x) = a_s x^s + a_{s-1}x^{s-1} + \cdots + a_1 x + a_0$ with $a_s \neq 0$ be a polynomial with real coefficients a_j. We say that $p(M)$ is partially separated if $f \in H$, $M^s f \in H$ together imply that $M^r f \in H$, for $r = 1, 2, \cdots, s-1$. (Note that $p(M)$ is partially separated if and only if M^s is partially separated.)

THEOREM 4.3.4. *Let the hypotheses and notation of Corollary 4.3.1 hold and assume that M^s is partially separated. Then*

a: *If $s = 2r$, then*

$$d^+(M^s) = d^-(M^s) = r\,(d^+(M) + d^-(M)).$$

b: *If $s = 2r+1$, $r > 0$, then*

$$d^+(M^s) = (r+1)d^+(M) + rd^-(M),$$
$$d^-(M^s) = rd^+(M) + (r+1)d^-(M).$$

Conversely,

c: *If, for $s = 2r$, even, either $d^+(M^s) = (r+1)d^+(M) + rd^-(M)$ or $d^-(M^s) = rd^+(M) + (r+1)d^-(M)$, then M^s is partially separated.*

d: *If, for $s = 2r+1$, $r > 0$, odd, either $d^+(M^s) = (r+1)d^+(M) + rd^-(M)$ or $d^-(M^s) = rd^+(M) + (r+1)d^-(M)$, then M^s is partially separated.*

PROOF. See [**608**], [**607**]; see also [**326**]. □

Next we state an important Corollary.

COROLLARY 4.3.2. *Let $Q \in Z_n(J)$ be L-symmetric, $M = M_Q$, and assume that the endpoint a is regular. Let $M^s = M_{Q^{[s]}}$, $s \in \mathbb{N}_2$ be constructed as above. If $d^+(M) = d^-(M) = d(M)$, as is always the case when $Q \in Z_n(J,\mathbb{R})$, then $d(M^s) = s\,d(M)$ if and only if M^s is partially separated.*

What are the possible sequences $d^+(M^s)$, $d^-(M^s)$, $s = 1, 2, 3, \cdots$? The next theorem answers this question for $Q \in Z_{2k}(J,\mathbb{R})$.

THEOREM 4.3.5. *Assume $Q \in Z_n(J,\mathbb{R})$, $n = 2k$, $M = M_Q$, Q is Lagrange symmetric and a is a regular endpoint. Let $M^m = M_{Q^{[m]}}$, $r_0 = 0$, $r_m = d(M^m)$, $m = 1, 2, 3, \cdots$. Then*

a:

$$k\,m \leq r_m \leq 2\,k\,m, \quad m = 0, 1, 2, 3, \cdots.$$

b: *The sequence*

$$\{s_m = r_m - r_{m-1}\}_{m=1}^{\infty}$$

is nondecreasing.

c: *Given any sequence of integers* $\{r_m\}_{m=0}^{\infty}$ *satisfying* (*a*) *and* (*b*) *there exists an* $M = M_Q$ *with* $Q \in Z_{2k}(J, \mathbb{R})$ *such that* $r_m = d(M^m)$, $m = 1, 2, 3, \cdots$.

PROOF. Part (a) follows from Theorem 4.3.3 applied to M^m. The proof for part (b) given in [**326**] for the classical case with smooth coefficients readily adapts to the quasi-differential expressions discussed here; it is based on a method of Kauffman [**325**]. Examples for part (c) are also constructed in [**326**]. □

Next we give two well known examples to illustrate some of the above results and make some comments about some of their other interesting features. See [**326**] for many other examples.

Let

$$My = -y'' - qy, \quad q(t) = t^{\alpha}, \quad t \in J = (1, \infty). \tag{4.3.1}$$

We state the next results as theorems even though they are Corollaries of either well known theorems or of theorems from above, then comment on some of their interesting features.

THEOREM 4.3.6. *Consider equation* (4.3.1).

(1) *If* $\alpha \leq 2$ *then* $d(M) = 1$ *in* $L^2(J, 1)$, *i.e.* M *is LP at* ∞.
(2) *If* $\alpha > 2$, *then* $d(M) = 2$ *in* $L^2(J, 1)$, *i.e.* M *is LC at* ∞.
(3) *If* $\alpha \leq 2$ *then* $d(M^s) = s$ *in* $L^2(J, 1)$, *i.e.* M^s *is LP at* ∞ *for all* $s = 1, 2, 3, \cdots$.
(4) *If* $\alpha > 2$, *then* $d(M^s) = 2s$ *in* $L^2(J, 1)$, *i.e.* M^s *is LC at* ∞ *for all* $s = 1, 2, 3, \cdots$.

PROOF. We give references for the proofs and make comments for each part.

(1) This is a special case of the well known Levinson LP condition. Together with part (2) it shows that $\alpha = 2$ is the critical exponent which separates the LP and LC cases. Evans and Zettl in [**139**] extended the well known Levinson condition for $d(M) = 1$ and showed that the extended version is sufficient for all powers of M to be in the LP case.
(2) This is well known; see [**326**] for more information.
(3) This is established by Evans and Zettl in [**139**]; see also [**326**].
(4) This follows from Theorem 4.3.4.

□

Consider

$$My = y^{(4)} + qy, \quad q(t) = t^{\alpha}, \quad t \in J = (1, \infty). \tag{4.3.2}$$

Hinton [**293**] found that $\alpha = 4/3$ is the critical constant which separates the LP and non LP cases for M in (4.3.2) and Evans and Zettl [**144**] extended this to M^s for all $s \in \mathbb{N}_2$.

THEOREM 4.3.7 (Hinton). *For equation* (4.3.2):

(1) $d(M) = 2$ *in* $L^2(J, 1)$, *i.e.* M *is LP at* ∞ *if and only if* $\alpha \leq \frac{4}{3}$.
(2) *If* $\alpha \leq \frac{4}{3}$, *then all powers of* M *are LP at* ∞.

PROOF. We give references for the proofs and make comments for each part.

(1) The critical constant was found by Hinton [**293**], see [**326**] for a general discussion including extensions of various kinds.
(2) This is established by Evans and Zettl [**144**], see [**326**] for a general discussion including various other extensions of this kind.

□

4. Complex Parameter Decompositions of the Maximal Domain

The von Neumann formula for the domain of the adjoint of a symmetric operator in Hilbert space is fundamental for the study of self-adjoint and symmetric operators.

THEOREM 4.4.1 (von Neumann). *Let T be a closed densely defined symmetric operator on a complex Hilbert space H, and let N_+ and N_- be the deficiency spaces of T. Then we have*

$$D(T^*) = D(T) \dot{+} N_+ \dot{+} N_- \tag{4.4.1}$$

An operator S is a closed symmetric extension of T if and only if there exist closed subspaces F_+ of N_+ and F_- of N_- and an isometric mapping V of F_+ onto F_- such that

$$D(S) = D(T) + \{g + Vg : \ g \in F_+\}.$$

Furthermore, S is self-adjoint if and only if $F_+ = N_+$ and $F_- = N_-$.

PROOF. For the definition of deficiency spaces and a proof of this theorem see Dunford and Schwartz [**111**], Naimark [**440**], or Weidmann [**573**]. □

The von Neumann formula can be applied to get the following decomposition of $D(S_{\max})$.

THEOREM 4.4.2. *Suppose $Q \in Z_n(J)$ is Lagrange symmetric, $M = M_Q$ and let $S_{\min} = S_{\min}(Q)$. Assume the deficiency index of M is d, i.e. $d^+ = d^- = d$. Then*

$$D(S_{\max}) = D(S_{\min}) \dot{+} N_\lambda \dot{+} N_{\overline{\lambda}}, \quad \mathrm{Im}(\lambda) \neq 0, \tag{4.4.2}$$

where

$$N_\lambda = \{y \in D(S_{\max}) : M_Q\, y = \lambda w\, y, \ \ \mathrm{Im}(\lambda) \neq 0\}. \tag{4.4.3}$$

Since the deficiency index is d the equation $My = M_Q\, y = \lambda w\, y$ has exactly d linearly independent solutions on $J = (a, b)$ for every λ with $\mathrm{Im}(\lambda) \neq 0$. Thus it is clear from (4.4.2) that $D_{\max}$ is a $2d$ dimensional extension of $D_{\min}$. Therefore $S_{\min}$ has self-adjoint extensions and every self-adjoint extension is a d dimensional extension. Furthermore, every d dimensional symmetric extension of $S_{\min}$ is self-adjoint. Moreover, every symmetric extension of $S_{\min}$ is an m dimensional extension with

$$0 \leq m \leq d$$

and an $l = 2d - m$ dimensional restriction of $S_{\max}$ with

$$d \leq l \leq 2d.$$

PROOF. This decomposition of $D_{\max}$ is well known [**573**], [**621**], [**440**] and the furthermore and moreover statements follow from the von Neumann Theorem. □

The deficiency spaces in the von Neumann formula (4.4.2) consist of solutions of the equation $My = \lambda w y$, $\mathrm{Im}(\lambda) \neq 0$, on the whole interval $J = (a, b)$. These solutions may be very different near the two endpoints a, b of the interval J. To take this different behavior at the endpoints into account we develop two different decompositions of $D_{\max}$ each of which replaces the direct sum $N_\lambda \dotplus N_{\overline{\lambda}}$ in formula (4.4.2) with

$$\mathrm{span}\{u_1, \cdots, u_{m_a}\} \dotplus \mathrm{span}\{v_1, \cdots, v_{m_b}\}$$

where - in this chapter - the u_i are maximal domain functions which are solutions for some complex number λ_a, $\mathrm{Im}(\lambda_a) \neq 0$, near the endpoint a and are identically 0 near b while v_i are maximal domain functions which are solutions for some complex number λ_b, $\mathrm{Im}(\lambda_b) \neq 0$, near the endpoint b and are identically 0 near a.

The next theorem gives the first of these two decompositions of $D_{\max}$. The second decomposition will be given in Part II. In the second decomposition the u_i, v_i will also be maximal domain functions which are solutions near a, and b but for some *real* value λ_a, λ_b. This is an important difference which will be used to obtain information about the spectrum of self-adjoint operators. (We will continue to use the notation u_i, v_i for these real λ solutions in Part II.)

NOTATION 4.4.1. Let $a < c < b$. Consider equation $M_Q y = \lambda w y$. Note that if $Q \in Z_n(J)$, then it follows that $Q \in Z_n(a, c)$, $Q \in Z_n(c, b)$ and we can study this equation on (a, c) and (c, b) as well as on $J = (a, b)$. Note that from the definition of $Z_n(a, c)$ and $Q \in Z_n(c, b)$ it follows that c is a regular endpoint for both intervals. Also the minimal and maximal operators are defined for these two subintervals and we can study the operator theory generated by M in the Hilbert spaces $L^2((a, c), w)$ and $L^2((c, b), w)$. Below we will use the notation $S_{\min}(I)$, $S_{\max}(I)$ for the minimal and maximal operators on the interval I for $I = (a, c)$, $I = (c, b)$, $I = (a, b) = J$. The interval I may be omitted from this notation when it is clear from the context. So we make the following definition.

DEFINITION 4.4.1. Let $a < c < b$. Let d_a^+, d_b^+ denote the dimension of the solution space of $My = i w y$ lying in $L^2((a, c), w)$ and $L^2((c, b), w)$, respectively, and let d_a^-, d_b^- denote the dimension of the solution space of $My = -i w y$ lying in $L^2((a, c), w)$ and $L^2((c, b), w)$, respectively. Then d_a^+ and d_a^- are called the positive deficiency index and the negative deficiency index of $S_{\min}(a, c)$, respectively. Similarly for d_b^+ and d_b^-. Also d^+, d^- denote the deficiency indices of $S_{\min}(a, b)$; these are the dimensions of the solution spaces of $My = iwy$, $My = -iwy$ lying in $L^2((a, b), w)$. If $d_a^+ = d_a^-$, then the common value is denoted by d_a and is called the deficiency index of $S_{\min}(a, c)$, or the deficiency index at a. Similarly for d_b. Note that d_a, d_b are independent of c. If $d^+ = d^-$, then we denote the common value by d and call it the deficiency index of $S_{\min}(a, b)$ or just of $S_{\min}$. Below when we speak of 'the deficiency index' d or d_a, d_b it is automatically assumed that $d^+ = d^- = d$, $d_a^+ = d_a^- = d_a$, $d_b^+ = d_b^- = d_b$.

REMARK 4.4.1. It is well known that $S_{\min}(I)$ has a self-adjoint extension if and only if $d^+(I) = d^-(I) = d(I)$, the deficiency index. In this case $My = \lambda w y$ has exactly $d(I)$ linearly independent solutions in $L^2(I, w)$ for any λ with $\mathrm{Im}(\lambda) \neq 0$.

The relationships between d_a, d_b and d are well known and summarized in the next lemma along with some additional information.

LEMMA 4.4.1. *For $d_a^+, d_b^+, d_a^-, d_b^-, d^+, d^-, d_a, d_b$ defined as in Definition 4.4.1, we have*

(1) $d^+ = d_a^+ + d_b^+ - n$, $d^- = d_a^- + d_b^- - n$;
(2) *if* $d_a^+ = d_a^- = d_a, d_b^+ = d_b^- = d_b$, *then* $[\frac{n+1}{2}] \leq d_a, d_b \leq n$;
(3) $m_a = 2d_a - n \leq d_a$, $m_b = 2d_b - n \leq d_b$, $m_a + m_b = 2d$.
(4) *the minimal operator $S_{\min}$ has self-adjoint extensions in H if and only if $d^+ = d^-$, in this case we let $d = d^+ = d^-$. If $d = 0$ then $S_{\min}$ is self-adjoint with no proper self-adjoint extension. In all other cases $S_{\min}$ has an uncountable number of symmetric extensions, i.e. there are an uncountable number of symmetric operators S in H satisfying*

$$S_{\min} \subset S \subset S^* \subset S_{\max}.$$

PROOF. This is well known, e.g. see the book [**574**]. □

In the following, we will state the above mentioned decomposition of $D_{\max}$ when both endpoints are singular; this reduces to the cases when one or both are regular. Also, it reduces to the well known Sun [**528**] result when M is a special symmetric expression considered by Sun and one endpoint is regular.

Here we first extend the Sun decomposition theorem [**528**] to the general symmetric expression with one regular endpoint and one singular endpoint. Let a be regular, b singular and $a < c < b$. By the Naimark Patching Lemma, we may choose functions $z_j \in D_{\max}$, $j = 1, \cdots, n$ such that $z_j(t) = 0$ for $t \geq c$ and $z_j^{[k-1]}(a) = \delta_{jk}$, $j, k = 1, \cdots, n$, where δ_{jk} is the Kronecker δ.

THEOREM 4.4.3 (Sun). *Let the endpoint a be regular, b singular, d denote the deficiency index and let $m = 2d - n$. Let $\lambda \in \mathbb{C}$, $\mathrm{Im}(\lambda) \neq 0$. Then there exist solutions ϕ_j, $j = 1, \cdots, m$ of $My = \lambda w y$ such that the $m \times m$ matrix $([\phi_i, \phi_j](b))_{1 \leq i, j \leq m}$ is nonsingular and*

$$D_{\max} = D_{\min} \dotplus \mathrm{span}\{z_1, z_2, \cdots, z_n\} \dotplus \mathrm{span}\{\phi_1, \phi_2, \cdots, \phi_m\}. \tag{4.4.4}$$

PROOF. The proof given in Sun [**528**] can be adapted to fit our hypotheses. Here, we just give a brief outline of the proof. Let $\phi_1, \cdots, \phi_d$ be d linearly independent solutions of $My = \lambda w y$, and $\phi_{d+1}, \phi_{d+2}, \cdots, \phi_{2d}$ be d linearly independent solutions of $My = \overline{\lambda} w y$.

By Sun's method, we know that there exist m linearly independent solutions, say $\phi_1, \cdots, \phi_m$, such that the $m \times m$ matrix $([\phi_i, \phi_j](b))_{1 \leq i, j \leq m}$ is nonsingular and each ϕ_j $(j = m+1, \cdots, d, d+1, \cdots, 2d)$ has a unique representation:

$$\phi_j = y_j + \sum_{k=1}^{n} a_{jk} z_k + \sum_{s=1}^{m} b_{js} \phi_s, \quad j = m+1, \cdots, 2d, \tag{4.4.5}$$

where $y_j \in D_{\min}$, $a_{jk}, b_{js} \in \mathbb{C}$. By Theorem 4.4.2, every $y \in D_{\max}$ can be uniquely written as

$$y = y_0 + \sum_{j=1}^{2d} c_j \phi_j, \tag{4.4.6}$$

where $y_0 \in D_{\min}$, $c_j \in \mathbb{C}$. Substitute (4.4.5) into (4.4.6), and note that the uniqueness of the representations of y and $\phi_j (j = m+1, \cdots, 2d)$, imply that

$$D_{\max} \subset D_{\min} \dotplus \mathrm{span}\{z_1, z_2, \cdots, z_n\} \dotplus \mathrm{span}\{\phi_1, \phi_2, \cdots, \phi_m\}.$$

Therefore (4.4.4) holds since $D_{\min}$ and $\operatorname{span}\{z_1, z_2, \cdots, z_n\}$, $\operatorname{span}\{\phi_1, \phi_2, \cdots, \phi_m\}$ are contained in $D_{\max}$. □

The next theorem states the decomposition of $D_{\max}$ when both endpoints are singular.

THEOREM 4.4.4. *Let $Q \in Z_n(J)$, $J = (a, b)$, $-\infty \leq a < b \leq \infty$, be Lagrange symmetric, w a weight function, and let $My = M_Q y = \lambda w y$ be the corresponding symmetric differential equation. Let $a < c < b$. Then $Q \in Z_n((a, c))$, $Q \in Z_n((c, b))$. Let d_a, d_b and d be the deficiency indices of $My = \lambda w y$ on (a, c), (c, b) and (a, b); let $m_a = 2d_a - n$, $m_b = 2d_b - n$. Fix $\lambda_a \in \mathbb{C}$, $\lambda_b \in \mathbb{C}$ with $\operatorname{Im}(\lambda_a) \neq 0 \neq \operatorname{Im}(\lambda_b)$. Then*

a:

(1) *there exist linearly independent solutions $u_1, \cdots, u_{m_a}$ of $My = \lambda_a w y$ on (a, c) such that the $m_a \times m_a$ matrix*

$$E_{m_a} = \begin{pmatrix} [u_1, u_1] & \cdots & [u_{m_a}, u_1] \\ \cdots & \cdots & \cdots \\ [u_1, u_{m_a}] & \cdots & [u_{m_a}, u_{m_a}] \end{pmatrix} (a)$$

is nonsingular;

(2) *$u_1, \cdots, u_{m_a}$ can be extended to (a, b) such that the extended functions, still denoted by $u_1, \cdots, u_{m_a}$, are in $D_{\max}(a, b)$ and are identically 0 near b;*

(3) *$u_1, \cdots, u_{m_a}$ are linearly independent modulo $D_{\min}$;*

b:

(1) *there exist linearly independent solutions $v_1, \cdots, v_{m_b}$ of $My = \lambda_b w y$ on (c, b) such that the $m_b \times m_b$ matrix*

$$E_{m_b} = \begin{pmatrix} [v_1, v_1] & \cdots & [v_{m_b}, v_1] \\ \cdots & \cdots & \cdots \\ [v_1, v_{m_b}] & \cdots & [v_{m_b}, v_{m_b}] \end{pmatrix} (b)$$

is nonsingular;

(2) *$v_1, \cdots, v_{m_b}$ can be extended to (a, b) such that the extended functions, still denoted by $v_1, \cdots, v_{m_b}$, are in $D_{\max}(a, b)$ and are identically 0 near a;*

(3) *$v_1, \cdots, v_{m_b}$ are linearly independent modulo $D_{\min}$;*

c:

The maximal domain can has the following representation:

$$(4.4.7) \qquad D_{\max}(a, b) = D_{\min}(a, b) \dot{+} \operatorname{span}\{u_1, \cdots, u_{m_a}\} \dot{+} \operatorname{span}\{v_1, \cdots, v_{m_b}\}.$$

PROOF. Part (1) can be obtained by Theorem 4.4.3. Part (2) follows from (1) and the Naimark Patching Lemma; part (3) follows from (1) and (2). It follows from parts (a) and (b) that $u_1, \cdots, u_{m_a}, v_1, \cdots, v_{m_b} \in D_{\max}$ and are linearly independent modulo $D_{\min}$. From this and the von Neumann formula we obtain

$$\dim(D_{\max}/D_{\min}) = 2d = m_a + m_b.$$

This completes the proof. □

The next two Corollaries are immediate consequences of Theorem 4.4.4. Note that if a is either a regular or LC endpoint then $d_a = n$. Similarly, $d_b = n$ if b is either a regular or LC endpoint. Recall that a singular endpoint a is LC if all solutions on (a,c) are in $L^2((a,c),w)$. Similarly for the endpoint b. This for any $c \in (a,b)$.

COROLLARY 4.4.1. *Let the hypotheses and notation of Theorem 4.4.4 hold and assume that* $d_a = n$. *Then* (4.4.7) *holds for any linearly independent solutions* $u_1, \cdots, u_n$ *on* (a,c) *and for* $v_1, \cdots, v_{m_b}$, $m_b = 2d_b - n$ *solutions on* (c,b), *for any* c, $a < c < b$, *chosen and extended to be in* $D_{\max}(a,b)$ *as in Theorem* 4.4.4.

COROLLARY 4.4.2. *Let the hypotheses and notation of Theorem 4.4.4 hold and assume that* $d_b = n$. *Then* (4.4.7) *holds for any linearly independent solutions* $v_1, \cdots, v_n$ *on* (c,b) *and for* $u_1, \cdots, u_{m_a}$, $m_a = 2d_a - n$ *solutions on* (a,c), *for any* c, $a < c < b$, *chosen and extended to be in* $D_{\max}(a,b)$ *as in Theorem* 4.4.4.

The next theorem extends the well known characterization of the minimal operator

$$D_{\min} = \{y \in D_{\max} : y^{[i]}(a) = 0 = y^{[i]}(b),\ i = 0, \cdots, n-1\}$$

from regular endpoints to singular endpoints.

THEOREM 4.4.5. *Let the notation and hypotheses of Theorem 4.4.4 hold. Then*

$$\begin{aligned} D_{\min} = \{y \in D_{\max} : {} & [y,u_j](a) = 0, \quad for\ j = 1, \cdots, m_a; \\ & [y,v_j](b) = 0, \quad for\ j = 1, \cdots, m_b\}. \end{aligned} \tag{4.4.8}$$

PROOF. Recall that in the decomposition of $D_{\max}$ given by (4.4.7) the u_j are identically 0 in a neighborhood of b and the v_j are identically zero in a neighborhood of a. Note the characterization of the minimal domain $D(S_{\min})$ given by Theorem 3.4.2, i.e.

$$D(S_{\min}) = \{y \in D_{\max} : [y,z](a) = 0 = [y,z](b),\ for\ all\ z \in D_{\max}\}. \tag{4.4.9}$$

In the following, we prove (4.4.8) is equivalent to (4.4.9). Let $y \in D_{\min}$, i.e. $[y,u_j](a) = 0$, $j = 1, \cdots, m_a$; and $[y,v_j](b) = 0$, $j = 1, \cdots, m_b$. For any $z \in D_{\max}$, let $z = z_0 + c_1u_1 + \cdots + c_{m_a}u_{m_a} + h_1v_1 + \cdots + h_{m_b}v_{m_b}$. Then

$$[y,z](b) = \sum_{j=1}^{m_b} \overline{h}_j[y,v_j](b) = 0, \qquad [y,z](a) = \sum_{j=1}^{m_a} \overline{c}_j[y,u_j](a) = 0$$

and hence $y \in D(S_{\min})$.

Conversely, we assume that $y \in D(S_{\min})$, then for all $z \in D_{\max}$, $[y,z](b) = [y,z](a) = 0$. Therefore for the functions $u_j, j = 1, 2, \cdots, m_a$, we have $[y,u_j](a) = 0$. Similarly, $[y,v_j](b) = 0$, *for* $j = 1, \cdots, m_b$. This shows that $y \in D_{\min}$. This completes the proof. □

The Singular Patching Theorem established next is used below in the proof of the symmetric domain characterization.

THEOREM 4.4.6 (Singular Patching Theorem). *Let the hypotheses and notation of Theorem 4.4.4 hold. For any complex numbers* $\alpha_1, \alpha_2, \cdots, \alpha_{m_a}, \beta_1, \beta_2, \cdots, \beta_{m_b}$, *there exists* $y \in D_{\max}(a,b)$ *such that*

$$\begin{aligned} &[y,u_1](a) = \alpha_1, \quad [y,u_2](a) = \alpha_2, \quad \cdots, \quad [y,u_{m_a}](a) = \alpha_{m_a}, \\ &[y,v_1](b) = \beta_1, \quad [y,v_2](b) = \beta_2, \quad \cdots, \quad [y,v_{m_b}](b) = \beta_{m_b}. \end{aligned}$$

PROOF. Consider the equation

$$\begin{pmatrix} [u_1,u_1](a) & \cdots & [u_{m_a},u_1](a) \\ \cdots & \cdots & \cdots \\ [u_1,u_{m_a}](a) & \cdots & [u_{m_a},u_{m_a}](a) \end{pmatrix} \begin{pmatrix} c_1 \\ \vdots \\ c_{m_a} \end{pmatrix} = \begin{pmatrix} \alpha_1 \\ \vdots \\ \alpha_{m_a} \end{pmatrix}$$

i.e.

$$E_{m_a} \begin{pmatrix} c_1 \\ \vdots \\ c_{m_a} \end{pmatrix} = \begin{pmatrix} \alpha_1 \\ \vdots \\ \alpha_{m_a} \end{pmatrix}. \tag{4.4.10}$$

Since E_{m_a} is nonsingular, equation (4.4.10) has a unique solution

$$\begin{pmatrix} c_1 \\ \vdots \\ c_{m_a} \end{pmatrix} = E_{m_a}^{-1} \begin{pmatrix} \alpha_1 \\ \vdots \\ \alpha_{m_a} \end{pmatrix}.$$

Similarly, by the fact that E_{m_b} is nonsingular, the following equation

$$\begin{pmatrix} [v_1,v_1](b) & \cdots & [v_{m_b},v_1](b) \\ \cdots & \cdots & \cdots \\ [v_1,v_{m_b}] & \cdots & [v_{m_b},v_{m_b}](b) \end{pmatrix} \begin{pmatrix} h_1 \\ \vdots \\ h_{m_b} \end{pmatrix} = \begin{pmatrix} \beta_1 \\ \vdots \\ \beta_{m_b} \end{pmatrix},$$

has a unique solution

$$\begin{pmatrix} h_1 \\ \vdots \\ h_{m_b} \end{pmatrix} = E_{m_b}^{-1} \begin{pmatrix} \beta_1 \\ \vdots \\ \beta_{m_b} \end{pmatrix}.$$

Let

$$y = y_0 + c_1 u_1 + \cdots + c_{m_a} u_{m_a} + h_1 v_1 + \cdots + h_{m_b} v_{m_b},$$

where $y_0 \in D_{\min}$. By the decomposition (4.4.7), we have $y \in D_{\max}(a,b)$, and then

$$[y,u_1](a) = c_1[u_1,u_1](a) + c_2[u_2,u_1](a) + \cdots + c_{m_a}[u_{m_a},u_1](a) = \alpha_1,$$
$$[y,u_2](a) = c_1[u_1,u_2](a) + c_2[u_2,u_2](a) + \cdots + c_{m_a}[u_{m_a},u_2](a) = \alpha_2,$$
$$\cdots\cdots\cdots$$
$$[y,u_{m_a}](a) = c_1[u_1,u_{m_a}](a) + c_2[u_2,u_{m_a}](a) + \cdots + c_{m_a}[u_{m_a},u_{m_a}](a) = \alpha_{m_a}.$$

Similarly,

$$[y,v_1](b) = \beta_1, \quad [y,v_2](b) = \beta_2, \quad \cdots, \quad [y,v_{m_b}](b) = \beta_{m_b}.$$

This completes the proof. □

5. Comments

1. The deficiency index d determines the number of independent boundary conditions required to obtain a self-adjoint operator.

2. The proof of the Deficiency Index Conjecture is based on the 2018 paper [**560**] by the authors. This proof was discovered by the authors while writing this book. This came as a very pleasant surprise since many people worked on this Conjecture after the landmark results of Kogan and Rofe-Beketov [**338**], [**337**] in 1975, 1976, and Gilbert [**237**], [**238**] in 1978, 1979.

As mentioned in the Preface, Everitt and Markus in their book [**184**] say: "... the Deficiency Index Conjecture is essential for a satisfactory completion of the conclusions of Section V. (Section V is their Chapter V.) So their conclusions in Section V of [**184**] can now be considered 'completed'.

3. This section is based largely on the monograph by Kauffman-Read-Zettl [**326**] and on the paper [**615**] for constructing powers of differential expressions without smoothness assumptions.

4. The decomposition of the maximal domain in terms of solutions on the intervals (a,c) and (c,b) with $a < c < b$ given here is an extension of the well known paper of Sun [**528**] from one singular endpoint to two singular endpoints and from a special class of expressions M to the general class used here.

5. Theorems 4.4.5 and 4.4.6 were established by the authors in [**562**]. These results play a critical role in the characterization of symmetric domains given in Chapter 6 below. In particular, Theorem 4.4.6: the Singular Patching Theorem.

Part 2

Symmetric, Self-Adjoint, and Dissipative Operators

CHAPTER 5

Regular Symmetric Operators

1. Introduction

In this chapter we characterize the domains of the symmetric realizations S of the equation

$$My = \lambda wy \quad on \quad J = (a,b), \quad -\infty \leq a < b \leq \infty, \tag{5.1.1}$$

where $M = M_Q$, $Q \in Z_n(J)$, $Q = Q^+$ and both endpoints a and b are regular. These operators S in the Hilbert space $L^2(J,w)$ satisfy

$$S_{\min} \subset S \subset S^* \subset S_{\max}.$$

The self-adjoint operators $S = S^*$ are a special case.

2. Boundary Conditions and Boundary Matrices

For regular endpoints a, b let

$$Y(a) = \begin{pmatrix} y(a) \\ y^{[1]}(a) \\ \vdots \\ y^{[n-1]}(a) \end{pmatrix}, \quad Y(b) = \begin{pmatrix} y(b) \\ y^{[1]}(b) \\ \vdots \\ y^{[n-1]}(b) \end{pmatrix}, \quad Y_{a,b} = \begin{pmatrix} Y(a) \\ Y(b) \end{pmatrix}. \tag{5.2.1}$$

Recall from Chapter 2 that at a regular endpoint the quasi-derivatives are well defined.

DEFINITION 5.2.1. A matrix $U \in M_{l,2n}$ with rank l, $0 \leq l \leq 2n$, is called a boundary condition matrix of (5.1.1), and for any $y \in D_{\max}$ the equation

$$U\,Y_{a,b} = 0 \tag{5.2.2}$$

is called a boundary condition of (5.1.1). If $l = 0$ then $U = 0$ and the boundary condition (5.2.2) is vacuous, i.e. there is no boundary condition.

REMARK 5.2.1. The hypothesis that rank$(U) = l$ is not significant since the number of equations of the linear algebra system (5.2.2) can be reduced to the number of linearly independent rows of U by elementary matrix operations.

Next we define operators $S(U)$ in $L^2(J,w)$.

DEFINITION 5.2.2. Suppose $U \in M_{l,2n}$ is a boundary condition matrix. Define an operator $S(U)$ in $L^2(J,w)$ by

$$\begin{aligned} D(S(U)) &= \{y \in D_{\max} : U\,Y_{a,b} = 0\}, \\ S(U)y &= S_{\max}y \quad for \quad y \in D(S(U)). \end{aligned} \tag{5.2.3}$$

REMARK 5.2.2. If $l = 0$ the matrix $U = 0$ and $S(U) = S_{\max}$. If $l = 2n$ and $U = I_{2n}$, the identity matrix, then $S(I_{2n}) = S_{\min}$ since $D_{\min} = \{y \in D_{\max} : Y(a) = 0 = Y(b)\}$.

LEMMA 5.2.1. *For any boundary condition matrix U, $D(S(U))$ is a linear submanifold of $D_{\max}$ and*

$$S_{\min} \subset S(U) \subset S_{\max}. \tag{5.2.4}$$

Consequently, since $S_{\max}$ is a closed finite dimensional extension of $S_{\min}$, it follows that every operator $S(U)$ is a closed finite dimensional extension of $S_{\min}$ and a closed finite dimensional restriction of $S_{\max}$.

PROOF. That $D(S(U))$ is a linear submanifold of $D_{\max}$ follows directly from its definition, $S_{\max} = S_{\min}^*$ and from the fact that every adjoint operator is closed. Hence $S_{\max}$ is a closed finite dimensional extension of $S_{\min}$ and therefore from (5.2.4) it follows that every operator $S(U)$ is a closed finite dimensional extension of $S_{\min}$ and a closed finite dimensional restriction of $S_{\max}$. □

3. Characterization of Symmetric Domains

The next theorem characterizes the boundary condition matrices U which determine all symmetric operators $S(U)$ satisfying (5.2.4). The proof is long and technical and is given with the help of several lemmas. The self-adjoint operators are a special case.

NOTATION 5.3.1. Given a matrix $U \in M_{l,2m}$ recall the notation $U = (A : B)$ where $A \in M_{l,m}$ consists of the first m columns of U in the same order as they are in U and $B \in M_{l,m}$ consists of the next m columns of U in the same order as they are in U.

THEOREM 5.3.1. *Assume $M = M_Q$, $Q \in Z_n(J)$, $Q = Q^+$, $J = (a, b)$, $-\infty \leq a < b \leq \infty$, and both a and b are regular. Suppose U is a boundary matrix with* $\text{rank}(U) = l$, $0 \leq l \leq 2n$. *Let $U = (A : B)$, $A \in M_{l,n}$, $B \in M_{l,n}$. Define the operator $S(U)$ in $L^2(J, w)$ by (5.2.3) and let*

$$C = C(A, B) = AE_nA^* - BE_nB^*, \quad E_n = ((-1)^r \delta_{r,n+1-s})_{r,s=1}^n, \text{ and let } r = \text{rank}\, C.$$

Then we have:

1. *If $l < n$, then $S(U)$ is not symmetric.*
2. *If $l = n$, then $S(U)$ is self-adjoint (and hence also symmetric) if and only if $r = 0$.*
3. *Let $l = n + s$, $0 < s < n$. Then $S(U)$ is symmetric and not self-adjoint if and only if $r = 2s$.*
4. *If $l = 2n$, then $S_{\min}$ is symmetric and has no proper symmetric extension.*

PROOF. Part (1) follows from Theorem 4.4.2. For part (4), if $l = 2n$, then $d = 0$ and it is well known that $S_{\min}$ is self-adjoint and has no proper self-adjoint extension. If $S_{\min}$ had a proper symmetric extension T then it has a self-adjoint extension as can be seen from the equivalence of (1) and (3) in Lemma 5.3.7 below.

The proof of parts (2) and (3) is long and given with the help of several lemmas, some of these may be of independent interest. These lemmas are established next. □

For the benefit of the reader the next two lemmas recall some basic facts from linear algebra which are used below. We do not have specific references for these but the discussions on pages 7-17 of Horn and Johnson [**305**] are helpful, and so is Chapter 1 of the book by Kato [**323**].

LEMMA 5.3.1. *If S is a subset of $\mathbb{C}^n$, $n \in \mathbb{N}_2$, then*

(1) *$S^\perp$ is a subspace of $\mathbb{C}^n$.*
(2) *$(S^\perp)^\perp =$ span of S.*
(3) *$(S^\perp)^\perp = S$, if S is a subspace.*
(4) *$n = \dim S^\perp + \dim(S^\perp)^\perp$.*
(5) *Suppose $A \in M_{l,m}$. Then $\mathcal{R}(A) = (\mathcal{N}(A^*))^\perp$ i.e. $Ax = y$ has a solution (not necessarily unique) if and only if $y^*z = 0$ for all $z \in \mathbb{C}^l$ such that $A^*z = 0$.*

LEMMA 5.3.2. *Let G be any invertible $p \times p$ matrix and F an $l \times p$ matrix with* rank $F = l$. *Then the following assertions are equivalent:*

(i) *$\mathcal{N}(F) \subset \mathcal{R}(GF^*)$;*
(ii) $\operatorname{rank}(FGF^*) \leq 2l - p$;
(iii) $\operatorname{rank}(FGF^*) = 2l - p$;
(iv) $\mathcal{N}(F) = GF^*\big(\mathcal{N}(FGF^*)\big)$.

The next lemma 'connects' the Lagrange Identity with the boundary condition (5.2.2).

LEMMA 5.3.3. *Assume that $U \in M_{l,2n}$,* rank $U = l$, *$n \leq l \leq 2n$. Let $y, z \in D_{\max}$ and define $Y_{a,b}$, $Z_{a,b}$ by (5.2.1). Let*

$$P = i^n \begin{pmatrix} E_n & 0 \\ 0 & -E_n \end{pmatrix} \tag{5.3.1}$$

and note that $P^{-1} = -P = P^*$. Then $S(U)$ is symmetric if and only if

$$Z_{a,b}^* P Y_{a,b} = 0, \quad \textit{for all } y, z \in D(S(U)). \tag{5.3.2}$$

PROOF. From the Lagrange Identity it follows that for any $y, z \in D_{\max}$,

$$\int_a^b \{\overline{z}My - y\overline{Mz}\} = [y, z](b) - [y, z](a).$$

Therefore, it follows from the definition of $S(U)$ that $S(U)$ is symmetric if and only if for all $y, z \in D(S(U))$,

$$(S(U)y, z) - (y, S(U)z) = \int_a^b \{\overline{z}My - y\overline{Mz}\} = [y, z](b) - [y, z](a) = 0,$$

i.e.

$$Z_{a,b}^* P Y_{a,b} = 0 \ \textit{ for all } y, z \in D(S(U)).$$

□

LEMMA 5.3.4. *Each of the following statements is equivalent to (5.3.2):*

(1) *For all $Y, Z \in \mathcal{N}(U)$, $Z^*PY = 0$;*
(2) *$\mathcal{N}(U) \perp P(\mathcal{N}(U))$;*
(3) *$P(\mathcal{N}(U)) \subset \mathcal{N}(U)^\perp = \mathcal{R}(U^*)$;*
(4) *$\mathcal{N}(U) \subset \mathcal{R}(P^{-1}U^*) = \mathcal{R}(PU^*)$.*

PROOF. First we prove (1) is equivalent to (5.3.2). For any $Y, Z \in \mathcal{N}(U)$, by the Naimark Patching Lemma 3.4.1 there exist $y, z \in D_{\max}$ such that

$$Y_{a,b} = \begin{pmatrix} y(a) \\ \vdots \\ y^{[n-1]}(a) \\ y(b) \\ \vdots \\ y^{[n-1]}(b) \end{pmatrix} = Y, \quad Z_{a,b} = \begin{pmatrix} z(a) \\ \vdots \\ z^{[n-1]}(a) \\ z(b) \\ \vdots \\ z^{[n-1]}(b) \end{pmatrix} = Z.$$

Therefore $UY_{a,b} = UY = 0$, $UZ_{a,b} = UZ = 0$. This shows $y, z \in D(S(U))$. It follows from (5.3.2) that

$$Z^*PY = Z^*_{a,b}PY_{a,b} = 0.$$

Hence, for all $Y, Z \in \mathcal{N}(U)$, $Z^*PY = 0$.

On the other hand, for any $y, z \in D(S(U))$, $UY_{a,b} = UZ_{a,b} = 0$. This shows that $Y_{a,b}, Z_{a,b} \in \mathcal{N}(U)$. If (1) holds, then $Z^*_{a,b}PY_{a,b} = 0$. Therefore, when (1) holds, then for all $y, z \in D(S(U))$, $Z^*_{a,b}PY_{a,b} = 0$.

Statements (1) and (2) are basically the same statements. The equivalence of (2) and (3) follows from Lemma 5.3.1. The equivalence of (3) and (4) can be easily obtained from the property $P^{-1} = -P$. □

THEOREM 5.3.2. *Let U be an $l \times 2n$ matrix with* rank $U = l$, *where $n \leq l \leq 2n$. Then the operator $S(U)$ is symmetric if and only if*

$$\mathcal{N}(U) \subset \mathcal{R}(PU^*),$$

where P is defined by (5.3.1).

PROOF. This follows from the previous two lemmas. □

LEMMA 5.3.5. *Suppose $A, B \in M_{l,n}$. Let $U = (A : B)$ and assume that* rank $U = l$. *Then the operator $S(U)$ is self-adjoint if and only if*

$$l = n \quad \text{and} \quad AE_nA^* = BE_nB^*.$$

PROOF. It follows from Theorem 4.4.2 and Theorem 5.3.2 that $S(U)$ is self-adjoint if and only if $S(U)$ is a n dimensional symmetric extension of the minimal operator $S_{\min}$, i.e. if and only if $l = n$ and $\mathcal{N}(U) \subset \mathcal{R}(PU^*)$. When $l = n$, then $\dim(\mathcal{N}(U)) = n$ and $\dim(\mathcal{R}(PU^*)) = n$. Hence $\mathcal{N}(U) \subset \mathcal{R}(PU^*)$ is equivalent to $\mathcal{R}(PU^*) \subset \mathcal{N}(U)$, and this is equivalent to $UPU^* = 0$, i.e. $AE_nA^* = BE_nB^*$. □

Next we study matrices U such that $(S(U))^*$ is symmetric.

THEOREM 5.3.3. *Let $U \in M_{l,2n}$, $0 \leq l \leq 2n$ and assume that* rank $U = l$. *Then*

$$D((S(U))^*) = \{z \in D_{\max} : Z_{a,b} = \begin{pmatrix} Z(a) \\ Z(b) \end{pmatrix} \in \mathcal{R}(PU^*)\}.$$

PROOF. Let $z \in D_{\max}$. Then by definition of the adjoint operator we have $z \in D((S(U))^*)$ if and only if

$$(S_{\max}y, z) = (y, S_{\max}z), \quad \textit{for all} \quad y \in D(S(U)).$$

From this and the Lagrange Identity it follows that $Z^*_{a,b}PY_{a,b} = 0$ for all $y \in D(S(U))$. Therefore $z \in D((S(U))^*)$ if and only if $Y^*_{a,b}P^*Z_{a,b} = 0$, i.e. $P^*Z_{a,b} \in \mathcal{N}(U)^\perp = \mathcal{R}(U^*)$, and the conclusion follows directly from $P^{-1} = P^*$. □

LEMMA 5.3.6. *Let* $U \in M_{l,2n}$ *and assume* rank $U = l$ *and* $0 \leq l \leq n$. *Then the following assertions are equivalent:*

(1) $(S(U))^*$ *is symmetric;*
(2) $\mathcal{N}(U) \supset \mathcal{R}(PU^*)$*;*
(3) $UPU^* = 0$.

PROOF. From Lemma 5.3.3 and Theorem 5.3.3 it follows that $(S(U))^*$ is symmetric if and only if

$$Z_{a,b}^* P Y_{a,b} = 0, \quad \text{for all } y, z \in D((S(U))^*), \tag{5.3.3}$$

where $Y_{a,b}$, $Z_{a,b} \in \mathcal{R}(PU^*)$. By Lemma 3.4.1 and Theorem 5.3.3, we obtain that (5.3.3) is equivalent to

$$Z^* P Y = 0 \quad \text{for all } Y, Z \in \mathcal{R}(PU^*). \tag{5.3.4}$$

Note that $P^{-1} = -P$. Therefore (5.3.4) is equivalent to $\mathcal{R}(PU^*) \perp \mathcal{R}(U^*)$. Since $\mathcal{R}(U^*) = (\mathcal{N}(U))^\perp$, it follows that $\mathcal{R}(PU^*) \perp (\mathcal{N}(U))^\perp$ which is equivalent to $\mathcal{N}(U) \supset \mathcal{R}(PU^*)$. This completes the proof of the equivalence of (1) and (2). The equivalence of (2) and (3) can be obtained immediately. □

LEMMA 5.3.7. *Let* $U \in M_{l,2n}$ *and assume that* rank $U = l$ *and* $n \leq l \leq 2n$. *Then the following statements are equivalent:*

(1) $S(U)$ *is a symmetric extension of the minimal operator* $S_{\min}$*;*
(2) $\mathcal{N}(U) \subset \mathcal{R}(PU^*)$*;*
(3) *There exists a* $n \times 2n$ *matrix* $\widetilde{U}$ *satisfying* rank $\widetilde{U} = n$, $\mathcal{N}(U) \subset \mathcal{N}(\widetilde{U})$ *and* $\widetilde{U} P \widetilde{U}^* = 0$*;*
(4) *There exists a* $n \times l$ *matrix* $\widetilde{V}$ *satisfying* rank $\widetilde{V} = n$ *and* $\widetilde{V} UPU^* \widetilde{V}^* = 0$*;*
(5) $\operatorname{rank}(UPU^*) = 2(l-n)$*;*
(6) $\operatorname{rank}(UPU^*) \leq 2(l-n)$*;*
(7) $\mathcal{N}(U) = PU^*(\mathcal{N}(UPU^*))$.

PROOF. The equivalence of (1) and (2) is given in Theorem 5.3.2.

(1) ⇒ (3): Note that every symmetric extension of $S_{\min}$ is a restriction of a self-adjoint extension of $S_{\min}$. By (2), $S(U)$ is a symmetric extension of $S_{\min}$, and by Lemma 5.3.5, $S(\widetilde{U})$ is self-adjoint. Therefore (3) holds.

(3) ⇒ (2): By Lemma 5.3.5 and condition (3), we obtain that $\mathcal{N}(U) \subset \mathcal{N}(\widetilde{U}) = \mathcal{R}(P\widetilde{U}^*)$. It follows from $\mathcal{N}(U) \subset \mathcal{N}(\widetilde{U})$ that

$$\mathcal{R}(\widetilde{U}^*) = \mathcal{N}(\widetilde{U})^\perp \subset \mathcal{N}(U)^\perp = \mathcal{R}(U^*).$$

Thus $\mathcal{R}(P\widetilde{U}^*) \subset \mathcal{R}(PU^*)$, and then it follows that $\mathcal{N}(U) \subset \mathcal{R}(PU^*)$. This shows that (2) holds.

(3) ⇒ (4): Since $\mathcal{N}(U) \subset \mathcal{N}(\widetilde{U})$, we have $\mathcal{R}(U^*) \supset \mathcal{R}(\widetilde{U}^*)$. Therefore there exists a $n \times l$ matrix $\widetilde{V}$ such that $\widetilde{U}^* = U^* \widetilde{V}^*$, i.e. $\widetilde{U} = \widetilde{V} U$. From $\widetilde{U} P \widetilde{U}^* = 0$, it follows that $\widetilde{V} UPU^* \widetilde{V}^* = \widetilde{U} P \widetilde{U}^* = 0$. By $\operatorname{rank} U = l$, one has $\operatorname{rank} \widetilde{V} = \operatorname{rank}(\widetilde{V} U) = \operatorname{rank} \widetilde{U} = n$.

(4) ⇒ (3): Set $\widetilde{U} = \widetilde{V} U$. Then $\widetilde{U} P \widetilde{U}^* = \widetilde{V} UPU^* \widetilde{V}^* = 0$. It follows from $\operatorname{rank} U = l$ that $\operatorname{rank} \widetilde{U} = \operatorname{rank}(\widetilde{V} U) = \operatorname{rank} \widetilde{V} = n$. For any $Y \in N(U)$, $\widetilde{U} Y = \widetilde{V} UY = 0$ which shows that $\mathcal{N}(U) \subset \mathcal{N}(\widetilde{U})$.

The equivalence of (2), (5), (6) and (7) is obvious from Lemma 5.3.2. □

Based on the above lemmas we now re-state the characterization of the symmetric domains and complete the proof of Theorem 5.3.1.

THEOREM 5.3.4. *Suppose M is a symmetric differential expression on the interval (a,b), $-\infty \leq a < b \leq \infty$, of order $n \in \mathbb{N}_2$. Define $Y_{a,b}$ by (5.2.1). Let $A, B \in M_{l,n}$ and let $U = (A : B)$. Assume U has rank l, $0 \leq l \leq 2n$. Define the operator $S(U)$ in $L^2(J,w)$ by (5.2.3) and let*

$$C = C(A,B) = AE_nA^* - BE_nB^*, \text{ and let } r = \operatorname{rank} C.$$

Then we have

(1) *If $l < n$, then $S(U)$ is not symmetric.*
(2) *If $l = n$, then $S(U)$ is self-adjoint (and hence also symmetric) if and only if $r = 0$.*
(3) *Let $l = n + s$, $0 < s \leq n$. Then $S(U)$ is symmetric if and only if $r = 2s$.*

PROOF. Part (1) follows from the abstract von Neumann formula stated by Theorem 4.4.1 and Theorem 4.4.2.

Part (2) is given by Lemma 5.3.5.

Part (3): $n < l \leq 2n$. From Lemma 5.3.7 it follows that $S(U)$ is symmetric if and only if $\operatorname{rank} C = \operatorname{rank} UPU^* = 2(l-n) = 2s$. □

4. Examples of Symmetric Operators

In this section, we give examples of regular symmetric operators.

THEOREM 5.4.1. *Let $n = 2k$ and let A and B be $k \times n$ matrices satisfying*

$$\operatorname{rank} A = \operatorname{rank} B = k,$$

$$AE_nA^* = BE_nB^* = 0,$$

where $E_n = ((-1)^r \delta_{r,n+1-s})_{r,s=1}^n$. Define

$$U_A = \begin{pmatrix} A & 0 \\ 0 & I_n \end{pmatrix}, \quad U_B = \begin{pmatrix} 0 & B \\ I_n & 0 \end{pmatrix}.$$

Then $S(U_A)$ and $S(U_B)$ are symmetric but not self-adjoint extensions of the minimal operator $S_{\min}$.

PROOF. It is clear that $\operatorname{rank} U_A = \operatorname{rank} U_B = n + k$. For U_A, let

$$\widetilde{A}_1 = \begin{pmatrix} A \\ 0 \end{pmatrix}, \quad \widetilde{B}_1 = \begin{pmatrix} 0 \\ I_n \end{pmatrix}.$$

By computation, we have

$$\begin{aligned} C_1 &= \widetilde{A}_1 E_n \widetilde{A}_1^* - \widetilde{B}_1 E_n \widetilde{B}_1^* \\ &= \begin{pmatrix} AE_nA^* & 0 \\ 0 & -E_n \end{pmatrix} \\ &= \begin{pmatrix} 0 & 0 \\ 0 & -E_n \end{pmatrix}. \end{aligned}$$

So $\operatorname{rank} C_1 = n = 2k$. Therefore by Theorem 5.3.4 the boundary conditions

$$AY(a) = 0 \text{ and } Y(b) = 0,$$

where $Y(a)$ and $Y(b)$ are defined in (5.2.1) determine the symmetric operator $S(U_A)$. For U_B, let

$$\widetilde{A}_2 = \begin{pmatrix} 0 \\ I_n \end{pmatrix}, \quad \widetilde{B}_2 = \begin{pmatrix} B \\ 0 \end{pmatrix}.$$

It is easy to show that

$$\begin{aligned} C_2 &= \widetilde{A}_2 E_n \widetilde{A}_2^* - \widetilde{B}_2 E_n \widetilde{B}_2^* \\ &= \begin{pmatrix} -BE_nB^* & 0 \\ 0 & E_n \end{pmatrix} = \begin{pmatrix} 0 & 0 \\ 0 & E_n \end{pmatrix} \end{aligned}$$

and rank $C_2 = n = 2k$. Therefore the operator $S(U_B)$ determined by the boundary conditions

$$Y(a) = 0 \;\; and \;\; BY(b) = 0$$

is symmetric. □

THEOREM 5.4.2. *Let A_1 and B_1 be $s \times n$ complex matrices where $1 \leq s < n$. Let* $\operatorname{rank}(A_1 + B_1) = s$. *Define*

$$U = \begin{pmatrix} I_n & -I_n \\ A_1 & B_1 \end{pmatrix}.$$

Then U is a boundary matrix and the operator $S(U)$ is a symmetric extension of $S_{\min}$.

PROOF. Clearly rank $U = n + s$. Let

$$A = \begin{pmatrix} I_n \\ A_1 \end{pmatrix}, \quad B = \begin{pmatrix} -I_n \\ B_1 \end{pmatrix}.$$

Compute

$$C = AE_nA^* - BE_nB^* = \begin{pmatrix} 0 & E_n(A_1 + B_1)^* \\ (A_1 + B_1)E_n & A_1E_nA_1^* - B_1E_nB_1^* \end{pmatrix}.$$

It follows from $\operatorname{rank}(A_1 + B_1) = s$ that rank $C = 2s$. Therefore by Theorem 5.3.4 the operator $S(U)$ determined by the boundary conditions

$$Y(a) = Y(b) \quad and \quad A_1Y(a) + B_1Y(b) = 0$$

is symmetric. □

5. Comments

This chapter is based largely on the Möller-Zettl papers [**428**], [**427**]. Using a completely different method Niessen and Zettl [**453**] characterized the self-adjoint boundary conditions which determine the Friedrichs extension of the minimal operator $S_{\min}$. These authors also found the boundary conditions which determine self-adjoint extensions of some other symmetric operators $S(U)$.

CHAPTER 6

Singular Symmetric Operators

1. Introduction

In this chapter we characterize the domains of the symmetric realizations S of the equation

$$My = \lambda wy \quad on\ J = (a,b),\ -\infty \le a < b \le \infty,$$

where $M = M_Q$, $Q \in Z_n(J)$, $n \in \mathbb{N}_2$, $Q = Q^+$ and the two endpoints a and b are *singular*. These operators S in the Hilbert space $L^2(J,w)$ satisfy

$$S_{\min} \subset S \subset S^* \subset S_{\max}. \tag{6.1.1}$$

Our characterization reduces to the cases when one or both endpoints are regular. And we characterize the symmetric domains in terms of maximal domain functions.

Both of these characterizations are based on the maximal domain decomposition

$$D_{\max}(a,b) = D_{\min}(a,b) \dot{+} \operatorname{span}\{u_1, \cdots, u_{m_a}\} \dot{+} \operatorname{span}\{v_1, \cdots, v_{m_b}\} \tag{6.1.2}$$

given by (4.4.7) of Theorem 4.4.4. We will use the hypotheses and notation of Theorem 4.4.4 in this chapter.

Note that decomposition (6.1.2) is very different from the decomposition (4.4.2) which is based on the abstract von Neumann formula. In (4.4.2) the solutions are defined on the whole interval (a,b). The behavior of solutions may be very different near the two endpoints; our symmetric domain characterization in this chapter will make no restriction on the behavior of the solutions near the endpoints a,b. In particular, no restriction on their asymptotic or oscillatory behavior.

2. Singular Boundary Conditions

At a singular endpoint the quasi-derivatives $y^{[i]}$ are, in general, not defined at that endpoint. From the Lagrange Identity in Chapter 3 it follows that for any $y,z \in D_{\max}$ the Lagrange brackets $[y,z]$ are well defined at each singular endpoint. These brackets can be used to replace the quasi-derivatives as we will see below.

We start with the introduction of boundary matrices and singular boundary conditions.

DEFINITION 6.2.1. For any $y \in D_{\max}$ define

$$Y_{a,b} = \begin{pmatrix} Y(a) \\ Y(b) \end{pmatrix},\ Y(a) = \begin{pmatrix} [y,u_1](a) \\ \vdots \\ [y,u_{m_a}](a) \end{pmatrix},\ Y(b) = \begin{pmatrix} [y,v_1](b) \\ \vdots \\ [y,v_{m_b}](b) \end{pmatrix} \tag{6.2.1}$$

and recall that the Lagrange brackets $[y,u_j](a)$ and $[y,v_j](b)$ exist as finite limits.

DEFINITION 6.2.2. A matrix $U \in M_{l,2d}$ with rank l, $0 \leq l \leq 2d$, $2d = m_a + m_b$, is called a boundary condition matrix. And for $y \in D_{\max}$ and $Y_{a,b}$ given by (6.2.1) the equation

$$UY_{a,b} = 0 \tag{6.2.2}$$

is called a boundary condition. The null space of U is denoted by $\mathcal{N}(U)$, $\mathcal{R}(U)$ denotes its range and U^* is the conjugate transpose of U.

Note that any boundary condition (6.2.2) can be reduced by elementary matrix operations to the case that the rank of U is the number of its rows.

DEFINITION 6.2.3. Suppose $U \in M_{l,2d}$ is a boundary condition matrix. Define an operator $S(U)$ in $L^2(J, w)$ by

$$\begin{aligned} D(S(U)) &= \{y \in D_{\max} : UY_{a,b} = 0\}, \\ S(U)y &= S_{\max}y \quad \textit{for} \quad y \in D(S(U)). \end{aligned} \tag{6.2.3}$$

REMARK 6.2.1. From (6.2.3) for any boundary condition matrix U, $D(S(U))$ is a linear submanifold of $D_{\max}$ and we have

$$S_{\min} \subset S(U) \subset S_{\max}.$$

Consequently, since $S_{\max}$ is a closed finite dimensional extension of $S_{\min}$, it follows that every operator $S(U)$ is a closed finite dimensional extension of $S_{\min}$. For which matrices U is $S(U)$ a symmetric operator in $L^2(J, w)$? This is the question answered below.

From the von Neumann Theorem it follows that the operator $S(U)$ is not symmetric if $l < d$. But its adjoint operator $(S(U))^*$ may be symmetric. For example, when $d \neq 0$, $S_{\max}$ is not symmetric but its adjoint $S_{\min} = S_{\max}^*$ is symmetric. When $d = 0$, then $S_{\min} = S_{\max}$ and $S_{\max}$ is symmetric and self-adjoint. So we will continue to study $S(U)$ for $U \in M_{l,2d}$ with rank l for $0 \leq l \leq 2d$.

3. Symmetric Domains and Proofs

The next lemma 'connects' the Lagrange Identity with the boundary condition (6.2.2).

Here we will use E_{m_a} and E_{m_b} defined in Theorem 4.4.4. Note that E_{m_a} and E_{m_b} are nonsingular and $E_{m_a}^* = -E_{m_a}$, $(E_{m_a}^{-1})^* = -E_{m_a}^{-1}$, $E_{m_b}^* = -E_{m_b}$, $(E_{m_b}^{-1})^* = -E_{m_b}^{-1}$, $m_a + m_b = 2d$.

LEMMA 6.3.1. *Assume that* $U \in M_{l,2d}$, rank $U = l$, $d \leq l \leq 2d$. *Let* $y, z \in D_{\max}$ *and define* $Y_{a,b}$, $Z_{a,b}$ *by* (6.2.1). *Let*

$$P = \begin{pmatrix} E_{m_a}^{-1} & 0 \\ 0 & -E_{m_b}^{-1} \end{pmatrix} \tag{6.3.1}$$

and note that $P^* = -P$, $(P^{-1})^* = -P^{-1}$. *Then* $S(U)$ *is symmetric if and only if*

$$Z_{a,b}^* P Y_{a,b} = 0, \quad \textit{for all } y, z \in D(S(U)). \tag{6.3.2}$$

PROOF. By the Lagrange Identity, for any $y, z \in D_{\max}$,

$$\int_a^b \{\overline{z}My - y\overline{Mz}\} = [y, z](b) - [y, z](a).$$

Therefore, it follows from the definition of $S(U)$ given in (6.2.3) that $S(U)$ is symmetric if and only if for all $y, z \in D(S(U))$,

$$(S(U)y, z) - (y, S(U)z) = \int_a^b \{\overline{z}My - y\overline{Mz}\} = [y, z](b) - [y, z](a) = 0.$$

By the decomposition (4.4.7), functions $y, z \in D_{\max}$ can be represented as

$$y = y_0 + c_1 u_1 + \cdots + c_{m_a} u_{m_a} + h_1 v_1 + \cdots + h_{m_b} v_{m_b},$$

$$z = z_0 + \widehat{c}_1 u_1 + \cdots + \widehat{c}_{m_a} u_{m_a} + \widehat{h}_1 v_1 + \cdots + \widehat{h}_{m_b} v_{m_b},$$

where $y_0, z_0 \in D_{\min}$ and $c_j, \widehat{c}_j \in \mathbb{C}$, $j = 1, \cdots, m_a$; $h_j, \widehat{h}_j \in \mathbb{C}$, $j = 1, \cdots, m_b$.

Since

$$\begin{pmatrix} [y, v_1](b) \\ \vdots \\ [y, v_{m_b}](b) \end{pmatrix} = E_{m_b} \begin{pmatrix} h_1 \\ \vdots \\ h_{m_b} \end{pmatrix},$$

we have

$$\begin{pmatrix} h_1 \\ \vdots \\ h_{m_b} \end{pmatrix} = E_{m_b}^{-1} \begin{pmatrix} [y, v_1](b) \\ \vdots \\ [y, v_{m_b}](b) \end{pmatrix}.$$

Similarly,

$$\begin{pmatrix} \widehat{h}_1 \\ \vdots \\ \widehat{h}_{m_b} \end{pmatrix} = E_{m_b}^{-1} \begin{pmatrix} [z, v_1](b) \\ \vdots \\ [z, v_{m_b}](b) \end{pmatrix}.$$

Therefore

$$\begin{aligned} [y, z](b) =& (\overline{\widehat{h}}_1,\ \overline{\widehat{h}}_2,\ \cdots, \overline{\widehat{h}}_{m_b}) E_{m_b} \begin{pmatrix} h_1 \\ h_2 \\ \vdots \\ h_{m_b} \end{pmatrix} \\ =& -\left(\overline{[z, v_1](b)},\ \cdots,\ \overline{[z, v_{m_b}](b)} \right) E_{m_b}^{-1} \begin{pmatrix} [y, v_1](b) \\ \vdots \\ [y, v_{m_b}](b) \end{pmatrix}. \end{aligned}$$

Similarly,

$$[y, z](a) = -\left(\overline{[z, u_1](a)},\ \cdots,\ \overline{[z, u_{m_a}](a)} \right) E_{m_a}^{-1} \begin{pmatrix} [y, u_1](a) \\ \vdots \\ [y, u_{m_a}](a) \end{pmatrix}.$$

So

$$[y,z](b)-[y,z](a)=\left(\overline{[z,u_1](a)},\cdots,\overline{[z,u_{m_a}](a)},\overline{[z,v_1](b)},\cdots,\overline{[z,v_{m_b}](b)}\right)$$

$$\cdot\begin{pmatrix} E_{m_a}^{-1} & 0 \\ 0 & -E_{m_b}^{-1}\end{pmatrix}\begin{pmatrix}[y,u_1](a)\\ \vdots \\ [y,u_{m_a}](a)\\ [y,v_1](b)\\ \vdots \\ [y,v_{m_b}](b)\end{pmatrix}.$$

Hence, the operator $S(U)$ is symmetric if and only if

$$[y,z](b)-[y,z](a)=0 \quad for\ all\ y,z\in D(S(U)),$$

i.e.

$$Z_{a,b}^*PY_{a,b}=0 \quad for\ all\ y,z\in D(S(U)).$$

□

LEMMA 6.3.2. *Each of the following statements is equivalent to* (6.3.2)*:*

(1) *For all* $Y,Z\in\mathcal{N}(U)$, $Z^*PY=0$*;*
(2) $\mathcal{N}(U)\perp P(\mathcal{N}(U))$*;*
(3) $P(\mathcal{N}(U))\subset\mathcal{N}(U)^{\perp}=\mathcal{R}(U^*)$*;*
(4) $\mathcal{N}(U)\subset\mathcal{R}(P^{-1}U^*)$.

PROOF. Now we prove that the equivalence of (6.3.2) and (1). For all $Y,Z\in\mathcal{N}(U)$, let

$$Y=\begin{pmatrix}y_1\\ y_2\\ \vdots\\ y_{2d}\end{pmatrix},\quad Z=\begin{pmatrix}z_1\\ z_2\\ \vdots\\ z_{2d}\end{pmatrix}.$$

Then $UY=0$ and $UZ=0$. By the Singular Patching Theorem 4.4.6, there exist $y,z\in D_{\max}$ such that

$$Y_{a,b}=\begin{pmatrix}[y,u_1](a)\\ \vdots \\ [y,u_{m_a}](a)\\ [y,v_1](b)\\ \vdots \\ [y,v_{m_b}](b)\end{pmatrix}=Y,\quad Z_{a,b}=\begin{pmatrix}[z,u_1](a)\\ \vdots \\ [z,u_{m_a}](a)\\ [z,v_1](b)\\ \vdots \\ [z,v_{m_b}](b)\end{pmatrix}=Z,$$

and $UY_{a,b}=UZ_{a,b}=0$. Therefore $y,z\in D(S(U))$, and by (6.3.2) we have

$$Z^*PY=Z_{a,b}^*PY_{a,b}=0.$$

Hence, for all $Y,Z\in\mathcal{N}(U)$, $Z^*PY=0$.

On the other hand, for all $y,z\in D(S(U))$, $UY_{a,b}=UZ_{a,b}=0$. This shows that $Y_{a,b},Z_{a,b}\in\mathcal{N}(U)$. If (1) holds, then $Z_{a,b}^*PY_{a,b}=0$. Therefore, when (1) holds, then for all $y,z\in D(S(U))$, $Z_{a,b}^*PY_{a,b}=0$.

Statements (1) and (2) are essentially the same statements, just written differently. The equivalence of (2) and (3) follows from Lemma 5.3.1. Whereas the equivalence of (3) and (4) immediately follows from the fact that P is an invertible matrix. □

THEOREM 6.3.1. *Let U be an $l \times 2d$ matrix with* rank $U = l$, *where* $d \leq l \leq 2d$, $d = d_a + d_b - n$. *Then the operator $S(U)$ is symmetric if and only if*

$$\mathcal{N}(U) \subset \mathcal{R}(P^{-1}U^*),$$

where P is defined by (6.3.1).

PROOF. This follows from the Singular Patching Theorem 4.4.6, Lemma 6.3.1 and Lemma 6.3.2. □

LEMMA 6.3.3. *Suppose $U \in M_{l,2d}$. Let $U = (A : B)$, where $A \in M_{l,m_a}$, $B \in M_{l,m_b}$, and recall that $m_a + m_b = 2d$. Assume that* rank $U = l$. *Then the operator $S(U)$ is self-adjoint if and only if*

$$l = d \quad \textit{and} \quad UP^{-1}U^* = 0, \ \textit{i.e.}\ AE_{m_a}A^* - BE_{m_b}B^* = 0.$$

PROOF. It follows from Theorem 4.4.2 and Theorem 6.3.1 that $S(U)$ is self-adjoint if and only if $S(U)$ is a d dimensional symmetric extension of the minimal operator $S_{\min}$, i.e. if and only if $l = d$ and $\mathcal{N}(U) \subset \mathcal{R}(P^{-1}U^*)$. When $l = d$, we have $\dim(\mathcal{N}(U)) = d$ and $\dim(\mathcal{R}(P^{-1}U^*)) = d$. Hence $\mathcal{N}(U) \subset \mathcal{R}(P^{-1}U^*)$ is equivalent to $\mathcal{R}(P^{-1}U^*) \subset \mathcal{N}(U)$, and this is equivalent to $UP^{-1}U^* = 0$, i.e. $AE_{m_a}A^* - BE_{m_b}B^* = 0$. □

Next we study matrices U such that $(S(U))^*$ is symmetric.

THEOREM 6.3.2. *Let $U \in M_{l,2d}$,* $0 \leq l \leq 2d$ *and assume that* rank $U = l$. *Then*

$$D((S(U))^*) = \{z \in D_{\max} : Z_{a,b} = \begin{pmatrix} [z, u_1](a) \\ \vdots \\ [z, u_{m_a}](a) \\ [z, v_1](b) \\ \vdots \\ [z, v_{m_b}](b) \end{pmatrix} \in \mathcal{R}(P^{-1}U^*)\}.$$

PROOF. Let $z \in D_{\max}$. Then $z \in D((S(U))^*)$ if and only if

$$(S_{\max}y, z) = (y, S_{\max}z), \quad \textit{for all} \ \ y \in D(S(U)).$$

This is equivalent to $Z^*_{a,b}PY_{a,b} = 0$ for all $y \in D(S(U))$. Therefore $z \in D((S(U))^*)$ if and only if $Y^*_{a,b}P^*Z_{a,b} = 0$, i.e. $P^*Z_{a,b} \in \mathcal{N}(U)^{\perp} = \mathcal{R}(U^*)$. Hence $Z_{a,b} \in \mathcal{R}(P^{-1}U^*)$. This completes the proof. □

LEMMA 6.3.4. *Let $U \in M_{l,2d}$ and assume* rank $U = l$ *and* $0 \leq l \leq d$. *Then the following statements are equivalent:*

(1) $(S(U))^*$ *is symmetric;*
(2) $\mathcal{N}(U) \supset \mathcal{R}(P^{-1}U^*)$;
(3) $UP^{-1}U^* = 0$.

PROOF. From Lemma 6.3.1 and Theorem 6.3.2, it follows that $(S(U))^*$ is symmetric if and only if

$$Z^*_{a,b}PY_{a,b} = 0, \quad \textit{for all}\ y, z \in D((S(U))^*), \tag{6.3.3}$$

where $Y_{a,b}$, $Z_{a,b} \in \mathcal{R}(P^{-1}U^*)$ are defined as in (6.2.1). It follows from the Singular Patching Theorem 4.4.6 and Theorem 6.3.2 that (6.3.3) is equivalent to

$$Z^*PY = 0, \quad \textit{for all}\ Y, Z \in \mathcal{R}(P^{-1}U^*). \tag{6.3.4}$$

Since P is invertible, (6.3.4) is equivalent to $\mathcal{R}(P^{-1}U^*) \perp \mathcal{R}(U^*)$. By Lemma 5.3.1, this is equivalent to $\mathcal{R}(P^{-1}U^*) \perp (\mathcal{N}(U))^{\perp}$. Therefore (1) and (2) are equivalent. The equivalence of (2) and (3) can be obtained immediately. □

LEMMA 6.3.5. *Let $U \in M_{l,2d}$ and assume that* rank $U = l$ *and* $d \leq l \leq 2d = m_a + m_b$. *Then the following statements are equivalent:*

(1) *$S(U)$ is a symmetric extension of the minimal operator $S_{\min}$;*
(2) $\mathcal{N}(U) \subset \mathcal{R}(P^{-1}U^*)$*;*
(3) *There exists a $d \times 2d$ matrix $\widetilde{U}$ satisfying* rank $\widetilde{U} = d$, $\mathcal{N}(U) \subset \mathcal{N}(\widetilde{U})$ *and* $\widetilde{U}P^{-1}\widetilde{U}^* = 0$*;*
(4) *There exists a $d\times l$ matrix $\widetilde{V}$ satisfying* rank $\widetilde{V} = d$ *and* $\widetilde{V}UP^{-1}U^*\widetilde{V}^* = 0$*;*
(5) $\operatorname{rank}(UP^{-1}U^*) = 2l - (m_a + m_b) = 2(l-d)$*;*
(6) $\operatorname{rank}(UP^{-1}U^*) \leq 2l - (m_a + m_b) = 2(l-d)$*;*
(7) $\mathcal{N}(U) = P^{-1}U^*(\mathcal{N}(UP^{-1}U^*))$.

PROOF. The equivalence of (1) and (2) is given in Theorem 6.3.1.

(1) $\Rightarrow$ (3): Note that every symmetric extension of $S_{\min}$ is a restriction of a self-adjoint extension of $S_{\min}$. By (1), $S(U)$ is a symmetric extension of $S_{\min}$, and by Lemma 6.3.3, $S(\widetilde{U})$ is self-adjoint. Therefore (3) holds.

(3) $\Rightarrow$ (2): By Lemma 6.3.3 and condition (3), we obtain that $\mathcal{N}(U) \subset \mathcal{N}(\widetilde{U}) = \mathcal{R}(P^{-1}\widetilde{U}^*)$. It follows from $\mathcal{N}(U) \subset \mathcal{N}(\widetilde{U})$ that

$$\mathcal{R}(\widetilde{U}^*) = \mathcal{N}(\widetilde{U})^{\perp} \subset \mathcal{N}(U)^{\perp} = \mathcal{R}(U^*).$$

Thus $\mathcal{R}(P^{-1}\widetilde{U}^*) \subset \mathcal{R}(P^{-1}U^*)$, and then it follows that $\mathcal{N}(U) \subset \mathcal{R}(P^{-1}U^*)$. This shows that (2) holds.

(3) $\Rightarrow$ (4): Since $\mathcal{N}(U) \subset \mathcal{N}(\widetilde{U})$, we have $\mathcal{R}(U^*) \supset \mathcal{R}(\widetilde{U}^*)$. Therefore there exists a $d \times l$ matrix $\widetilde{V}$ such that $\widetilde{U}^* = U^*\widetilde{V}^*$, i.e. $\widetilde{U} = \widetilde{V}U$. From $\widetilde{U}P^{-1}\widetilde{U}^* = 0$, it follows that $\widetilde{V}UP^{-1}U^*\widetilde{V}^* = \widetilde{U}P^{-1}\widetilde{U}^* = 0$. Therefore $\operatorname{rank}\widetilde{V} = \operatorname{rank}(\widetilde{V}U) = \operatorname{rank}\widetilde{U} = d$.

(4) $\Rightarrow$ (3): Set $\widetilde{U} = \widetilde{V}U$. Then $\widetilde{U}P^{-1}\widetilde{U}^* = \widetilde{V}UP^{-1}U^*\widetilde{V}^* = 0$. It follows from $\operatorname{rank} U = l$ that $\operatorname{rank}\widetilde{U} = \operatorname{rank}(\widetilde{V}U) = \operatorname{rank}\widetilde{V} = d$. For any $Y \in \mathcal{N}(U)$, $\widetilde{U}Y = \widetilde{V}UY = 0$ which shows that $\mathcal{N}(U) \subset \mathcal{N}(\widetilde{U})$.

The equivalence of (2), (5), (6) and (7) can be obtained from the Linear Algebra Lemma 5.3.2. □

Based on the above lemmas and theorems we now obtain our main result: the characterization of symmetric operators $S(U)$ in the Hilbert space $L^2(J,w)$ determined by two-point boundary conditions.

THEOREM 6.3.3. *Let $M = M_Q$, $Q \in Z_n(J)$, $J = (a,b)$, $-\infty \leq a < b \leq \infty$, be Lagrange symmetric, w a weight function. Let d_a, d_b and d be the deficiency indices of $My = \lambda wy$ on (a,c), (c,b) and (a,b), respectively. Recall that $d = d_a + d_b - n$. Let $m_a = 2d_a - n$, $m_b = 2d_b - n$, $Y_{a,b}$ be defined by (6.2.1). Assume $U \in M_{l,2d}$ has rank l, $0 \leq l \leq m_a + m_b = 2d$ and let $U = (A : B)$ with $A \in M_{l,m_a}$ consisting of the first m_a columns of U in the same order as they are in U and $B \in M_{l,m_b}$ consisting of the next m_b columns of U in the same order as they are in U. Define the operator $S(U)$ in $L^2(J,w)$ by (6.2.3) and let*

$$C = C(A,B) = UP^{-1}U^* = AE_{m_a}A^* - BE_{m_b}B^*, \textit{ and let } r = \operatorname{rank} C.$$

Then we have

(1) *If $l < d$, then $S(U)$ is not symmetric.*
(2) *If $l = d$, then $S(U)$ is self-adjoint (and hence also symmetric) if and only if $r = 0$.*
(3) *Let $l = d + s$, $0 < s \leq d$. Then $S(U)$ is symmetric if and only if $r = 2s$.*

PROOF. Part (1) follows from the von Neumann formula.

Part (2) is given by Lemma 6.3.3.

Part (3): $d < l \leq 2d$. From Lemma 6.3.5 it follows that $S(U)$ is symmetric if and only if rank $C =$ rank $UP^{-1}U^* = 2(l - d) = 2s$. □

4. Symmetric Domain Characterization with Maximal Domain Functions

In this section we characterize the linear manifolds $D(S)$ of $D_{\max}$ which are the domains of symmetric operators S, $S_{\min} \subset S \subset S^* \subset S_{\max}$, in $L^2(J, w)$. The self-adjoint characterization is a special case.

THEOREM 6.4.1. *Let $Q \in Z_n(J)$ be a Lagrange symmetric matrix and $M = M_Q$ the corresponding symmetric differential expression of order n, even or odd, with complex coefficients and let w be a weight function. Let $a < c < b$, and denote the equal deficiency indices of M on (a, c), (c, b) by d_a, d_b, respectively. Then the deficiency index of M on (a, b) is $d = d_a + d_b - n$ by the Kodaira formula. A linear submanifold $D(S)$ of $D_{\max}$ is the domain of a symmetric extension S of $S_{\min}$ if and only if there exist functions $w_1, w_2, \cdots, w_l$ $(d \leq l \leq 2d)$ in $D_{\max}$ such that the following conditions hold:*

(i) *$w_1, w_2, \cdots, w_l$ are linearly independent modulo $D_{\min}$;*
(ii) *rank $W = 2(l - d)$, where*

$$W = \begin{pmatrix} [w_1, w_1]_a^b & [w_2, w_1]_a^b & \cdots & [w_l, w_1]_a^b \\ \cdots & \cdots & \cdots & \cdots \\ [w_1, w_l]_a^b & [w_2, w_l]_a^b & \cdots & [w_l, w_l]_a^b \end{pmatrix};$$

(iii) *$D(S)$ is given by*

$$D(S) = \{y \in D_{\max} : \ [y, w_j]_a^b = 0, \quad j = 1, 2, \cdots, l\}. \tag{6.4.1}$$

PROOF. *Sufficiency.* Assume that there exist $w_1, w_2, \cdots, w_l$ $(d \leq l \leq 2d)$ in $D_{\max}$ satisfying conditions (i) and (ii). Now we prove that $D(S)$ defined by (6.4.1) is the domain of a symmetric extension S of $S_{\min}$ satisfying (6.1.1). By (4.4.7) w_j can be represented by

(6.4.2)

$$w_j = w_{j0} + a_{j1}u_1 + a_{j2}u_2 + \cdots + a_{jm_a}u_{m_a} + b_{j1}v_1 + \cdots + b_{jm_b}v_{m_b}, \quad j = 1, 2, \cdots, l,$$

where $w_{j0} \in D_{\min}$ and $a_{j1}, \cdots a_{jm_a}$, $b_{j1}, \cdots b_{jm_b} \in \mathbb{C}$. For any $y \in D_{\max}$ and $j = 1, 2, \cdots, l$,

$$[y, w_j](a) = \overline{a}_{j1}[y, u_1](a) + \overline{a}_{j2}[y, u_2](a) + \cdots + \overline{a}_{jm_a}[y, u_{m_a}](a),$$

$$[y, w_j](b) = \overline{b}_{j1}[y, v_1](b) + \overline{b}_{j2}[y, v_2](b) + \cdots + \overline{b}_{jm_b}[y, v_{m_b}](b).$$

Let $y \in D(S)$, then $[y, w_j]_a^b = 0, \quad j = 1, 2, \cdots, l$, i.e.

$$\begin{pmatrix} [y, w_1](b) \\ \vdots \\ [y, w_l](b) \end{pmatrix} - \begin{pmatrix} [y, w_1](a) \\ \vdots \\ [y, w_l](a) \end{pmatrix}$$
$$= B \begin{pmatrix} [y, v_1](b) \\ \vdots \\ [y, v_{m_b}](b) \end{pmatrix} + A \begin{pmatrix} [y, u_1](a) \\ \vdots \\ [y, u_{m_a}](a) \end{pmatrix} = 0,$$

where

$$A = -\begin{pmatrix} \overline{a}_{11} & \cdots & \overline{a}_{1m_a} \\ \cdots & \cdots & \cdots \\ \overline{a}_{l1} & \cdots & \overline{a}_{lm_a} \end{pmatrix}, \quad B = \begin{pmatrix} \overline{b}_{11} & \cdots & \overline{b}_{1m_b} \\ \cdots & \cdots & \cdots \\ \overline{b}_{l1} & \cdots & \overline{b}_{lm_b} \end{pmatrix}.$$

Therefore the boundary condition (6.4.1) of $D(S)$ is equivalent to $(A : B)Y_{a,b} = UY_{a,b} = 0$, where $Y_{a,b}$ is defined by (6.2.1). Now we prove that the matrices A and B satisfy the following two conditions:

(1) rank $U =$ rank $(A : B) = l$;

(2) rank $C =$ rank $C(A, B) =$ rank $(AE_{m_a}A^* - BE_{m_b}B^*) = 2(l - d)$.

It is clear that rank $U \leq l$. If rank $U < l$, then there exist constants $h_1, h_2, \cdots, h_l$, not all zero, such that $(h_1, h_2, \cdots, h_l)\, U = 0$, i.e.

$$(h_1, h_2, \cdots, h_l)A = 0 \ \text{ and } \ (h_1, h_2, \cdots, h_l)B = 0. \tag{6.4.3}$$

Let $z = \overline{h}_1 w_1 + \overline{h}_2 w_2 + \cdots + \overline{h}_l w_l$. It follows from (6.4.2) that

$$z = \sum_{k=1}^{l} \overline{h}_k w_{k0} + \sum_{j=1}^{m_a} \sum_{k=1}^{l} \overline{h}_k a_{kj} u_j + \sum_{j=1}^{m_b} \sum_{k=1}^{l} \overline{h}_k b_{kj} v_j.$$

By (6.4.3), we obtain that $\sum_{j=1}^{m_a} \sum_{k=1}^{l} \overline{h}_k a_{kj} u_j = 0$ and $\sum_{j=1}^{m_b} \sum_{k=1}^{l} \overline{h}_k b_{kj} v_j = 0$. Hence $z = \sum_{k=1}^{l} \overline{h}_k w_{k0} \in D_{\min}$ contradicting that $w_1, w_2, \cdots, w_l$ are linearly independent modulo $D_{\min}$. Therefore rank $U = l$.

By (6.4.2), we have for all $k, j = 1, 2, \cdots, l$,

$$[w_k, w_j](b) = (b_{k1}, b_{k2}, \cdots, b_{km_b}) \begin{pmatrix} [v_1, v_1](b) & \cdots & [v_1, v_{m_b}](b) \\ \cdots & \cdots & \cdots \\ [v_{m_b}, v_1](b) & \cdots & [v_{m_b}, v_{m_b}](b) \end{pmatrix} \begin{pmatrix} \overline{b}_{j1} \\ \overline{b}_{j2} \\ \vdots \\ \overline{b}_{j,m_b} \end{pmatrix}.$$

Therefore

$$\begin{pmatrix} [w_1, w_1](b) & [w_2, w_1](b) & \cdots & [w_l, w_1](b) \\ \cdots & \cdots & \cdots & \cdots \\ [w_1, w_l](b) & [w_2, w_l](b) & \cdots & [w_l, w_l](b) \end{pmatrix}$$
$$= B \begin{pmatrix} [v_1, v_1](b) & \cdots & [v_{m_b}, v_1](b) \\ \cdots & \cdots & \cdots \\ [v_1, v_{m_b}](b) & \cdots & [v_{m_b}, v_{m_b}](b) \end{pmatrix} B^*$$
$$= BE_{m_b}B^*.$$

Similarly, we have

$$\begin{aligned}&\begin{pmatrix} [w_1,w_1](a) & \cdots & [w_l,w_1](a) \\ \cdots & \cdots & \cdots \\ [w_1,w_l](a) & \cdots & [w_l,w_l](a) \end{pmatrix} \\ =&A\begin{pmatrix} [u_1,u_1](a) & \cdots & [u_{m_a},u_1](a) \\ \cdots & \cdots & \cdots \\ [u_1,u_{m_a}](a) & \cdots & [u_{m_a},u_{m_a}](a) \end{pmatrix} A^* \\ =&AE_{m_a}A^*. \end{aligned}$$

Therefore

$$W = \begin{pmatrix} [w_1,w_1]_a^b & \cdots & [w_l,w_1]_a^b \\ \cdots & \cdots & \cdots \\ [w_1,w_l]_a^b & \cdots & [w_l,w_l]_a^b \end{pmatrix} = BE_{m_b}B^* - AE_{m_a}A^*.$$

Since $w_1, w_2, \cdots, w_l$ satisfy condition (ii), we have that

$$\text{rank}\,(AE_{m_a}A^* - BE_{m_b}B^*) = \text{rank}\,W = 2(l-d).$$

By Theorem 6.3.3, the operator $S(U)$ determined by the boundary condition $(A : B)Y_{a,b} = 0$ is symmetric. By the equivalence of (6.4.1) and $(A : B)Y_{a,b} = 0$, it is clear that $D(S)$ defined by (6.4.1) is the domain of a symmetric extension S of $S_{\min}$.

Necessity. Let a linear submanifold $D(S)$ of $D_{\max}$ be the domain of a symmetric extension S of $S_{\min}$. Now we prove that there exist functions $w_1, w_2, \cdots, w_l$ $(d \le l \le 2d)$ in $D_{\max}$ satisfying conditions (i), (ii) such that (iii) holds, i.e. $D(S)$ can be represented by (6.4.1).

Since S is symmetric, by Theorem 6.3.3, there exist matrices A, B of order $l \times m_a$, $l \times m_b$, respectively, such that $\text{rank}\,U = l$ and $\text{rank}\,(AE_{m_a}A^* - BE_{m_b}B^*) = 2(l-d)$, where $d \le l \le 2d$, $U = (A : B)$ and $D(S)$ can be characterized by

$$D(S) = \{y \in D_{\max} : UY_{a,b} = 0\}. \tag{6.4.4}$$

Let

$$A = -\begin{pmatrix} \bar{a}_{11} & \cdots & \bar{a}_{1m_a} \\ \cdots & \cdots & \cdots \\ \bar{a}_{l1} & \cdots & \bar{a}_{lm_a} \end{pmatrix}, \quad B = \begin{pmatrix} \bar{b}_{11} & \cdots & \bar{b}_{1m_b} \\ \cdots & \cdots & \cdots \\ \bar{b}_{l1} & \cdots & \bar{b}_{lm_b} \end{pmatrix}$$

and

$$w_k = \sum_{j=1}^{m_a} a_{kj}u_j + \sum_{j=1}^{m_b} b_{kj}v_j, \quad k = 1, 2, \cdots, l. \tag{6.4.5}$$

By direct computation, we get

$$\begin{pmatrix} [y,w_1](a) \\ \vdots \\ [y,w_l](a) \end{pmatrix} = -A\begin{pmatrix} [y,u_1](a) \\ \vdots \\ [y,u_{m_a}](a) \end{pmatrix} = -AY(a),$$

$$\begin{pmatrix} [y,w_1](b) \\ \vdots \\ [y,w_l](b) \end{pmatrix} = B\begin{pmatrix} [y,v_1](b) \\ \vdots \\ [y,v_{m_b}](b) \end{pmatrix} = BY(b).$$

Therefore

$$AY(a)+BY(b)=\begin{pmatrix}[y,w_1]_a^b\\ \vdots\\ [y,w_l]_a^b\end{pmatrix}.$$

This shows that (6.4.1) is equivalent to (6.4.4). Next we show $w_1, w_2, \cdots, w_l$ are linearly independent modulo $D_{\min}$. If not, there exist constants $c_1, c_2, \cdots, c_l$, not all zero, such that

$$\gamma = c_1w_1 + c_2w_2 + \cdots + c_lw_l \in D_{\min}.$$

Hence

$$[u_j, \gamma](a) = 0, \quad j = 1, 2, \cdots, m_a,$$

and

$$[v_j, \gamma](b) = 0, \quad j = 1, 2, \cdots, m_b.$$

From (6.4.5), we obtain

$$\gamma = \sum_{k=1}^{l} c_k(\sum_{j=1}^{m_a} a_{kj}u_j + \sum_{j=1}^{m_b} b_{kj}v_j).$$

By direct computation, we can obtain that

$$\begin{pmatrix}[u_1,\gamma](a)\\ [u_2,\gamma](a)\\ \vdots\\ [u_{m_a},\gamma](a)\end{pmatrix}=0$$

is equivalent to

$$(\bar{c}_1, \bar{c}_2, \cdots, \bar{c}_l)(-A)E_{m_a} = 0.$$

Hence $(\bar{c}_1, \bar{c}_2, \cdots, \bar{c}_l)A = 0$. Similarly

$$\begin{pmatrix}[v_1,\gamma](b)\\ [v_2,\gamma](b)\\ \vdots\\ [v_{m_b},\gamma](b)\end{pmatrix}=0$$

is equivalent to

$$(\bar{c}_1, \bar{c}_2, \cdots, \bar{c}_l)BE_{m_b} = 0.$$

Hence $(\bar{c}_1, \bar{c}_2, \cdots, \bar{c}_l)B = 0$. Therefore $(\bar{c}_1, \bar{c}_2, \cdots, \bar{c}_l)(A : B) = (\bar{c}_1, \bar{c}_2, \cdots, \bar{c}_l)U = 0$ contradicting that $\operatorname{rank} U = l$. Therefore $w_1, w_2, \cdots, w_l$ are linearly independent modulo $D_{\min}$. It follows from (6.4.5) that

$$\begin{pmatrix}[w_1,w_1](b) & \cdots & [w_l,w_1](b)\\ \cdots & \cdots & \cdots\\ [w_1,w_l](b) & \cdots & [w_l,w_l](b)\end{pmatrix} = BE_{m_b}B^*$$

and

$$\begin{pmatrix}[w_1,w_1](a) & \cdots & [w_l,w_1](a)\\ \cdots & \cdots & \cdots\\ [w_1,w_l](a) & \cdots & [w_l,w_l](a)\end{pmatrix} = AE_{m_a}A^*.$$

Therefore

$$W = \begin{pmatrix} [w_1, w_1]_a^b & \cdots & [w_l, w_1]_a^b \\ \cdots & \cdots & \cdots \\ [w_1, w_l]_a^b & \cdots & [w_l, w_l]_a^b \end{pmatrix} = BE_{m_b}B^* - AE_{m_a}A^*.$$

It follows from rank $(AE_{m_a}A^* - BE_{m_b}B^*) = 2(l-d)$ that rank $W = 2(l-d)$. This completes the proof. □

The next theorem is the self-adjoint special case of Theorem 6.4.1.

THEOREM 6.4.2. *A linear submanifold $D(S)$ of $D_{\max}$ is the domain of a self-adjoint extension S of $S_{\min}$ if and only if there exist functions $w_1, w_2, \cdots, w_d$ in $D_{\max}$ such that the following conditions hold:*

(i) *$w_1, w_2, \cdots, w_d$ are linearly independent modulo $D_{\min}$;*
(ii) *$[w_j, w_k]_a^b = 0,\ j, k = 1, 2, \cdots, d$;*
(iii) *$D(S) = \{y \in D_{\max} :\ [y, w_j]_a^b = 0, \quad j = 1, 2, \cdots, d\}$.*

PROOF. This is the special case $l = d$ of Theorem 6.4.1. □

REMARK 6.4.1. Theorem 6.4.2 is the well known GKN theorem. It characterizes all self-adjoint extensions of $S_{\min}$ for two-point boundary conditions. As mentioned above this GKN theorem has been extended in many directions: to n-point boundary conditions for any n finite or infinite [**199**], difference operators, Hamiltonian systems, etc. We expect that similar extensions will be found for the symmetric Theorem 6.4.1.

5. Comments

Theorem 6.3.3 extends Theorem 5.3.1 from regular to singular endpoints. This generalization is based on the recent paper by the authors [**557**] and uses the maximal domain representation of Section 4.4, the Singular Patching Theorem 4.4.6, and an adaptation of the Linear Algebra proof of the regular case in Chapter 5.

It is interesting to note that the proof of the singular symmetric theorem in Section 6.4 uses the singular symmetric domain characterization of Theorem 6.3.3 in contrast to the self-adjoint case where the GKN Theorem is used to prove the self-adjoint domain characterization.

Consider the equation

$$My = \lambda w y \quad on \quad J = (a, b), \quad -\infty \leq a < b \leq \infty. \tag{6.5.1}$$

As mentioned in the Preface, for the self-adjoint case it can be argued that our results in Chapters 5 and 6 culminates the program started by Naimark with Theorem 4 and Theorem 5 in 1968. A landmark paper in this period is the 1986 paper of Sun [**528**] in which Theorems 4 and 5 are extended to problems with one regular and one singular endpoint and any deficiency index. Another landmark result is the extension of the class of symmetric expressions M by several dimensions achieved, before the publication of Naimark's book [**440**], by Shin in a series of papers [**513**], [**514**], [**515**], [**516**]. These are mentioned in a footnote in [**440**] but were not used there. They were rediscovered, in a slightly different but equivalent form, in 1975 by Zettl [**607**] who showed that the Hilbert space methods of Naimark, based largely on the work of Glazman, can be applied to these much more general expressions M.

Theorem 6.4.1 extends the GKN Theorem from self-adjoint operators to symmetric operators and contains the self-adjoint theorem GKN theorem a special case. This theorem is based on the authors 2017 and 2018 papers [**559**], [**562**]. Our proofs of these theorems were influenced by the method used by Sun [**528**] for self-adjoint operators and by the method of Möller-Zettl [**428**] for symmetric operators. The method of Sun is based on a decomposition of the maximal domain using a non-real value of the spectral parameter. Recall that von Neumann's formula for the adjoint of a symmetric operator in Hilbert space depends critically on a nonreal value of the spectral parameter. The method of Möller-Zettl [**428**] for the characterization of symmetric boundary conditions uses a heavy dose of linear algebra.

CHAPTER 7

Self-Adjoint Operators

1. Introduction

In this chapter we give a new characterization of all self-adjoint realizations of the equation

(7.1.1) $$My = \lambda wy \text{ on } J = (a,b), \; -\infty \leq a < b \leq \infty$$

in the Hilbert space $H = L^2(J, w)$. A self-adjoint realization of equation (7.1.1) is an operator S which satisfies

(7.1.2) $$S_{\min} \subset S = S^* \subset S_{\max},$$

where $S_{\min}$ and $S_{\max}$ are the minimal and maximal operators of (7.1.1) defined in Chapter 3. Clearly each such operator S is an extension of $S_{\min}$ and a restriction of $S_{\max}$. These operators S are generally referred to as self-adjoint extensions of the minimal operator $S_{\min}$ but are characterized as restrictions of the maximal operator $S_{\max}$.

This characterization is different than the one given by Theorem 6.3.3 in four important respects:

1. It is based on solutions $u_1, \cdots, u_{m_a}$ and $v_1, \cdots, v_{m_b}$, where (although we use the same notation) u_j, v_j are solutions on (a,c) and on (c,b) for some (not any) *real* values of λ_a, λ_b, respectively.

2. The matrices E_{m_a}, E_{m_b} are simple symplectic matrices E_h, $h = m_a, m_b$.

3. $M = M_Q$, $Q = Q^+$ and $Q \in Z_n(J, \mathbb{R})$, $n = 2k$, $k > 1$. (For $k = 1$ see the book [**620**].)

4. The hypothesis **EH** holds:

DEFINITION 7.1.1. **EH**: Let $a < c < b$. Assume that the equation (7.1.1) has d_a linearly independent solutions, denoted by $u_1, u_2, \cdots, u_{d_a}$, in $L^2((a,c), w)$ for some real $\lambda = \lambda_a$ and on (c,b) has d_b linearly independent solutions, denoted by $v_1, v_2, \cdots, v_{d_b}$, in $L^2((c,b), w)$ for some real $\lambda = \lambda_b$.

DEFINITION 7.1.2. Recall that for $Q \in Z_n(J, \mathbb{R})$, $Q = Q^+$, we have $d_a^+ = d_a^- = d_a$, $d_b^+ = d_b^- = d_b$, and $d = d_a + d_b - n$. Also recall that for any $\lambda \in \mathbb{C}$ with $\text{Im}(\lambda) \neq 0$, the number of linearly independent solutions of (7.1.1) in $L^2((a,c), w)$ is d_a. Similarly for d_b. But, as we will see below, for *some* real values of λ the number of linearly independent solutions of (7.1.1) in $L^2((a,c), w)$ may be less than d_a. Similarly for d_b.

REMARK 7.1.1. We comment on hypothesis **EH**. If, for some interval (a,c) or (c,b), this hypothesis does not hold then the essential spectrum of every self-adjoint extension of the minimal operator $S_{\min}(a,b)$ covers the whole real line. In this case any eigenvalue of any self-adjoint extension, if there is one, is imbedded in the essential spectrum and nothing, other than examples, seems to be known about such eigenvalues and their boundary conditions.

2. LC Solutions and Real Parameter Decompositions of the Maximal Domain

Both the abstract von Neumann representation (4.4.1) and the ordinary differential equation representation (4.4.2) hold for any nonreal complex number; but the assumption that $\mathrm{Im}(\lambda) \neq 0$ is critical. For the even order case $n = 2k$, $k > 1$ with real valued coefficients we establish a representation of $D_{\max}$ using real solutions for certain *real* values of λ. These solutions are called LC solutions in analogy with the Weyl limit circle solutions in the second order case since they play a similar role in the characterization of singular self-adjoint domains. However, in contrast to the second order case, not all solutions in $L^2(J, w)$ are LC solutions. To guarantee the existence of such LC solutions for real λ we need the additional hypothesis **EH** (extra hypothesis). In Chapter 9 we will use this real λ decomposition to obtain information about the spectrum of self-adjoint operators.

LEMMA 7.2.1. *Assume $M = M_Q$, where $Q = Q^+ \in Z_n(J, \mathbb{R})$, $n = 2k$, $k > 1$, $J = (a, b)$, w is a weight function, and the endpoint a is regular. The number d of linearly independent solutions of*

$$My = \lambda wy \text{ on } J \tag{7.2.1}$$

lying in $H = L^2(J, w)$ is independent of $\lambda \in \mathbb{C}$, provided $\mathrm{Im}(\lambda) \neq 0$. The inequalities

$$k \leq d \leq 2k = n \tag{7.2.2}$$

hold. For $\lambda \in \mathbb{R}$, the number of linearly independent solutions of (7.2.1) is less than or equal to d.

PROOF. The fact that the number d of linearly independent solutions of (7.2.1) is independent of $\lambda \in \mathbb{C}$, $\mathrm{Im}(\lambda) \neq 0$, follows from Theorem 4.1.1. The inequalities (7.2.2) follow from Theorem 4.1.2.

Now we prove that for $\lambda \in \mathbb{R}$, the number of linearly independent solutions of (7.2.1) is less than or equal to d. First observe that if there are r linearly independent solutions of $My = \lambda wy$ in H for some real λ, then there exist r linearly independent real-valued solutions in H for this λ. This follows from the fact that the real and imaginary parts of a complex solution are real solutions. Let χ_j, $j = 1, \cdots, r$ be linearly independent real valued solutions of (7.2.1) in H. Suppose that $r > d$ and define

$$D = D_{\min} \dotplus \mathrm{span}\{\chi_1, \cdots, \chi_d\},$$
$$D_r = D_{\min} \dotplus \mathrm{span}\{\chi_1, \cdots, \chi_r\}.$$

Then D is a self-adjoint domain by Corollary 3.3.1 and Theorem 6.4.2. Since, for all $u, v \in D_r$, $[u, v](b) - [u, v](a) = 0$, we have that D_r is a symmetric domain. Note that the dimensions mod $D_{\min}$ of D and D_r are d and r, respectively. Let S and T be the restrictions of the maximal operator $S_{\max}$ to D and D_r, respectively. Then T is a proper symmetric extension of the self-adjoint operator S. But this is impossible since

$$S \subset T \subset T^* \subset S^* = S$$

implies that $T = S$. This contradiction proves that the number of linearly independent solutions of $My = \lambda wy$ lying in H, for any *real* λ is less than or equal to the deficiency index d. $\square$

THEOREM 7.2.1. *Assume $M = M_Q$, where $Q \in Z_n(J, \mathbb{R})$, $n = 2k$, $k > 1$, $J = (a, b)$, is Lagrange symmetric, w is a weight function, and the endpoint a is regular. Let d denote the deficiency index and let $m = 2d - n = 2d - 2k$. Assume there exists $\lambda \in \mathbb{R}$ such that (7.2.1) has d linearly independent real valued solutions lying in H. Then there exist real solutions u_j, $j = 1, \cdots, m$ of (7.2.1) lying in $H = L^2(J, w)$ such that the $m \times m$ matrix*

$$U = \begin{pmatrix} [u_1, u_1](a) & \cdots & [u_1, u_m](a) \\ \cdots & \cdots & \cdots \\ [u_m, u_1](a) & \cdots & [u_m, u_m](a) \end{pmatrix} = ([u_i, u_j](a))_{1 \le i, j \le m} \tag{7.2.3}$$

is nonsingular and

$$D_{\max} = D_{\min} \dotplus \operatorname{span}\{z_1, z_2, \cdots, z_n\} \dotplus \operatorname{span}\{u_1, u_2, \cdots, u_m\}, \tag{7.2.4}$$

where z_j, $j = 1, 2, \cdots, n$ are defined in Theorem 4.4.3.

PROOF. Let $\theta_1, \cdots, \theta_d$ be d linearly independent real solutions of (7.2.1) for some real λ. By (4.4.4) there exist $y_i \in D_{\min}$ and d_{is}, $f_{ij} \in \mathbb{C}$ such that

$$\theta_i = y_i + \sum_{s=1}^{n} d_{is} z_s + \sum_{j=1}^{m} f_{ij} \phi_j, \quad i = 1, \cdots, d, \tag{7.2.5}$$

where ϕ_j, $j = 1, 2, \cdots, m$ are defined in Theorem 4.4.3. From this it follows that

$$\begin{aligned} &([\theta_h, \theta_l](b))_{1 \le h,l \le d} \\ &= \begin{pmatrix} [\theta_1, \theta_1](b) & \cdots & [\theta_1, \theta_d](b) \\ \cdots & \cdots & \cdots \\ [\theta_d, \theta_1](b) & \cdots & [\theta_d, \theta_d](b) \end{pmatrix} \\ &= \left([\sum_{j=1}^{m} f_{hj} \phi_j, \sum_{j=1}^{m} f_{lj} \phi_j](b) \right)_{1 \le h,l \le d} \\ &= F \left([\phi_i, \phi_j](b) \right)_{1 \le i,j \le m} F^*, \qquad F = (f_{ij})_{d \times m}. \end{aligned} \tag{7.2.6}$$

Hence

$$\operatorname{rank} \left([\theta_h, \theta_l](b) \right)_{1 \le h,l \le d} \le \operatorname{rank} \left([\phi_i, \phi_j](b) \right)_{1 \le i,j \le m} = m.$$

Since $\theta_1, \cdots, \theta_d$ are real valued solutions of (7.2.1) for the same real λ, we have

$$\left([\theta_h, \theta_l](b) \right)_{d \times d}^T = \left([\theta_h, \theta_l](a) \right)_{d \times d}^T = (-1)^{k+1} G^T E_n G,$$

where $G = \begin{pmatrix} \theta_1(a) & \cdots & \theta_d(a) \\ \cdots & \cdots & \cdots \\ \theta_1^{[n-1]}(a) & \cdots & \theta_d^{[n-1]}(a) \end{pmatrix}$.

Since rank $E_n = n$ and rank $G = d$, we may conclude that

$$\operatorname{rank} \left([\theta_h, \theta_l](b) \right)_{d \times d} \ge m.$$

Hence

$$\operatorname{rank} \left([\theta_h, \theta_l](b) \right)_{d \times d} = m.$$

Note that

$$([\theta_h,\theta_l](b))_{1\le h,l\le d}^T = -\left([\theta_h,\theta_l](b)\right)_{1\le h,l\le d}.$$

Therefore there exists a non-singular real matrix $P=(p_{ij})_{d\times d}$ such that

$$P^T\left([\theta_h,\theta_l](b)\right)_{1\le h,l\le d}P=(-1)^k\begin{pmatrix} & & & -1 & \\ & & 1 & & \\ & \cdots & & & 0_{m\times(n-d)} \\ -1 & & & & \\ 1 & & & & \\ & 0_{(n-d)\times m} & & & 0_{(n-d)\times(n-d)}\end{pmatrix}.$$

Let

$$\begin{pmatrix}u_1\\ \vdots\\ u_d\end{pmatrix}=P^T\begin{pmatrix}\theta_1\\ \vdots\\ \theta_d\end{pmatrix}. \tag{7.2.7}$$

Then $u_1,\cdots,u_d$ are linearly independently solutions of (7.2.1) satisfying

$$\begin{aligned}&([u_i,u_j](b))_{1\le i,j\le d}\\ =&\begin{pmatrix}[u_1,u_1](b) & \cdots & [u_1,u_d](b)\\ \cdots & \cdots & \cdots\\ [u_d,u_1](b) & \cdots & [u_d,u_d](b)\end{pmatrix}\\ =&(-1)^k\begin{pmatrix} & & & -1 & \\ & & 1 & & \\ & \cdots & & & 0_{m\times(n-d)} \\ -1 & & & & \\ 1 & & & & \\ & 0_{(n-d)\times m} & & & 0_{(n-d)\times(n-d)}\end{pmatrix}.\end{aligned} \tag{7.2.8}$$

Clearly the $m\times m$ matrix $U=([u_i,u_j](a))_{m\times m}=([u_i,u_j](b))_{m\times m}=(-1)^kE_m$ is nonsingular.

By (7.2.6) and (7.2.7), we have

$$([u_h,u_l](b))_{1\le h,l\le m}=(P_1F)\left([\phi_i,\phi_j](b)\right)_{1\le i,j\le m}(P_1F)^*,$$

where $P_1=((p_{ij})_{1\le i\le d,1\le j\le m})^T$. Hence $P_1F=M=(m_{ij})_{m\times m}$ is nonsingular.

By (7.2.5) and (7.2.7) , we get

$$\begin{aligned}u_j&=\sum_{i=1}^d p_{ij}\theta_i\\ &=\sum_{i=1}^d p_{ij}(y_i+\sum_{s=1}^n d_{is}z_s+\sum_{s=1}^m f_{is}\phi_s)\\ &=\sum_{i=1}^d p_{ij}y_i+\sum_{i=1}^d\sum_{s=1}^n p_{ij}d_{is}z_s+\sum_{i=1}^d\sum_{s=1}^m p_{ij}f_{is}\phi_s,\\ &=\sum_{i=1}^d p_{ij}y_i+\sum_{i=1}^d\sum_{s=1}^n p_{ij}d_{is}z_s+\sum_{s=1}^m m_{js}\phi_s,\quad j=1,\cdots,m.\end{aligned}$$

Therefore we have unique solutions

$$\phi_j = \widetilde{y}_j + \sum_{i=1}^{n} \widetilde{b}_{ji} z_i + \sum_{s=1}^{m} \widetilde{c}_{js} u_s, \quad j = 1, \cdots, m, \tag{7.2.9}$$

where $\widetilde{y}_j \in D_{\min}$, $\widetilde{b}_{ji}, \widetilde{c}_{js} \in \mathbb{C}$.

It follows from (7.2.9) and Theorem 4.4.3 that

$$D_{\max} = D_{\min} \dotplus \operatorname{span}\{z_1, z_2, \cdots, z_n\} \dotplus \operatorname{span}\{u_1, u_2, \cdots, u_m\}.$$

Thus in terms of the solutions of (7.2.1) for some real λ we have established the decomposition (7.2.4) of the maximal domain. □

REMARK 7.2.1. If $d = k$, $m = 2d - 2k = 0$, then the term involving the u_j drops out in (7.2.4) and there is no boundary condition at the singular endpoint b required or allowed. This is the higher order analogue of the Weyl limit-point case at b for second order Sturm-Liouville problems. In this case all self-adjoint extensions of $S_{\min}$ are determined by boundary conditions at the regular endpoint a only.

From the proof of Theorem 7.2.1 we have the following Corollary.

COROLLARY 7.2.1. *Let d denote the deficiency index and let $m = 2d - n = 2d - 2k$. Assume there exists $\lambda \in \mathbb{R}$ such that (7.2.1) has d linearly independent solutions lying in H. Then there exist d linearly independent real solutions $u_1, \cdots, u_d$ of (7.2.1) for this λ satisfying the following three conditions:*

(1) *The $m \times m$ matrix $U = ([u_i, u_j](a))_{1 \le i, j \le m}$ is given by*

$$U = (-1)^k \begin{pmatrix} 0 & 0 & 0 & 0 & -1 \\ 0 & 0 & 0 & 1 & 0 \\ \cdots & \cdots & \cdots & \cdots & \cdots \\ 0 & -1 & 0 & 0 & 0 \\ 1 & 0 & 0 & 0 & 0 \end{pmatrix}_{m \times m} = (-1)^k E_m,$$

and is therefore nonsingular.

(2) *For every $y \in D_{\max}$ we have*

$$[u_j, y](b) = 0, \quad for \ j = m+1, \cdots, d.$$

(3)

$$[u_i, u_j](a) = [u_i, u_j](b) = 0, \quad for \ j = m+1, \cdots, d.$$

PROOF. Part (1) follows from (7.2.8).

For part (2), by the decomposition Theorem 7.2.1, any $y \in D_{max}$, can be uniquely written as:

$$y = \widetilde{y}_0 + \sum_{s=1}^{n} \widetilde{d}_s z_s + \sum_{j=1}^{m} c_j u_j,$$

where $\widetilde{y}_0 \in D_{min}$, $\widetilde{d}_s, c_i \in \mathbb{C}$. By (7.2.8) we have

$$[u_i, u_j](b) = 0, \quad i = 1, \cdots m, \quad j = m+1, \cdots d.$$

and the conclusion follows.

Part (3) follows from (7.2.8) and the fact that all u_j are real solutions of (7.2.1) for the same real λ. □

NOTATION 7.2.1. Given d linearly independent real solutions of (7.2.1) lying in H, Theorem 7.2.1 selects m of these such that the $m \times m$ matrix U is nonsingular. We will see below that these solutions can be used to characterize the self-adjoint boundary conditions at the singular endpoint. In the intermediate cases when $k < d < 2k$, since $m = 2d - 2k$ implies that $d - m = 2k - d > 0$, Corollary 7.2.1 shows that the remaining $d - m$ solutions play no role in determining the singular boundary conditions.

DEFINITION 7.2.1 (LC and LP Solutions). We say that the solutions $u_1, \cdots, u_m$ are of limit-circle (or LC) type at b and $u_{m+1}, \cdots, u_d$ are of limit-point (or LP) type at b.

REMARK 7.2.2. Definition 7.2.1 is motivated by the fact that, just as in the Sturm-Liouville case, the LC solutions contribute to the singular self-adjoint boundary conditions while the LP solutions do not. If $d = n$ then all solutions u_j, $j = 1, \cdots, d$ are LC solutions; whereas if $d = k$ then all solutions u_j, $j = 1, \cdots, d$ are LP solutions.

Note that in the intermediate deficiency cases, $k < d < 2k$, when the solutions $u_1, \cdots, u_m, u_{m+1}, \cdots, u_d$ which lie in H are completed to a full bases of solutions of $My = \lambda wy$: $u_1, \cdots, u_m, u_{m+1}, \cdots, u_d$, $u_{d+1}, \cdots, u_n$ the solutions $u_{d+1}, \cdots, u_n$ are not in H. Thus there are three classes of solutions: those not in H and those in H are categorized into two disjoint classes, LP and LC.

Based on Theorem 7.2.1 we now give a characterization of all self-adjoint domains in terms of real-valued solutions of (7.2.1) for real λ.

THEOREM 7.2.2. *Let the hypotheses and notation of Theorem* 7.2.1 *hold. Then a linear submanifold $D(S)$ of $D_{\max}$ is the domain of a self-adjoint extension S of $S_{\min}$ if and only if there exist a complex $d \times n$ matrix A and a complex $d \times m$ matrix B such that the following three conditions hold:*

(1) *The* $\operatorname{rank}(A : B) = d$;
(2) $AE_nA^* = BE_mB^*$;
(3) $D(S) = \{y \in D_{\max}$:

$$A\begin{pmatrix} y(a) \\ \vdots \\ y^{[n-1]}(a) \end{pmatrix} + B\begin{pmatrix} [y, u_1](b) \\ \vdots \\ [y, u_m](b) \end{pmatrix} = \begin{pmatrix} 0 \\ \vdots \\ 0 \end{pmatrix}\}.$$

In (2) E_j is the symplectic matrix (3.2.1) *of order j.*

PROOF. See Theorem 4 of [**553**]. □

Next, and also based on Theorem 7.2.1, in the following we will discuss the real-parameter decomposition of $D_{\max}$ when both endpoints a and b are singular. See [**253**].

THEOREM 7.2.3. *Let $M = M_Q$, where $Q \in Z_n(J, \mathbb{R})$, $n = 2k$, $k > 1$, $J = (a, b)$, be Lagrange symmetric and let $c \in (a, b)$. Consider the equation*

$$My = \lambda wy. \tag{7.2.10}$$

Let d_a denote the deficiency index of (7.2.10) *on (a, c) and d_b the deficiency index of* (7.2.10) *on (c, b). Assume that for some $\lambda = \lambda_a \in \mathbb{R}$* (7.2.10) *has d_a linearly independent solutions on (a, c) which lie in $L^2((a, c), w)$ and that for some $\lambda = \lambda_b \in$*

$\mathbb{R}$ (7.2.10) *has* d_b *linearly independent solutions on* (c,b) *which lie in* $L^2((c,b),w)$. *Then*

(1) *There exist* d_a *linearly independent real-valued solutions* $u_1,\cdots,u_{d_a}$ *of* (7.2.10) *with* $\lambda=\lambda_a$ *on* (a,c) *which lie in* $L^2((a,c),w)$.
(2) *There exist* d_b *linearly independent real-valued solutions* $v_1,\cdots,v_{d_b}$ *of* (7.2.10) *with* $\lambda=\lambda_b$ *on* (c,b) *which lie in* $L^2((c,b),w)$.
(3) *For* $m_a=2d_a-2k$ *the solutions* $u_1,\cdots,u_{d_a}$ *can be ordered such that the* $m_a\times m_a$ *matrix* $U=([u_i,u_j](c))_{1\leq i,j\leq m_a}$ *is given by*
$$U=(-1)^kE_{m_a},$$
where E_{m_a} *is a symplectic matrix defined in* (3.2.1).
(4) *For* $m_b=2d_b-2k$ *the solutions* $v_1,\cdots,v_{d_b}$ *on* (c,b) *can be ordered such that the* $m_b\times m_b$ *matrix* $V=([v_i,v_j](c))$, $1\leq i,j\leq m_b$, *is given by*
$$V=(-1)^kE_{m_b},$$
where E_{m_b} *is a symplectic matrix defined in* (3.2.1).
(5) *For every* $y\in D_{\max}(a,b)$ *we have*
$$[y,u_j](a)=0,\ \text{for } j=m_a+1,\cdots,d_a. \tag{7.2.11}$$
(6) *For every* $y\in D_{\max}(a,b)$ *we have*
$$[y,v_j](b)=0,\ \text{for } j=m_b+1,\cdots,d_b. \tag{7.2.12}$$
(7) *For* $1\leq i,j\leq d_a$, *we have*
$$[u_i,u_j](a)=[u_i,u_j](c)$$
(8) *For* $1\leq i,j\leq d_b$, *we have*
$$[v_i,v_j](b)=[v_i,v_j](c)$$
(9) *The solutions* $u_1,\cdots,u_{d_a}$ *can be extended to* (a,b) *such that the extended functions, also denoted by* $u_1,\cdots,u_{d_a}$, *satisfy* $u_j\in D_{\max}(a,b)$ *and* u_j *is identically zero in a left neighborhood of* b, $j=1,\cdots,d_a$.
(10) *The solutions* $v_1,\cdots,v_{d_b}$ *can be extended to* (a,b) *such that the extended functions, also denoted by* $v_1,\cdots,v_{d_b}$, *satisfy* $v_j\in D_{\max}(a,b)$ *and* v_j *is identically zero in a right neighborhood of* a, $j=1,\cdots,d_b$.

PROOF. Parts (1) and (2) follow from the fact that the real and imaginary parts of a complex solution are real solutions. Parts (3), (4), (5) and (6) follow from Corollary 7.2.1. Parts (7) and (8) follow from Corollary 3.3.1. By Lemma 3.4.1 the solutions $u_1,\cdots,u_{d_a}$ can be 'patched' at c to obtain maximal domain functions in $D_{\max}(a,b)$. By another application of Lemma 3.4.1 these extended functions can be modified to be identically zero in a left neighborhood of b. So (9) holds and (10) follows similarly. □

DEFINITION 7.2.2. Let the hypothesis and Notation of Theorem 7.2.3 hold. The solutions $u_1,\cdots,u_{m_a}$ and $v_1,\cdots,v_{m_b}$ are called LC solutions of the endpoints a and b, respectively. The solutions $u_{m_a+1},\cdots,u_{d_a}$, and $v_{m_b+1},\cdots,v_{d_m}$ are called LP solutions at a and b, respectively.

REMARK 7.2.3. Since the Lagrange brackets in (7.2.11) and (7.2.12) are zero for all maximal domain functions y the LP solutions play no role in the determination of the self-adjoint boundary conditions. Nevertheless, the LP solutions play an important role in the study of the continuous spectrum (see [**531**]) and in the approximation of singular problems with regular ones.

The next theorem gives the representation of the maximal domain in terms of real-valued solutions for real values of the spectral parameter λ.

THEOREM 7.2.4. *Let the notation and hypotheses of Theorem 7.2.3 hold. Then*

$$D_{\max}(a,b) = D_{\min}(a,b) \dot{+} \operatorname{span}\{u_1, \cdots, u_{m_a}\} \dot{+} \operatorname{span}\{v_1, \cdots, v_{m_b}\}. \tag{7.2.13}$$

PROOF. By von Neumann's formula, $\dim(D_{\max}(a,b)/D_{\min}(a,b)) \leq 2d$. From Theorem 7.2.3 parts (7), (8), (9), (10) and the fact that the matrices U and V are nonsingular it follows that $u_1, \cdots, u_{m_a}$ and $v_1, \cdots, v_{m_b}$ are linearly independent mod$(D_{\min}(a,b))$ and therefore $\dim(D_{\max}(a,b)/D_{\min}(a,b)) \geq m_a + m_b = 2d$, completing the proof. □

Although Theorem 7.2.4 is stated for the case when both endpoints are singular, it can be specialized to the case when one or both are regular.

THEOREM 7.2.5. *Let the notation and hypotheses of Theorem 7.2.3 hold.*

(1) *If the endpoint a is regular, then there exist $z_j \in D_{\max}(a,b)$, $j = 1, \cdots, n$ which are identically zero in a neighborhood of b such that $u_1, \cdots, u_{m_a}$ in (7.2.13) can be replaced by z_j, $j = 1, \cdots, n$, i.e. we have*

$$D_{\max}(a,b) = D_{\min}(a,b) \dot{+} \operatorname{span}\{z_1, z_2, \cdots, z_n\} \dot{+} \operatorname{span}\{v_1, \cdots, v_{m_b}\}.$$

(2) *If the endpoint b is regular, then there exist $z_j \in D_{\max}(a,b)$, $j = 1, \cdots, n$ which are identically zero in a neighborhood of a such that $v_1, \cdots, v_{m_b}$ in (7.2.13) can be replaced by z_j, $j = 1, \cdots, n$, i.e. we have*

$$D_{\max}(a,b) = D_{\min}(a,b) \dot{+} \operatorname{span}\{u_1, \cdots, u_{m_a}\} \dot{+} \operatorname{span}\{z_1, z_2, \cdots, z_n\}.$$

PROOF. This follows from Theorem 7.2.1; also the proof of Theorem 7.2.4 can be specialized to these cases. □

REMARK 7.2.4. Note that, although we use the same notation $u_1, \cdots, u_{m_a}$ and $v_1, \cdots, v_{m_b}$ in (7.2.13) as in (4.4.7) these solutions are very different. All are maximal domain functions. The u_i are solutions on the interval (a,c) which are identically 0 near b and the v_i are solutions on (c,b) which are identically 0 near a. But in (4.4.7) the u_i are solutions for a nonreal λ and in (7.2.13) the u_i are real solutions for a certain real λ_a. Similarly for the v_i. Another important difference is that E_{m_a}, E_{m_b} in Theorem 7.2.3 are the very simple symplectic matrices E_k, $k = m_a, m_b$ compared with E_{m_a}, E_{m_b} of Theorem 4.4.4.

REMARK 7.2.5. We will see below that only the LC solutions are used in the construction of the boundary conditions which characterize the self-adjoint operators in the Hilbert space $L^2(J,w)$. The LP solutions and the solutions not in the corresponding Hilbert space make no contribution to the construction of the self-adjoint boundary conditions.

In Chapter 8 the LC solutions will be used to characterize the symmetric domains and to construct examples of symmetric domains.

3. A Real Parameter Characterization of Self-Adjoint Domains

In this section we give another characterization of the domains of the self-adjoint operators S satisfying (7.1.2). This real λ characterization is based on the decomposition Theorem 7.2.4 of the maximal domain which is characterized by LC solutions.

In Chapter 8 these LC solutions will be used to classify the self-adjoint boundary conditions into three mutually exclusive classes when $n > 2$: separated, coupled, mixed, and to characterize the symmetric domains. In Chapter 9, the LC representations of self-adjoint domains will be used to obtain information about the spectrum of self-adjoint operators. We find that the LC and LP solutions constructed in Section 7.2 have an important role in the investigation of the spectrum.

THEOREM 7.3.1. *Let $M = M_Q$, where $Q \in Z_n(J,\mathbb{R})$, $n = 2k$, $k > 1$, $J = (a,b)$, be Lagrange symmetric, and let w be a weight function on J. Assume hypothesis* ***EH*** *holds. Let the deficiency index of $S_{\min}(a,b)$ be d, $d = d_a + d_b - n$, and let $m_a = 2d_a - n$, $m_b = 2d_b - n$. Let $u_1, \cdots, u_{m_a}$ be real valued LC solutions of equation (7.1.1) on (a,c) for some real λ_a; and let $v_1, \cdots, v_{m_b}$ be real valued LC solutions of equation (7.1.1) on (c,b) for some real λ_b as constructed in Section 7.2. Then a linear manifold $D(S)$ of $D_{\max}$ is the domain of a self-adjoint extension S satisfying (7.1.2) if and only if there exists a complex $d \times m_a$ matrix A and a complex $d \times m_b$ matrix B such that the following three conditions hold:*

(1)
$$\operatorname{rank}(A : B) = d,$$

(2)
$$AE_{m_a}A^* = BE_{m_b}B^*,$$

(3)
$$D(S) = \{y \in D_{\max} : A\begin{pmatrix} [y,u_1](a) \\ \vdots \\ [y,u_{m_a}](a) \end{pmatrix} + B\begin{pmatrix} [y,v_1](b) \\ \vdots \\ [y,v_{m_b}](b) \end{pmatrix} = \begin{pmatrix} 0 \\ \vdots \\ 0 \end{pmatrix}\}. \tag{7.3.1}$$

The Lagrange brackets in (7.3.1) have finite limits.

PROOF. *Sufficiency.* Let the matrices A and B satisfy conditions (1) and (2) of Theorem 7.3.1. We prove that $D(S)$ defined by (7.3.1) is the domain of a self-adjoint extension S of $S_{\min}$. Let

$$A = -(\overline{a}_{ij})_{d\times m_a}, \qquad B = (\overline{b}_{ij})_{d\times m_b}.$$

$$w_i = \sum_{j=1}^{m_a} a_{ij}\, u_j + \sum_{j=1}^{m_b} b_{ij} v_j, \qquad i = 1, \cdots, d.$$

Then for $y \in D_{\max}(a,b)$ we have

$$-A\begin{pmatrix} [y,u_1](a) \\ \vdots \\ [y,u_{m_a}](a) \end{pmatrix} = \begin{pmatrix} [y,\sum_{j=1}^{m_a} a_{1j}u_j](a) \\ \vdots \\ [y,\sum_{j=1}^{m_a} a_{dj}u_j](a) \end{pmatrix} = \begin{pmatrix} [y,w_1](a) \\ \vdots \\ [y,w_d](a) \end{pmatrix}.$$

$$B\begin{pmatrix} [y,v_1](b) \\ \vdots \\ [y,v_{m_b}](b) \end{pmatrix} = \begin{pmatrix} [y,\sum_{j=1}^{m_b} b_{1j}v_j](b) \\ \vdots \\ [y,\sum_{j=1}^{m_b} b_{dj}v_j](b) \end{pmatrix} = \begin{pmatrix} [y,w_1](b) \\ \vdots \\ [y,w_d](b) \end{pmatrix}.$$

Therefore the boundary condition (3) of Theorem 7.3.1 becomes the boundary condition (iii) of the GKN Theorem 6.4.2, i.e.

$$[y, w_i](b) - [y, w_i](a) = 0, \qquad i = 1, \cdots, d.$$

It remains to show that $w_1, \cdots, w_d$ satisfy conditions (i) and (ii) of Theorem 6.4.2. To show that condition (i) holds, assume that it does not hold. Then there exist constants $c_1, \cdots, c_d$, not all zero, such that

$$\gamma = \sum_{j=1}^{d} c_j w_j \in D_{\min}.$$

Hence we have $[\gamma, y](a) = 0$ for all $y \in D_{\max}$ by the characterization of $D_{\min}$. Using the notation U from Theorem 7.2.3, then we have

$$\begin{aligned}(0, \cdots, 0) &= ([\sum_{j=1}^{d} c_j w_j, u_1](a), \cdots, [\sum_{j=1}^{d} c_j w_j, u_{m_a}](a)) \\ &= (c_1, \cdots, c_d)(a_{ij})_{d\times m_a} U.\end{aligned}$$

Since U is nonsingular, we have $(\overline{c}_1 \cdots \overline{c}_d)\, A = 0$. Similarly, we have $(\overline{c}_1 \cdots \overline{c}_d)\, B = 0$. Hence

$$(\overline{c}_1 \cdots \overline{c}_d)\, (A : B) = 0.$$

This contradicts the fact that $\operatorname{rank}(A : B) = d$.

Next we show that (ii) holds. Since

$$[w_i, w_j](a) = [\sum_{l=1}^{m_a} a_{il} u_l, \sum_{s=1}^{m_a} a_{js} u_s](a) = \sum_{l=1}^{m_a}\sum_{s=1}^{m_a} a_{il}\overline{a}_{js}[u_l, u_s](a),$$

by computation, we have

$$([w_i, w_j](a))^T_{d\times d} = AU^T A^* = (-1)^{k+1} A E_{m_a} A^*$$

and

$$(([w_i, w_j](b))^T_{d\times d} = BV^T B^* = (-1)^{k+1} B E_{m_b} B^*,$$

where U and V are defined in Theorem 7.2.3. Therefore

$$([w_i, w_j]_a^b)^T = (-1)^{k+1} B E_{m_b} B^* - (-1)^{k+1} A E_{m_a} A^* = 0.$$

It follows from Theorem 6.4.2 that $D(S)$ is a self-adjoint domain.

Necessity. Let $D(S)$ be the domain of a self-adjoint extension S of $S_{\min}$. By Theorem 6.4.2, there exist $w_1, \cdots, w_d \in D_{\max}$ satisfying conditions (i), (ii), (iii) of this theorem. By Theorem 7.2.4 each w_i can be uniquely written as:

$$w_i = \widehat{y}_{i0} + \sum_{j=1}^{m_a} a_{ij} u_j + \sum_{j=1}^{m_b} b_{ij} v_j, \tag{7.3.2}$$

where $\widehat{y}_{i0} \in D_{\min}$, $a_{ij}, b_{ij} \in \mathbb{C}$. Let

$$A = -(\overline{a}_{ij})_{d\times m_a}, \; B = (\overline{b}_{ij})_{d\times m_b}.$$

Then

$$\begin{pmatrix} [y, w_1](a) \\ \vdots \\ [y, w_d](a) \end{pmatrix} = \begin{pmatrix} [y, \sum_{j=1}^{m_a} a_{1j} u_j](a) \\ \vdots \\ [y, \sum_{j=1}^{m_a} a_{dj} u_j](a) \end{pmatrix} = -A \begin{pmatrix} [y, u_1](a) \\ \vdots \\ [y, u_{m_a}](a) \end{pmatrix},$$

$$\begin{pmatrix} [y,w_1](b) \\ \vdots \\ [y,w_d](b) \end{pmatrix} = \begin{pmatrix} [y,\sum_{j=1}^{m_b} b_{1j}v_j](b) \\ \vdots \\ [y,\sum_{j=1}^{m_b} b_{dj}v_j](b) \end{pmatrix} = B \begin{pmatrix} [y,v_1](b) \\ \vdots \\ [y,v_{m_b}](b) \end{pmatrix}.$$

Hence the boundary condition (iii) of Theorem 6.4.2 is equivalent to part (3) of Theorem 7.3.1. Next we prove that A, B satisfy conditions (1) and (2) of Theorem 7.3.1. Clearly $\text{rank}(A:B) \le d$. If $\text{rank}(A:B) < d$, then there exist constants $c_1, \cdots, c_d$, not all zero, such that

$$(c_1 \cdots c_d)(A:B) = 0. \tag{7.3.3}$$

Let $g = \sum_{i=1}^d \overline{c}_i w_i$, from (7.3.2), we obtain

$$g = \sum_{i=1}^{d} \overline{c}_i \widehat{y}_{i0} + \sum_{i=1}^{d}\sum_{j=1}^{m_a} \overline{c}_i a_{ij} u_j + \sum_{i=1}^{d}\sum_{j=1}^{m_b} \overline{c}_i b_{ij} v_j.$$

By (7.3.3), we have $(c_1 \cdots c_d)A = (c_1 \cdots c_d)B = 0$. Hence

$$g = \sum_{i=1}^{d} \overline{c}_i \widehat{y}_{i0}.$$

So $g \in D_{\min}$. This contradicts the fact that the functions $w_1, w_2, ..., w_d$ are linearly independent modulo $D_{\min}$. Therefore $\text{rank}(A:B) = d$. Now we verify A, B satisfy condition (2). By (7.3.2), we have

$$[w_i, w_j](a) = [\sum_{l=1}^{m_a} a_{il}u_l, \sum_{s=1}^{m_a} a_{js}u_s](a) = \sum_{l=1}^{m_a}\sum_{s=1}^{m_a} a_{il}\overline{a}_{js}[u_l, u_s](a), \quad i,j = 1,\cdots,d.$$

From Theorem 7.2.3 we obtain

$$([w_i, w_j](a))^T_{d\times d} = AU^T A^* = (-1)^{k+1} A E_{m_a} A^*.$$

and similarly, we have

$$([w_i, w_j](b))^T_{d\times d} = BV^T B^* = (-1)^{k+1} B E_{m_b} B^*.$$

Hence condition (ii) of Theorem 6.4.2 becomes

$$AE_{m_a}A^* = BE_{m_b}B^*.$$

This completes the proof. □

4. The Maximal Deficiency Cases

The maximal deficiency case $d = n$ occurs when each endpoint is either regular or LC singular. So we have the four cases for the endpoints: R/R, LC/LC, LC/R, R/LC. Note that the next theorem does not use hypothesis **EH**.

THEOREM 7.4.1. *Let $M = M_Q$, $Q \in Z_n(J,\mathbb{R})$, $n = 2k$, $k > 1$, be Lagrange symmetric and let w be a weight function on J. Assume that $d = n$. Then*

(1) *Each endpoint is either regular or LC singular.*
(2) *Choose any $\lambda \in \mathbb{R}$ and let $u_1, u_2, \cdots, u_n$ be real valued solutions of equation (7.1.1) on (a,b) which satisfy the condition $([u_i, u_j](a))_{n\times n} = (-1)^k E_n$. If $A, B \in M_n(\mathbb{C})$ satisfy conditions:*

$$\text{rank}(A:B) = n, \quad AE_nA^* = BE_nB^*,$$

then the following boundary condition

$$A\begin{pmatrix}[y,u_1](a)\\ \vdots\\ [y,u_n](a)\end{pmatrix}+B\begin{pmatrix}[y,u_1](b)\\ \vdots\\ [y,u_n](b)\end{pmatrix}=\begin{pmatrix}0\\ \vdots\\ 0\end{pmatrix} \tag{7.4.1}$$

determines a self-adjoint operator S satisfying (7.1.2) *and every such operator S is generated this way.*

(3) *If a is regular then* $\begin{pmatrix}[y,u_1](a)\\ \vdots\\ [y,u_n](a)\end{pmatrix}$ *in* (7.4.1) *can be replaced by*

$$\begin{pmatrix}y(a)\\ \vdots\\ y^{[n-1]}(a)\end{pmatrix}.$$

(4) *If b is regular then* $\begin{pmatrix}[y,u_1](b)\\ \vdots\\ [y,u_n](b)\end{pmatrix}$ *in* (7.4.1) *can be replaced by*

$$\begin{pmatrix}y(b)\\ \vdots\\ y^{[n-1]}(b)\end{pmatrix}.$$

(5) *If both a and b are regular then* (7.4.1) *can be replaced by*

$$A\begin{pmatrix}y(a)\\ \vdots\\ y^{[n-1]}(a)\end{pmatrix}+B\begin{pmatrix}y(b)\\ \vdots\\ y^{[n-1]}(b)\end{pmatrix}=\begin{pmatrix}0\\ \vdots\\ 0\end{pmatrix}.$$

PROOF. In this maximal deficiency case all solutions are LC solutions on (a,b) and therefore also on (a,c) and (c,b) so we do not need the hypothesis **EH** to construct LC solutions on (a,c) and (c,b). In particular, item (2) follows from the Naimark Patching Lemma 3.4.1, Theorems 7.2.3, 7.2.4 and 7.3.1. For (3), (4) and (5), recall that at a regular endpoint the Lagrange brackets in (7.4.1) which involve that endpoint can be replaced by quasi-derivatives evaluated at that point. □

THEOREM 7.4.2. *Let $Q \in Z_n(J,\mathbb{R})$, $n=2k$, $k>1$, be Lagrange symmetric; let $M=M_Q$, and let w be a weight function on J. Then*

(1) *If $d=0$ then the minimal operator $S_{\min}$ is self-adjoint and has no proper self-adjoint extension.*
(2) *Suppose $d_a=k$ and $d_b=n$. Then $d=k$, $m_a=0$, $m_b=n$. If $B\in M_{k,n}(\mathbb{C})$ satisfies* $\operatorname{rank}(B)=k$ *and $BE_nB^*=0$ then every operator S satisfying* (7.1.2) *is determined by a boundary condition*

$$B\begin{pmatrix}[y,v_1](b)\\ \vdots\\ [y,v_n](b)\end{pmatrix}=\begin{pmatrix}0\\ \vdots\\ 0\end{pmatrix} \tag{7.4.2}$$

and every boundary condition (7.4.2) *determines such an operator S.*

(3) *Suppose* $d_a = n$ *and* $d_b = k$. *Then* $d = k$, $m_a = n$, $m_b = 0$. *If* $A \in M_{k,n}(\mathbb{C})$ *satisfies* $\text{rank}(A) = k$ *and* $AE_nA^* = 0$, *then every operator* S *satisfying* (7.1.2) *is determined by a boundary condition*

$$A \begin{pmatrix} [y, u_1](a) \\ \vdots \\ [y, u_n](a) \end{pmatrix} = \begin{pmatrix} 0 \\ \vdots \\ 0 \end{pmatrix} \tag{7.4.3}$$

and every boundary condition (7.4.3) *determines such an operator* S.

PROOF. This follows directly from Theorem 7.3.1. □

5. Boundary Conditions for the Friedrichs Extension

In this section we do not use hypothesis **EH.**

For H a separable Hilbert space with inner product $(\cdot, \cdot)$ and S a closed, densely defined, bounded below, symmetric operator in H, in a celebrated paper Friedrichs [**212**] constructed a self-adjoint extension of S_F of $S_{\min}$ in H with the same lower bound as S without directly referring to any boundary condition. This has come to be known as the Friedrichs extension. In general there are other self-adjoint extensions of S which have the same lower bound as S. (See [**620**] for examples.)

In Section 3.5 we identified densely defined closed symmetric operators $S_{\min}(Q)$ which are bounded below. In general - as we will see below - these minimal operators have an uncountable number of self-adjoint extensions. Which boundary condition determines the Friedrichs extension? This question is discussed in this Section.

Friedrichs himself asked this question [**213**] and proved that for the expression $-y'' + qy$ on $J = [a, b]$ with regular finite endpoints and q real and continuous the

Dirichlet condition

$$y(a) = 0 = y(b)$$

determines the Friedrichs extension. This result has been generalized by many authors for regular and singular problems. In the regular case there is a very general result due to Niessen-Zettl [**453**] and Möller-Zettl [**428**]:

THEOREM 7.5.1. *Suppose* $Q = (q_{rs}) \in Z_n(J, \mathbb{R})$ *with* $n = 2k$, $k \geq 1$ *is Lagrange symmetric with positive leading coefficient and each endpoint is regular. Then for any weight function* w, *the Friedrichs extension of the minimal operator* $S_{\min}(Q)$ *in* $L^2(J, w)$ *is determined by the Dirichlet boundary condition:*

$$y^{[j]}(a) = 0 = y^{[j]}(b), \quad j = 0, 1, \cdots, k-1.$$

PROOF. See [**453**] or [**428**]. □

Next we discuss the second order singular case. Consider the equation

$$-(py')' + qy = \lambda wy \ \text{ on } J, \quad \frac{1}{p}, q, w \in L_{loc}(J, \mathbb{R}), \ p > 0, \ w > 0, \ \lambda \in \mathbb{R}.$$

Note that we have added the condition that $p > 0$ since the minimal operator is not bounded below when p changes sign. Let

$$Q = \begin{bmatrix} 0 & \frac{1}{p} \\ -q & 0 \end{bmatrix}$$

and let $M = M_Q$. Then $Q \in Z_2(J, \mathbb{R})$, M is Lagrange symmetric and $S_{\min} = S_{\min}(Q)$ is a closed, densely defined, symmetric operator, which is bounded below in $H = L^2(J, w)$ for any weight function w.

THEOREM 7.5.2. *Suppose each endpoint is either regular or LCNO (limit-circle non-oscillatory) and u_a is the principal solution at a and u_b is the principal solution at b. Then the closed densely defined minimal operator $S_{\min}$ is bounded below and its Friedrichs extension is given by the boundary condition:*

$$[y, u_a](a) = 0 = [y, u_b](b). \tag{7.5.1}$$

Recall that the principal solution at an endpoint is unique up to constant multiples so that (7.5.1) does not depend on the choice of principal solutions. Also recall that if a is regular $[y, u_a](a) = 0$ reduces to $y(a) = 0$. Similarly for b. And recall that $[y, z]$ is the Lagrange sesquilinear form $[y, z] = y(p\overline{z}') - \overline{z}(py')$.

PROOF. For a proof and definition of principal solution and LCNO endpoint see [**454**], also see [**620**]. □

The characterization (7.5.1) is a restriction of the maximal domain $D_{\max}$ determined by principal solutions u_a and u_b. These depend on the LCNO classification of the endpoints a, b. For a different proof which is closer in spirit to the Friedrichs proof see the paper by Yao et. al. [**589**]. It starts with the assumption that $S_{\min}$ is bounded below and does not directly use oscillation or nonoscillation theory.

For a very special class of singular even order problems the next theorem determines the boundary conditions of the Friedrichs extension explicitly. The proof of this theorem may be of more interest than the result. The method of proof is to 'regularize' these singular problems and then apply Theorem 7.5.1 for the regular case.

The next theorem also demonstrates that the representation $M = M_Q$ is not $1-1$: Q determines M uniquely but not conversely: we may have $M_A = M_B$ with $A \neq B$. (We use the notation M_A here instead of M_Q since we use Q for an anti-derivative of the coefficient q.)

THEOREM 7.5.3. *For $n = 2k$, $k > 1$ and w a weight function, consider*

$$My = (-1)^k y^{(n)} + qy = \lambda wy \quad \text{on } J = (a, b), \tag{7.5.2}$$

where q and w satisfy

$$q \in L_{loc}(J, \mathbb{R}), \quad Q = \int q \in L(J, \mathbb{R}), \quad w \in L(J) \tag{7.5.3}$$

and Q is an anti-derivative of q. Then

(1) *The minimal operator $S_{\min}(M)$ is bounded below in $H = L^2(J, w)$.*
(2) *For all $y \in D_{\max}(M)$ the limits*

$$\lim_{t \to a^+} y^{[j]}(t) \quad \text{and} \quad \lim_{t \to b^-} y^{[j]}(t) \tag{7.5.4}$$

exist and are finite for $j = 0, 1, \cdots, n-2$. (But not for $j = n-1$ in general.)

(3) *The domain D_F of the Friedrichs extension of $S_{\min}(M)$ is given by the self-adjoint boundary conditions*

$$y^{(j)}(a) = 0 = y^{(j)}(b), \quad j = 0, 1, \cdots, k-1.$$

PROOF. Under hypothesis (7.5.3) the endpoints may be singular. In general, the limits (7.5.4) do not exist at a singular endpoint, so the existence of these limits is part of the theorem. Let

$$A = \begin{bmatrix} & 1 & & & \\ & & 1 & & \\ & & & \ddots & \\ & & & & 1 \\ -q & & & & \end{bmatrix}, \quad B = \begin{bmatrix} & & 1 & & & \\ & & & 1 & & \\ & & & & \ddots & \\ -Q & & & & & 1 \\ & & Q & & & \end{bmatrix},$$

where all the nonspecified entries are zero. Note that A and B are in $Z_n(J,\mathbb{R})$ and both are Lagrange symmetric. We compute the quasi-derivatives and note that $y_A^{[j]} = y^{(j)} = y_B^{[j]}$, $j = 0, 1, \cdots, n-2$; and

$$y_A^{[n-1]} = y^{(n-1)} \quad but \quad y_B^{[n-1]} = (y^{(n-1)} + Qy).$$

Hence

$$M_A\, y = i^n y_A^{[n]} = (-1)^k y^{(n)} + qy$$

and

$$M_B\, y = i^n y_B^{[n]} = (-1)^k \{(y_B^{[n-1]}) + Qy\}' - Qy' = (-1)^k y^{(n)} + qy.$$

Therefore $D(A) = D(B)$ and $M_A = M_{B.}$ By hypothesis (7.5.3) M_B is regular on J and therefore the conclusions follows from Theorem 7.5.1. □

Next we give an example to illustrate Theorem 7.5.3.

EXAMPLE 7.5.1. For $n = 2k$, $k > 1$, consider

$$My = (-1)^k y^{(n)} + qy = \lambda wy \quad on\ J = (0,1),$$
$$q(t) = t^r, \quad t \in J, \quad r \in (-2, 1].$$

Note that $w = 1$, the endpoint $a = 0$ is singular, the endpoint $b = 1$ is regular and the hypothesis (7.5.3) holds. Hence the Friedrichs extension is given by the self-adjoint boundary conditions

$$y^{(j)}(0) = 0 = y^{(j)}(1), \quad j = 0, 1, \cdots, k-1.$$

REMARK 7.5.1. In this example, by Theorem 7.5.3, we have that for all $y \in D_{\max}(M)$ the limits

$$\lim_{t \to 0^+} y^{[j]}(t) \ and \ \lim_{t \to 1^-} y^{[j]}(t)$$

exist and are finite for $j = 0, 1, \cdots, n-2$. What about $j = n-1$? The limit $\lim_{t\to 1^-} y^{[n-1]}(t)$ exists since the endpoint $b = 1$ is regular. In this case the limit $\lim_{t\to 0^+} y^{[n-1]}(t)$ does not exist because if it did then that and the fact that $y_B^{[n-1]} = (y^{(n-1)} + Qy)$ has a finite limit at 0 would imply that Q has a finite limit at 0 which is not the case by inspection. For this reason the quasi-derivative $y_B^{[n-1]}$ is used in the formulation of the general self-adjoint boundary conditions for equation (7.5.2). Only those self-adjoint boundary conditions can be expressed solely in terms of classical derivatives which involve derivatives of order strictly less than $n-1$. The Friedrichs extension is one of these since it involves only derivatives up to order $k-1$ and these are all classical rather than quasi-derivatives in this case.

REMARK 7.5.2. Since $M = M_A = M_B$ we can give the characterization of the singular self-adjoint boundary conditions for M_A by using the regular representation M_B. (The use of matrices A, B for both the boundary conditions and the representations $M = M_A$ and $M = M_B$ should not be confusing here since the meaning is clear from the context.) Thus we have: All self-adjoint realizations of the problem (7.5.2), (7.5.3) are given by

$$A \begin{pmatrix} y(a) \\ y^{[1]}(a) \\ \vdots \\ y^{[n-1]}(a) \end{pmatrix} + B \begin{pmatrix} y(b) \\ y^{[1]}(b) \\ \vdots \\ y^{[n-1]}(b) \end{pmatrix} = \begin{pmatrix} 0 \\ \vdots \\ 0 \end{pmatrix},$$

where $y^{[j]} = y^{(j)}$, $j = 0, 1, \cdots, n-2$ and $y^{[n-1]} = (y^{(n-1)} + Qy)$ and the $n \times n$ complex matrices A, B satisfy

$$\operatorname{rank}(A : B) = n;$$

and

$$A\,E\,A^* = B\,E\,B^*, \quad E = ((-1)^r \delta_{r,n+1-s})_{r,s=1}^n.$$

6. Comments

1. This is just a general introduction to the chapter.

2. The proof of Theorem 7.2.1 is based on the recent Wang-Sun-Zettl [**553**] paper. This proof uses LC solutions which were originally constructed by Wang-Sun-Zettl [**553**] for problems with one regular endpoint and extended to problems with two singular endpoints by Hao-Sun-Wang-Zettl [**253**].

3. In the maximal deficiency index case $d = n$ of Theorem 7.4.1 all solutions are in $L^2(J, w)$ for every λ real or complex. This case is similar to the regular endpoint case and reduces to it. It has a much longer history dating back at least to the classic 1964 book of Atkinson [**23**] than the intermediate deficiency case when $0 < d < n$.

4. Theorem 7.5.1 is due to Niessen-Zettl [**453**] and Möller-Zettl [**428**]. Theorem 7.5.2 is due to Niessen-Zettl [**454**]; also see this paper for a detailed discussion of endpoints which are oscillatory/non-oscillatory, principal, etc.

Theorem 7.5.3 as well as the example and illustrations are based on Zettl [**612**].

CHAPTER 8

Self-Adjoint and Symmetric Boundary Conditions

1. Introduction

In this chapter we use LC solutions to:

1. Classify the self-adjoint boundary conditions given by Theorem 7.3.1 into three mutually exclusive classes when $n > 2$: separated, coupled, and mixed. Recall that for $n = 2$ there are only two classes: separated and coupled.
2. Construct examples of all three types.
3. Construct nonreal self-adjoint boundary conditions. (See the Everitt-Markus comment in the Preface.)
4. Characterize the symmetric domains.
5. Construct examples of symmetric operators.

DEFINITION 8.1.1. For given matrices A, B satisfying (1) and (2) of Theorem 7.3.1, at first glance it may seem that all d conditions of (7.3.1) 'connect' the two endpoints a, b with each other i.e. they are 'coupled'. This is not the case. We will see in this chapter that each of these d conditions is either 'separated' or 'coupled': separated when it involves only one of the two endpoints, coupled when it involves both endpoints. The set of d conditions in (7.3.1) is said to be mixed if it includes at least one separated and one coupled condition. We say that the boundary condition (7.3.1) is separated when every one of the d conditions is separated and coupled when all d of its conditions are coupled. If at least one of the d conditions of (7.3.1) is separated and at least one is coupled we say that the self-adjoint boundary condition (7.3.1) is mixed.

Throughout this chapter we assume that hypothesis EH holds. Let $a < c < b$, and assume that equation (7.1.1) on (a, c) has d_a linearly independent solutions in $L^2((a, c), w)$ for some real $\lambda = \lambda_a$ and on (c, b) has d_b linearly independent solutions in $L^2((c, b), w)$ for some real $\lambda = \lambda_b$.

REMARK 8.1.1. We recall the comment on this assumption made in Section 7.1. Let $a < c < b$. Let $r(\lambda)$ denote the number of linearly independent solutions of (7.1.1) on (a, c) which lie in $L^2((a, c), w)$. For any real λ it is known [**553**] that $r(\lambda) \leq d_a$ and that $r(\lambda) < d_a$ implies that λ is in the essential spectrum of every self-adjoint realization in $L^2((a, c), w)$. Thus, if there does not exist a real λ_a such that $r(\lambda_a) = d_a$ then the essential spectrum of every self-adjoint operator S in $L^2((a, c), w)$ covers the whole real line. If the essential spectrum of every self-adjoint realization in $L^2((a, c), w)$ covers the whole real line then this also holds for all self-adjoint realizations in $L^2((a, b), w)$. In this case every eigenvalue, if there are any, is embedded in the essential spectrum and the dependence of such eigenvalues on boundary conditions seems to be 'coincidental'. A similar comment applies to the interval (c, b) with respect to d_b and the interval (a, b) with respect to d.

2. Separated Conditions

In this section we construct separated self-adjoint boundary conditions for all values of $d > 0$. Recall that when $d = 0$, $S_{\min}$ is self-adjoint and has no proper self-adjoint extensions.

Next we establish two linear algebra lemmas which may be of independent interest. These will be used below.

LEMMA 8.2.1. *Let h be any even positive integer and let C be an $r \times h$ complex matrix with* $\text{rank}(C) = r$. *Assume that*

$$CE_hC^* = 0,$$

where E_h is the h-order symplectic matrix $E_h = ((-1)^j \delta_{j,h+1-s})_{j,s=1}^h$. Then $r \leq \frac{h}{2}$.

PROOF. Let $C = (\alpha_1, \alpha_2, \cdots, \alpha_r)^T$, where α_i are the row vectors of C. Then $\text{rank}(CE_h) = r$. The equation $CE_hC^* = 0$ is equivalent to the system

$$\begin{pmatrix} \alpha_1 E_h \alpha_1^* & \alpha_1 E_h \alpha_2^* & \cdots & \alpha_1 E_h \alpha_r^* \\ \cdots & \cdots & \cdots & \cdots \\ \alpha_r E_h \alpha_1^* & \alpha_r E_h \alpha_2^* & \cdots & \alpha_r E_h \alpha_r^* \end{pmatrix} = 0.$$

Hence

$$(\alpha_i, \alpha_j E_h) = 0, \quad i, j = 1, 2, \cdots, r,$$

where $(\cdot, \cdot)$ denote the usual inner product in $\mathbb{C}^h$.

This implies that

$$\alpha_i \in \{\alpha_1 E_h, \alpha_2 E_h, \cdots, \alpha_r E_h\}^\perp, \quad i = 1, 2, \cdots, r,$$

i.e.

$$\{\alpha_1, \alpha_2, \cdots, \alpha_r\} \perp \{\alpha_1 E_h, \alpha_2 E_h, \cdots, \alpha_r E_h\}.$$

Note that $\text{rank}(CE_h) = r$, and hence $\alpha_1, \cdots, \alpha_r, \alpha_1 E_h, \cdots, \alpha_r E_h$ are linearly independent in $\mathbb{C}^h$. Therefore $h \geq 2r$, i.e. $r \leq \frac{h}{2}$. □

LEMMA 8.2.2. *Let the hypotheses and notation of Lemma* 8.2.1 *hold. Then there exist $\alpha_1, \cdots, \alpha_{\frac{h}{2}}$ in $\mathbb{R}^h$ such that $CE_hC^* = 0$, where $C = (\alpha_1, \alpha_2, \cdots, \alpha_{\frac{h}{2}})^T$.*

PROOF. Let $\alpha_1 \in \mathbb{R}^h$ be a non-zero real vector, we have $(\alpha_1, \alpha_1 E_h) = 0$, i.e. $\alpha_1 \in \{\alpha_1 E_h\}^\perp$. Obviously $\{\alpha_1 E_h\}^\perp$ is an $h-1$ dimension subspace of $\mathbb{R}^h$. We can choose $\alpha_2 \in \{\alpha_1 E_h\}^\perp$ such that α_1, α_2 are linearly independent. Note that $(\alpha_2, \alpha_2 E_h) = 0$, and we have

$$\{\alpha_1, \alpha_2\} \perp \{\alpha_1 E_h, \alpha_2 E_h\}.$$

Let $r = \frac{h}{2}$. Following this procedure, we obtain linearly independent vectors $\{\alpha_1, \alpha_2, \cdots, \alpha_{r-1}\}$ such that

$$\{\alpha_1, \cdots, \alpha_{r-1}\} \perp \{\alpha_1 E_h, \cdots, \alpha_{r-1} E_h\}.$$

Since the dimension of the subspace $\{\alpha_1 E_h, \cdots, \alpha_{r-1} E_h\}^\perp$ is $r+1$, we can choose $\alpha_r \in \mathbb{R}^h$ such that $\alpha_1, \cdots, \alpha_{r-1}, \alpha_r$ are linearly independent and

$$\{\alpha_1, \cdots, \alpha_r\} \perp \{\alpha_1 E_h, \cdots, \alpha_r E_h\},$$

and the conclusion follows. □

REMARK 8.2.1. These two lemmas show that the rank of matrices C satisfying $CE_hC^* = 0$ is at most $\frac{h}{2}$ and that there are some matrices C for which it is $\frac{h}{2}$.

The next theorem gives a construction for separated self-adjoint boundary conditions.

THEOREM 8.2.1. *Let the hypotheses and notation of Theorem* 7.3.1 *hold. Suppose*

$$A = \begin{pmatrix} C_{l\times m_a} \\ 0_{(d-l)\times m_a} \end{pmatrix}, \quad B = \begin{pmatrix} 0_{l\times m_b} \\ D_{(d-l)\times m_b} \end{pmatrix},$$

and assume that $\operatorname{rank}(C) = l$ *and* $\operatorname{rank}(D) = d - l$. *Then* $\operatorname{rank}(A : B) = d$ *and the boundary conditions*

$$A \begin{pmatrix} [y, u_1](a) \\ \vdots \\ [y, u_{m_a}](a) \end{pmatrix} + B \begin{pmatrix} [y, v_1](b) \\ \vdots \\ [y, v_{m_b}](b) \end{pmatrix} = 0$$

are self-adjoint if and only if

$$CE_{m_a}C^* = 0 \text{ and } DE_{m_b}D^* = 0. \tag{8.2.1}$$

Furthermore $l = d_a - k$ *in this case, and so* $d - l = d_b - k$.

PROOF. It is clear that $\operatorname{rank}(A : B) = d$. Note that

$$AE_{m_a}A^* = \begin{pmatrix} CE_{m_a}C^* & 0_{l\times(d-l)} \\ 0_{(d-l)\times l} & 0_{(d-l)\times(d-l)} \end{pmatrix}, \quad BE_{m_b}B^* = \begin{pmatrix} 0_{l\times l} & 0_{l\times(d-l)} \\ 0_{(d-l)\times l} & DE_{m_b}D^* \end{pmatrix},$$

and (8.2.1) follows. The furthermore parts follow from Lemmas 8.2.1 and 8.2.2. By Lemma 8.2.1, we have $l \leq \frac{m_a}{2} = d_a - k$ and $d - l \leq \frac{m_b}{2} = d_b - k$, i.e. $l \geq d - d_b + k = d_a + d_b - 2k - d_b + k = d_a - k$. Therefore $l = d_a - k$, and it can be realized by Lemma 8.2.2 . □

REMARK 8.2.2. Although we do not give a rigorous technical definition of separated boundary conditions until Section 8.3, it is intuitively clear that the construction given by Theorem 8.2.1 yields exactly $d_a - k$ separated conditions at the endpoint a and exactly $d_b - k$ separated boundary conditions at the endpoint b. Moreover, if each of the d equations of a self-adjoint boundary condition (7.3.1) is specified at one endpoint only, then the boundary condition can be put into the form of Theorem 8.2.1 by elementary matrix transformations. We state this as the next Corollary.

COROLLARY 8.2.1. *Assume that* $d_b - k$ *rows of the matrix* A *are zero and* $\operatorname{rank}(A) = d_a - k$ *and suppose that the complementary* $d_a - k$ *rows of* B *are zero and* $\operatorname{rank}(B) = d_b - k$. *Let* C *denote the* $(d_a - k) \times m_a$ *submatrix of* A *consisting of the nonzero rows of* A. *Similarly, let* D *denote the submatrix of* B *consisting of the nonzero rows of* B. *Then* $\operatorname{rank}(A : B) = d$ *and the boundary conditions* (7.3.1) *are self-adjoint if and only if* (8.2.1) *holds.*

PROOF. This follows from the observation that the condition

$$AE_{m_a}A^* = BE_{m_b}B^*$$

is invariant under multiplication on the left by any nonsingular $d \times d$ matrix G. Note that

$$(GA)E_{m_a}(GA)^* = (GB)E_{m_b}(GB)^*$$

for any nonsingular $d \times d$ matrix G. By choosing elementary matrices G, the zero rows of A and of B can be interchanged. □

COROLLARY 8.2.2. *Assume that* $d = 2k = n$, *i.e. the maximal deficiency case holds. This occurs if and only if* $d_a = n$ *and* $d_b = n$. *Thus* $m_a = m_b = n$. *By Theorem* 8.2.1, *we have* $l = k$, $d - l = k$. *Let*

$$A = \begin{pmatrix} C_{k\times n} \\ 0_{k\times n} \end{pmatrix}, \quad B = \begin{pmatrix} 0_{k\times n} \\ D_{k\times n} \end{pmatrix}$$

and suppose that $\text{rank}(C) = \text{rank}(D) = k$. *Then the separated boundary conditions*

$$C \begin{pmatrix} [y, u_1](a) \\ \vdots \\ [y, u_n](a) \end{pmatrix} = 0 \quad \textit{and} \quad D \begin{pmatrix} [y, u_1](b) \\ \vdots \\ [y, u_n](b) \end{pmatrix} = 0 \tag{8.2.2}$$

are self-adjoint if and only if $CE_nC^* = DE_nD^* = 0$.

PROOF. It is clear that $\text{rank}(A : B) = n$ and by computation

$$A \begin{pmatrix} [y, u_1](a) \\ \vdots \\ [y, u_n](a) \end{pmatrix} + B \begin{pmatrix} [y, u_1](b) \\ \vdots \\ [y, u_n](b) \end{pmatrix} = \begin{pmatrix} 0 \\ \vdots \\ 0 \end{pmatrix}$$

is equivalent to (8.2.2). Note that

$$AE_nA^* = \begin{pmatrix} CE_nC^* & 0_{k\times k} \\ 0_{k\times k} & 0_{k\times k} \end{pmatrix} \quad \textit{and} \quad BE_nB^* = \begin{pmatrix} 0_{k\times k} & 0_{k\times k} \\ 0_{k\times k} & DE_nD^* \end{pmatrix}.$$

By Theorem 7.4.1, we obtain that (8.2.2) is a self-adjoint boundary condition if and only if $CE_nC^* = DE_nD^* = 0$. □

REMARK 8.2.3. Let $d_a = k$ and $d_b = n$. Then $d = d_a + d_b - n = k$, $m_a = 0$, $m_b = n$, $d_a - k = 0$ and $d_b - k = k$. In this case, by Theorem 8.2.1, all self-adjoint boundary conditions are separated. By Corollary 8.2.1 we may, without loss of generality, let

$$A = \begin{pmatrix} C_{(d_a-k)\times m_a} \\ 0_{(d_b-k)\times m_a} \end{pmatrix} = 0_{d\times 0}, \quad B = \begin{pmatrix} 0_{(d_a-k)\times m_b} \\ D_{(d_b-k)\times m_b} \end{pmatrix} = D_{k\times n}$$

with $\text{rank}(D) = k$. Then $\text{rank}(A : B) = k$ and

$$D \begin{pmatrix} [y, v_1](b) \\ \vdots \\ [y, v_n](b) \end{pmatrix} = 0$$

are self-adjoint if and only if $DE_nD^* = 0$.

Similarly, let $d_a = n$ and $d_b = k$. Then $d = k$, $m_a = n$, $m_b = 0$, $d_a - k = k$ and $d_b - k = 0$. In this case, by Theorem 8.2.1, all self-adjoint boundary conditions are separated. By Corollary 8.2.1 we may let

$$A = \begin{pmatrix} C_{(d_a-k)\times m_a} \\ 0_{(d_b-k)\times m_a} \end{pmatrix} = C_{k\times n}, \quad B = \begin{pmatrix} 0_{(d_a-k)\times m_b} \\ D_{(d_b-k)\times m_b} \end{pmatrix} = 0_{k\times 0}$$

with $\text{rank}(C) = k$. Then $\text{rank}(A : B) = k$ and

$$C \begin{pmatrix} [y, u_1](a) \\ \vdots \\ [y, u_n](a) \end{pmatrix} = 0$$

are self-adjoint if and only if $CE_nC^* = 0$.

3. Separated, Coupled, and Mixed Conditions

We classify the self-adjoint boundary conditions given by Theorem 7.3.1 into different types depending on how many of the conditions are coupled. Our classification depends on Theorem 8.3.1 given below in this section.

Although Definition 8.1.1 is intuitively clear we will give a rigorous technical definition of all three types of conditions in this section.

Recall that

$$d = d_a + d_b - n, \quad n = 2k,$$

and the following inequalities hold:

$$0 \leq d \leq n, \quad k \leq d_a, \, d_b \leq n = 2k.$$

Recall from Chapter 4 that all possible values within these ranges are realized.

THEOREM 8.3.1. *Let the notation and hypotheses of Theorem* 7.3.1 *hold and let* $k < d_a \leq n$ *and* $k < d_b \leq n$. *Then* $0 < d \leq n$, $d_a - k > 0$ *and* $d_b - k > 0$.

a: *Suppose* $d_a \geq d_b$. *Then*

(i)

$$d_a - k \leq \operatorname{rank}(A) \leq d, \quad d_b - k \leq \operatorname{rank}(B) \leq m_b = 2(d_b - k); \tag{8.3.1}$$

(ii) For any r *satisfying* $0 \leq r \leq d_b - k$, *if*

$$\operatorname{rank}(A) = d_a - k + r \tag{8.3.2}$$

then

$$\operatorname{rank}(B) = d_b - k + r. \tag{8.3.3}$$

Furthermore, for any r *satisfying* $0 \leq r \leq d_b - k$, *there exist matrices* A, B *satisfying conditions (1) and (2) of Theorem* 7.3.1 *such that* (8.3.1), (8.3.2) *and* (8.3.3) *hold.*

b: *Suppose* $d_a < d_b$. *Then*

(i)

$$d_a - k \leq \operatorname{rank}(A) \leq m_a = 2(d_a - k), \quad d_b - k \leq \operatorname{rank}(B) \leq d; \tag{8.3.4}$$

(ii) For any r *satisfying* $0 \leq r \leq d_a - k$, *if*

$$\operatorname{rank}(B) = d_b - k + r \tag{8.3.5}$$

then

$$\operatorname{rank}(A) = d_a - k + r. \tag{8.3.6}$$

Furthermore, for any r *satisfying* $0 \leq r \leq d_a - k$, *there exist matrices* A, B *satisfying conditions (1) and (2) of Theorem* 7.3.1 *such that* (8.3.4), (8.3.5) *and* (8.3.6) *hold.*

PROOF. We only prove part (a) since the proof of part (b) is similar.

(i) we first prove that

$$d_a - k = \frac{m_a}{2} \leq \operatorname{rank}(A) \leq d.$$

If not, we let $\operatorname{rank}(A) = h < d_a - k$. Then by multiplying the boundary conditions (7.3.1) on the left by a nonsingular matrix, (Note that the self-adjointness conditions

(1) and (2) of Theorem 7.3.1 are invariant.) we may assume that the first h rows of A are linearly independent and all other rows are identically zero. Let

$$A = (\xi_1, \cdots, \xi_h, 0, \cdots, 0)^T,$$

where $\xi_i, i = 1, \cdots, h$, are the row vectors of A. And let

$$B = (\gamma_1, \cdots, \gamma_h, \gamma_{h+1}, \cdots, \gamma_d)^T,$$

where $\gamma_i, i = 1, \cdots, d$, are the row vectors of B. Then we compute

$$AE_{m_a}A^* = \begin{pmatrix} \xi_1 E_{m_a}\xi_1^* & \cdots & \xi_1 E_{m_a}\xi_h^* & 0 & \cdots & 0 \\ \cdots & \cdots & \cdots & \cdots & \cdots & \cdots \\ \xi_h E_{m_a}\xi_1^* & \cdots & \xi_h E_{m_a}\xi_h^* & 0 & \cdots & 0 \\ 0 & \cdots & 0 & 0 & \cdots & 0 \\ \cdots & \cdots & \cdots & \cdots & \cdots & \cdots \\ 0 & \cdots & 0 & 0 & \cdots & 0 \end{pmatrix},$$

$$BE_{m_b}B^* = \begin{pmatrix} \gamma_1 E_{m_b}\gamma_1^* & \cdots & \gamma_1 E_{m_b}\gamma_h^* & \gamma_1 E_{m_b}\gamma_{h+1}^* & \cdots & \gamma_1 E_{m_b}\gamma_d^* \\ \cdots & \cdots & \cdots & \cdots & \cdots & \cdots \\ \gamma_h E_{m_b}\gamma_1^* & \cdots & \gamma_h E_{m_b}\gamma_h^* & \gamma_h E_{m_b}\gamma_{h+1}^* & \cdots & \gamma_h E_{m_b}\gamma_d^* \\ \gamma_{h+1} E_{m_b}\gamma_1^* & \cdots & \gamma_{h+1} E_{m_b}\gamma_h^* & \gamma_{h+1} E_{m_b}\gamma_{h+1}^* & \cdots & \gamma_{h+1} E_{m_b}\gamma_d^* \\ \vdots & \vdots & \vdots & \vdots & \vdots & \vdots \\ \gamma_d E_{m_b}\gamma_1^* & \cdots & \gamma_d E_{m_b}\gamma_h^* & \gamma_d E_{m_b}\gamma_{h+1}^* & \cdots & \gamma_d E_{m_b}\gamma_d^* \end{pmatrix}.$$

Then from $AE_{m_a}A^* = BE_{m_b}B^*$, we get

$$\begin{pmatrix} \gamma_{h+1} E_{m_b}\gamma_{h+1}^* & \cdots & \gamma_{h+1} E_{m_b}\gamma_d^* \\ \vdots & \vdots & \vdots \\ \gamma_d E_{m_b}\gamma_{h+1}^* & \cdots & \gamma_d E_{m_b}\gamma_d^* \end{pmatrix} = 0_{(d-r)\times(d-r)}.$$

Let $C = (\gamma_{h+1}, \cdots, \gamma_d)^T$. The above equation is equivalent to $CE_{m_b}C^* = 0$. By Lemma 8.2.1, we have $\text{rank}(C) \le \frac{m_b}{2} = d_b - k$. Note that we let $\text{rank}(A) = h < d_a - k$. Then we have $\text{rank}(A : B) < d$. This contradicts with condition (1) $\text{rank}(A : B) = d$ of Theorem 7.3.1. So $d_a - k \le \text{rank}(A)$. Moreover, note that $d \le m_a$. Hence $d_a - k \le \text{rank}(A) \le d$.

Similarly, we can prove $d_b - k \le \text{rank}(B) \le m_b = 2(d_b - k)$.

(ii) When $r = 0$, then $\text{rank}(A) = d_a - k$. By Theorem 8.2.1 we have $\text{rank}(B) = d_b - k$. In this case, the equations (7.3.1) determine separated self-adjoint boundary conditions.

Assume that $\text{rank}(A) = d_a - k + r$, $1 \le r \le d_b - k$, and $\text{rank}(B) = d_b - k + h$, $0 \le h \le d_b - k$. Then by multiplying the boundary conditions (7.3.1) by a nonsingular matrix and interchanging rows, if necessary, we may assume that the first $d_a - k + r$ rows of A are linearly independent and all other rows are identically zero. By condition (1) of Theorem 7.3.1 we may also assume that the last $d_b - k + h$ rows of B are linearly independent and all other rows are identically zero. For simplicity we set $s_1 = d_a - k$ and $s_2 = d_b - k$.

Let

$$A = (\ \alpha_1, \cdots, \alpha_{s_1}, \alpha_{s_1+1}, \cdots, \alpha_{s_1+r}, 0, \cdots, 0\ )^T,$$

and let

$$B = (\ 0, \cdots, 0, \beta_{s_2+h}, \cdots, \beta_{s_2+1}, \beta_{s_2}, \cdots, \beta_1\ )^T.$$

Then we compute

$$AE_{m_a}A^* = \begin{pmatrix} \alpha_1 E_{m_a}\alpha_1^* & \cdots & \alpha_1 E_{m_a}\alpha_{s_1+r}^* & 0 & \cdots & 0 \\ \cdots & \cdots & \cdots & \cdots & \cdots & \cdots \\ \alpha_{s_1+r}E_{m_a}\alpha_1^* & \cdots & \alpha_{s_1+r}E_{m_a}\alpha_{s_1+r}^* & 0 & \cdots & 0 \\ 0 & \cdots & 0 & 0 & \cdots & 0 \\ \cdots & \cdots & \cdots & \cdots & \cdots & \cdots \\ 0 & \cdots & 0 & 0 & \cdots & 0 \end{pmatrix},$$

$$BE_{m_b}B^* = \begin{pmatrix} 0 & \cdots & 0 & 0 & \cdots & 0 \\ \cdots & \cdots & \cdots & \cdots & \cdots & \cdots \\ 0 & \cdots & 0 & 0 & \cdots & 0 \\ 0 & \cdots & 0 & \beta_{s_2+h}E_{m_b}\beta_{s_2+h}^* & \cdots & \beta_{s_2+h}E_{m_b}\beta_1^* \\ \vdots & \vdots & \vdots & \vdots & \vdots & \vdots \\ 0 & \cdots & 0 & \beta_1 E_{m_b}\beta_{s_2+h}^* & \cdots & \beta_1 E_{m_b}\beta_1^* \end{pmatrix}.$$

It follows from $AE_{m_a}A^* = BE_{m_b}B^*$ that

$$\begin{pmatrix} \alpha_1 E_{m_a}\alpha_1^* & \cdots & \alpha_1 E_{m_a}\alpha_{s_1+r}^* \\ \cdots & \cdots & \cdots \\ \alpha_{d-(s_2+h)}E_{m_a}\alpha_1^* & \cdots & \alpha_{d-(s_2+h)}E_{m_a}\alpha_{s_1+r}^* \end{pmatrix} = 0_{(s_1-h)\times(s_1+r)}, \tag{8.3.7}$$

$$\begin{pmatrix} \beta_{d-s_1-r}E_{m_b}\beta_{s_2+h}^* & \cdots & \beta_{d-s_1-r}E_{m_b}\beta_1^* \\ \cdots & \cdots & \cdots \\ \beta_1 E_{m_b}\beta_{s_2+h}^* & \cdots & \beta_1 E_{m_b}\beta_1^* \end{pmatrix} = 0_{(s_2-r)\times(s_2+h)}. \tag{8.3.8}$$

Equations (8.3.7) are equivalent to

$$\alpha_i \in \{\alpha_1 E_{m_a}, \alpha_2 E_{m_a}, \cdots, \alpha_{s_1+r}E_{m_a}\}^\perp, \quad i = 1, 2, \cdots, s_1 - h, \tag{8.3.9}$$

and (8.3.8) is equivalent to

$$\beta_i \in \{\beta_1 E_{m_b}, \beta_2 E_{m_b}, \cdots, \beta_{s_2+h}E_{m_b}\}^\perp, \quad i = 1, 2, \cdots, s_2 - r. \tag{8.3.10}$$

Since

$$\dim\{\alpha_1 E_{m_a}, \alpha_2 E_{m_a}, \cdots, \alpha_{s_1+r}E_{m_a}\}^\perp = d_a - k - r$$

in $\mathbb{C}^{m_a}$, by (8.3.9) we have $s_1 - h \le d_a - k - r$, i.e. $h \ge r$.

Since

$$\dim\{\beta_1 E_{m_b}, \beta_2 E_{m_b}, \cdots, \beta_{s_2+h}E_{m_b}\}^\perp = d_b - k - h$$

in $\mathbb{C}^{m_b}$, by (8.3.10) we have $s_2 - r \le d_b - k - h$, i.e. $h \le r$. Therefore $h = r$.

This shows that if $0 \le r \le d_b - k$ and $\operatorname{rank}(A) = d_a - k + r$, then $\operatorname{rank}(B) = d_b - k + r$. For the furthermore part, the construction of matrices A, B satisfying these three conditions is routine and therefore omitted. □

The value of the parameter r in Theorem 8.3.1 determines the number of coupled boundary conditions. We expand on this point with a number of corollaries; these follow from Theorem 8.3.1 and its proof.

COROLLARY 8.3.1. *Let the notation and hypotheses of Theorem* 8.3.1 *hold. If* $r = 0$ *then the conditions* (7.3.1) *are separated with exactly* $d_a - k$ *conditions at* a *and exactly* $d_b - k$ *conditions at* b.

COROLLARY 8.3.2. *Let the notation and hypotheses of Theorem* 8.3.1 *(a) hold. If* $d_a > d_b$ *and* $r > 0$, *then* (7.3.1) *has exactly* $2r$ *coupled boundary conditions and exactly* $d_a - k - r$ *separated conditions at* a *and exactly* $d_b - k - r$ *separated conditions at* b.

COROLLARY 8.3.3. *Let the notation and hypotheses of Theorem* 8.3.1 *(a) hold. If* $d_a = d_b$ *and* $0 < r < d_b - k$, *then* (7.3.1) *has exactly* $2r$ *coupled boundary conditions and exactly* $d_a - k - r$ *separated conditions at* a *and exactly* $d_b - k - r$ *separated conditions at* b. *If* $d_a = d_b$ *and* $r = d_b - k$, *then all conditions are coupled. Note that* $d = 2r$ *in this case and that all conditions of* (7.3.1) *can be coupled only when* $d_a = d_b$.

COROLLARY 8.3.4. *Let the notation and hypotheses of Theorem* 8.3.1 *(b) hold. If* $r > 0$, *then* (7.3.1) *has exactly* $2r$ *coupled boundary conditions and exactly* $d_b - k - r$ *separated conditions at* b *and exactly* $d_a - k - r$ *separated conditions at* a.

REMARK 8.3.1. If $0 \leq r \leq \min\{d_a - k,\ d_b - k\}$, then there are exactly $2r$ coupled conditions. Thus we can classify the self-adjoint boundary conditions (7.3.1) into $\min\{d_a - k, d_b - k\} + 1$ 'types' in terms of the value of r. When $r = 0$ all conditions are separated, but note that all conditions can be coupled only when $d_a = d_b$. If $d_a = d_b$ and r assumes its maximum value $r = d_b - k$, then all conditions are coupled.

REMARK 8.3.2. We comment on the three remaining cases (For these cases we don't need Theorem 8.3.1 since they follow from Theorem 7.3.1 directly.):

(1) $d_a = k = d_b$. In this case the minimal operator S_{min} is itself a self-adjoint operator and has no proper self-adjoint extension in H. Thus there are no boundary conditions required or allowed i.e. in Theorem 7.3.1 $A = 0 = B$ and (7.3.1) is vacuous. This case occurs if and only if $d_a = k$ and $d_b = k$.

(2) $d_a = k < d_b$. Then $m_a = 0$ and $m_b = 2d_b - n$. In this case A is vacuous and all the self-adjoint boundary conditions are given by:

$$B \begin{pmatrix} [y, v_1](b) \\ \vdots \\ [y, v_{m_b}](b) \end{pmatrix} = \begin{pmatrix} 0 \\ \vdots \\ 0 \end{pmatrix}, \tag{8.3.11}$$

where the $d \times m_b$ complex matrix B satisfies $\text{rank}(B) = d$ and $BE_{m_b}B^* = 0$. In this case every one of the conditions (7.3.1) is specified at the endpoint b only. If $d_b = n$ the endpoint b is either LC or regular, and $d = k$, $m_b = n$. If b is LC Theorem 7.3.1 reduces to the self-adjoint boundary conditions at the endpoint b :

$$B \begin{pmatrix} [y, v_1](b) \\ \vdots \\ [y, v_n](b) \end{pmatrix} = \begin{pmatrix} 0 \\ \vdots \\ 0 \end{pmatrix}.$$

And if b is regular (8.3.11) reduces to

$$B \begin{pmatrix} y(b) \\ \vdots \\ y^{[n-1]}(b) \end{pmatrix} = \begin{pmatrix} 0 \\ \vdots \\ 0 \end{pmatrix}.$$

(3) $k < d_a \leq n$ and $d_b = k$. Then $m_a = 2d_a - n$ and $m_b = 0$. This case is the same as case (2) with the endpoints a, b interchanged: in case (2) replace b by a, m_b by m_a, v_j by u_j and B by A.

It is interesting to note that Theorem 8.3.1 can be used to give a completely rigorous definition of separated self-adjoint boundary conditions.

DEFINITION 8.3.1. Under the conditions of Theorem 8.3.1 we say that the self-adjoint boundary conditions (7.3.1) of Theorem 7.3.1 are separated if $\operatorname{rank}(A) = d_a - k$. In this case, by Theorem 8.2.1, $\operatorname{rank}(B) = d_b - k$. Note that this is the case $r = 0$ of Theorem 8.3.1.

REMARK 8.3.3. By Definition 8.3.1, we know that when all conditions in (7.3.1) are separated there must be $d_a - k$ at a and $d_b - k$ at b. The next Corollary shows that whenever all self-adjoint conditions are separated and there is only one condition at a given endpoint then this condition -if it is not real - can always be replaced by an equivalent *real* condition.

COROLLARY 8.3.5. *Let the hypotheses and notation of Theorem 7.3.1 hold. When all self-adjoint conditions are separated and there is only one condition at a given endpoint then this condition - if it is not real - can always be replaced by an equivalent real condition.*

PROOF. Let the self-adjoint boundary condition have $d_a - k > 1$ separated condition at a and $d_b - k = 1$ separated condition at b. Then in this case $m_b = 2(d_b - k) = 2$. Let B have the form

$$B = \begin{pmatrix} 0_{(d_a-k)\times 1} & 0_{(d_a-k)\times 1} \\ g & h \end{pmatrix}, \quad g, h \in \mathbb{C}.$$

In this case

$$E_{m_b} = E_2 = \begin{pmatrix} 0 & -1 \\ 1 & 0 \end{pmatrix}$$

and the condition $BE_2B^* = 0$ implies that

$$g\overline{h} = \overline{g}h. \tag{8.3.12}$$

Since $\operatorname{rank} B = 1$ at least one of h, g is not zero. If h is not zero, then multiplying the condition

$$g[y, v_1](b) + h[y, v_2](b) = 0$$

by $\overline{h}$ we may assume that h is real. This implies that g is real in (8.3.12). There is a similar argument when g is not zero. Clearly the same argument also applies regardless of which row g, h are in as long as all other rows of B are zero and the conclusion follows.

Similarly, if the self-adjoint conditions have $d_a - k = 1$ separated condition at a and $d_b - k > 1$ separated condition at b, we have a similar conclusion. □

4. Examples and Construction for All Types

In this section we will construct all types of self-adjoint boundary conditions determined by Theorem 8.3.1 and its corollaries. The construction shows that self-adjoint boundary conditions characterized by Theorem 8.3.1 in terms of the parameter r can be realized for any r, $0 \leq r \leq \min\{d_a - k,\, d_b - k\}$. Also we construct separated non-real self-adjoint boundary conditions. We start with an example to construct conditions for all values of the parameter r of Theorem 8.3.1 (a). The construction for part (b) is similar and hence omitted.

EXAMPLE 8.4.1. Let the notation and hypotheses of Theorem 8.3.1 (a) hold. If $0 \leq r \leq d_b - k$ and $\text{rank}(A) = d_a - k + r$. Then by Theorem 8.3.1, we have $\text{rank}(B) = d_b - k + r$.

Let

$$e_j = (0, 0, \cdots, 0, 1, 0, \cdots, 0)_{1\times m_a}.$$

where 1 is located at the j-th column. Let

$$h_j = (0, 0, \cdots, 0, 1, 0, \cdots, 0)_{1\times m_b},$$

where 1 is located at the j-th column.

The self-adjointness conditions $AE_{m_a}A^* = BE_{m_b}B^*$ given by Theorem 7.3.1 can be written as

$$(A\ \ B)\begin{pmatrix} E_{m_a} & 0 \\ 0 & -E_{m_b} \end{pmatrix}(A\ \ B)^* = 0.$$

(i) If $r = 0$, i.e. separated boundary conditions, we can choose

$$A_{d\times m_a} = (e_1^T, e_2^T, \cdots, e_{d_a-k}^T, 0_{1\times m_a}^T, \cdots, 0_{1\times m_a}^T)^T,$$

and

$$B_{d\times m_b} = (0_{1\times m_b}^T, \cdots, 0_{1\times m_b}^T, h_1^T, h_2^T, \cdots, h_{d_b-k}^T)^T.$$

By a computation, we get that (1) and (2) of Theorem 7.3.1 hold and therefore (7.3.1) is a self-adjoint boundary condition.

(ii) If $0 < r \leq d_b - k$, we can choose

$$A = \left(e_1^T, e_2^T, \ldots, e_{d_a-k}^T, (e_1E_{m_a})^T, \ldots, (e_rE_{m_a})^T, 0_{1\times m_a}^T, \ldots, 0_{1\times m_a}^T\right)^T,$$

$$B = \left(-(h_1E_{m_b})^T, -(h_2E_{m_b})^T, \ldots, -(h_rE_{m_b})^T, 0_{1\times m_b}^T, \ldots, 0_{1\times m_b}^T, h_1^T, \ldots, h_{d_b-k}^T\right)^T.$$

Then (1) and (2) of Theorem 7.3.1 hold and therefore (7.3.1) is a self-adjoint boundary condition.

(iii) Especially, when $r = d_b - k$ and $d_a = d_b$, the self-adjoint boundary conditions are coupled. In this case $d = 2(d_a - k) = m_a = m_b$, we can choose the matrices $A = B = I_{m_a}$, where I_{m_a} denotes the identity matrix.

It is well known [**620**] that in the second order case there are no separated complex boundary conditions in the sense that every such condition is equivalent to a separated real condition. For real coefficient differential expressions of any even order $n = 2k \geq 4$ which satisfy hypothesis **EH** we discuss all the cases $k \leq d_a, d_b \leq n$, and then illustrate whether they have separated non-real self-adjoint conditions. When they have, we construct some examples.

We start with the case $n = 4$.

EXAMPLE 8.4.2. Let

$$My = [(p_2y'')' + p_1y']' + qy = \lambda wy \quad on \quad J = (a, b), \quad -\infty \leq a < b \leq \infty,$$

where $\frac{1}{p_2}, p_1, q, w \in L_{loc}(J, \mathbb{R})$, $w > 0$ on J.

Here we assume that $d_a = k = 2$ and $2 < d_b \leq 4$. Then $m_a = 0$ and $m_b = 2d_b - n = 2d_b - 4$. In this case we have no condition at a and self-adjoint separated conditions at b.

(i) If $d_b = 3$, we have $m_b = 2d_b - 4 = 2$ and $d = d_a + d_b - n = 1$. Then the self-adjoint boundary condition is separated and there is only one condition at

b and this condition can be replaced by an equivalent real self-adjoint condition. This can be seen from Corollary 8.3.5.

(ii) If $d_b = 4$, we have $m_b = 4$ and $d = 2$. We choose

$$B_{2\times 4} = \begin{pmatrix} 1 & i & 0 & 0 \\ 0 & 0 & 1 & -i \end{pmatrix},$$

then $\operatorname{rank}(B) = 2$ and $BE_{m_b}B^* = 0$. By Theorem 7.3.1, these separated conditions are self-adjoint and thus we have the non-real self-adjoint boundary conditions:

$$[y, v_1](b) + i[y, v_2](b) = 0, \quad [y, v_3](b) - i[y, v_4](b) = 0. \tag{8.4.1}$$

EXAMPLE 8.4.3. For $n = 2k = 6$, we let

$$My = \{[(p_3 y''')' + (p_2 y'')']' + p_1 y'\}' + qy = \lambda wy \text{ on } J = (a, b), \ -\infty \le a < b \le \infty,$$

where $\frac{1}{p_3}$, p_2, p_1, q, $w \in L_{loc}(J, \mathbb{R})$, $w > 0$ on J.

We assume that $d_a = 3$ and $3 < d_b \le 6$, then $m_a = 0$, $m_b = 2d_b - 6$.

(i) When $d_b = 4$, then $m_b = 2$, $d = 1$ and every non-real self-adjoint condition is equivalent to a real self-adjoint condition by Corollary 8.3.5.

(ii) When $d_b = 5$, then $m_b = 4$, $d = 2$ and we can use (8.4.1).

(iii) When $d_b = 6$, then $m_b = 6$, $d = 3$. We choose

$$B_{3\times 6} = \begin{pmatrix} 1 & i & 0 & 0 & 0 & 0 \\ 0 & 0 & 0 & 0 & 1 & -i \\ 0 & 0 & 1 & 0 & 0 & 0 \end{pmatrix}.$$

Then $\operatorname{rank}(B) = 3$ and $BE_{m_b}B^* = 0$. By Theorem 7.3.1 the following condition is self-adjoint:

$$\begin{aligned} [y, v_1](b) + i[y, v_2](b) &= 0, \\ [y, v_5](b) - i[y, v_6](b) &= 0, \\ [y, v_3](b) &= 0. \end{aligned}$$

THEOREM 8.4.1. *Let $d_a = k$ and $k < d_b \le n$.*

(i) If $d_b = k + 1$, then $m_b = 2, d = 1$ and the self-adjoint boundary condition has only one condition at b and this condition can be replaced by an equivalent real self-adjoint condition by Corollary 8.3.5.

(ii) If $d_b = k + 2$, then $m_b = 4, d = 2$ and we may construct self-adjoint differential operators specified by non-real separated boundary conditions as in (8.4.1).

(iii) If $d_b = k + r$ $(2 < r \le k)$, then $d = r$, $m_b = 2r$. Let

$$e_j = (0, 0, \cdots, 0, 1, 0, \cdots, 0)_{1\times m_b},$$

where 1 is located at the j-th column. Then

$$\begin{aligned} e_1 E_{m_b} &= (0, 0, \cdots, 0, -1)_{1\times 2r}, \\ e_2 E_{m_b} &= (0, 0, \cdots, 0, 1, 0)_{1\times 2r}. \end{aligned}$$

Choose

$$B = \begin{pmatrix} e_1 + ie_2 \\ ie_1 E_{m_b} + e_2 E_{m_b} \\ e_3 \\ e_4 \\ \vdots \\ e_r \end{pmatrix}.$$

Then rank$(B) = d$ *and* $BE_{m_b}B^* = 0$. *By Theorem* 7.3.1, *the following are separated non-real self-adjoint boundary conditions:*

$$\begin{aligned}
[y, v_1](b) + i[y, v_2](b) &= 0, \\
[y, v_{m_b-1}](b) - i[y, v_{m_b}](b) &= 0, \\
[y, v_3](b) &= 0, \\
[y, v_4](b) &= 0, \\
&\cdots\cdots \\
[y, v_r](b) &= 0.
\end{aligned}$$

REMARK 8.4.1. Let $d_a = k$ and $d_b = k+1$. Then $d = 1$ and by Corollary 8.3.5 , any complex self-adjoint boundary condition at b can be replaced by an equivalent real self-adjoint condition and this can occur only when $d_b = k+1$.

Similar to Theorem 8.4.1, we can construct separated non-real self-adjoint boundary conditions for the case when $k < d_a \leq n$, $d_b = k$.

For the limit circle case, we have the following general result which illustrates the non-real separated self-adjoint boundary conditions.

THEOREM 8.4.2. *If* $d_a = n$ *and* $d_b = n$, *then* $d = n = 2k$, $d_a - k = k$ *and* $d_b - k = k$. *Let*

$$\begin{aligned}
e_1 &= (1, 0, 0, \cdots, 0, 0)_{1\times 2k}, \\
e_2 &= (0, 1, 0, \cdots, 0, 0)_{1\times 2k}, \\
&\vdots \\
e_k &= (0, 0, \cdots, 0, 1, 0, \cdots, 0)_{1\times 2k}, \quad \textit{where } 1 \textit{ is in the } k\textit{-th position.}
\end{aligned}$$

Then

$$\begin{aligned}
-e_1E_n &= (0, 0, \cdots, 0, 1)_{1\times 2k}, \\
e_2E_n &= (0, 0, \cdots, 0, 1, 0)_{1\times 2k}, \\
&\vdots \\
(-1)^k e_kE_n &= (0, 0, \cdots, 0, 1, 0, \cdots, 0)_{1\times 2k}, \quad \textit{where } 1 \textit{ is in the } (k+1)\textit{-th position.}
\end{aligned}$$

For k ***even:***

Let

$$A = \begin{pmatrix} \widetilde{A}_{k\times 2k} \\ 0_{k\times 2k} \end{pmatrix},$$

where

$$\widetilde{A} = \begin{pmatrix} e_1 + ie_2 \\ -e_1E_n + ie_2E_n \\ e_3 + ie_4 \\ -e_3E_n + ie_4E_n \\ \vdots \\ e_{k-1} + ie_k \\ -e_{k-1}E_n + ie_kE_n \end{pmatrix}.$$

Let

$$B = \begin{pmatrix} 0_{k\times 2k} \\ \widetilde{B}_{k\times 2k} \end{pmatrix},$$

where

$$\widetilde{B} = (I_{k\times k}, \quad 0_{k\times k}).$$

Then

$$\operatorname{rank}(A : B) = n$$

and

$$\widetilde{A}E_n\widetilde{A}^* = 0, \quad \widetilde{B}E_n\widetilde{B}^* = 0.$$

Therefore

$$AE_nA^* = 0, \quad BE_nB^* = 0.$$

By Theorem 7.4.1*, we have the separated non-real self-adjoint boundary conditions:*

$$\begin{gathered}
[y, u_1](a) + i[y, u_2](a) = 0, \\
i[y, u_{n-1}](a) + [y, u_n](a) = 0, \\
[y, u_3](a) + i[y, u_4](a) = 0, \\
\cdots\cdots \\
[y, u_{k-1}](a) + i[y, u_k](a) = 0, \\
i[y, u_{k+1}](a) + [y, u_{k+2}](a) = 0, \\
[y, v_1](b) = 0, \\
[y, v_2](b) = 0, \\
\cdots\cdots \\
[y, v_k](b) = 0.
\end{gathered}$$

For k ***odd:***

Set

$$A = \begin{pmatrix} \widetilde{A}_{k\times 2k} \\ 0_{k\times 2k} \end{pmatrix},$$

where

$$\widetilde{A} = \begin{pmatrix} e_1 + ie_2 \\ -e_1E_n + ie_2E_n \\ e_3 + ie_4 \\ -e_3E_n + ie_4E_n \\ \vdots \\ e_{k-2} + ie_{k-1} \\ -e_{k-2}E_n + ie_{k-1}E_n \\ e_k \end{pmatrix}.$$

Set

$$B = \begin{pmatrix} 0_{k\times 2k} \\ \widetilde{B}_{k\times 2k} \end{pmatrix},$$

where

$$\widetilde{B} = (I_{k\times k}, \quad 0_{k\times k}).$$

Then

$$\operatorname{rank}(A : B) = n$$

and

$$\widetilde{A}E_n\widetilde{A}^* = 0, \quad \widetilde{B}E_n\widetilde{B}^* = 0.$$

Therefore

$$AE_nA^* = 0, \quad BE_nB^* = 0.$$

By Theorem 7.4.1, we have the separated non-real self-adjoint boundary conditions:

$$\begin{aligned}
[y,u_1](a)+i[y,u_2](a)&=0,\\
i[y,u_{n-1}](a)+[y,u_n](a)&=0,\\
[y,u_3](a)+i[y,u_4](a)&=0,\\
&\cdots\cdots\\
[y,u_{k-2}](a)+i[y,u_{k-1}](a)&=0,\\
i[y,u_{k+2}](a)+[y,u_{k+3}](a)&=0,\\
[y,u_k](a)&=0,\\
[y,v_1](b)&=0,\\
[y,v_2](b)&=0,\\
&\cdots\cdots\\
[y,v_k](b)&=0.
\end{aligned}$$

REMARK 8.4.2. Using Theorem 8.4.2 we can construct non-real self-adjoint differential operators specified by separated boundary conditions for real limit-circle (LC) differential expressions of even order $n=2k\geq 4$. If the endpoint a is regular, we can simply replace $[y,u_1](a),[y,u_2](a),[y,u_3](a),\cdots$ with $y(a),y^{[1]}(a),y^{[2]}(a),\cdots$. Similarly for a regular endpoint b.

Next we construct non-real separated boundary conditions when neither deficiency index is minimal.

THEOREM 8.4.3. *Assume that $k<d_a\leq n$, $k<d_b\leq n$ and $d_a\geq d_b$.*

(i) Let $k<d_a\leq n$, $k<d_b<n$ and $d_a>d_b$. Then $d_a-k\geq 2$ and there are at least two separated conditions at the endpoint a. So, for this case, we can always construct non-real separated self-adjoint boundary conditions. Notice that when $d_b=k+1$, there is only one separated condition at b and this condition can always be replaced by an equivalent real condition.

Let $d_b=k+r$, $1\leq r\leq k-1$. Then for each fixed r, $k+r<d_a\leq n$, $d_b-k=r$, $m_b=2(d_b-k)=2r$, $m_a=2(d_a-k)$ and $d=d_a-k+r$.

(1) If $d_a-k=2$ and $d_a>d_b$, we have $d_b-k=1$, $m_a=4$, $m_b=2$ and $d=3$. Choose

$$A_{3\times 4}=\begin{pmatrix}1&i&0&0\\0&0&1&-i\\0&0&0&0\end{pmatrix},\quad B_{3\times 2}=\begin{pmatrix}0&0\\0&0\\1&0\end{pmatrix}.$$

Then

$$\operatorname{rank}(A:B)=3,$$

and

$$AE_4A^*=BE_2B^*=0.$$

Therefore we have the following separated non-real self-adjoint boundary conditions at a:

$$\begin{aligned}
[y,u_1](a)+i[y,u_2](a)&=0,\\
[y,u_3](a)-i[y,u_4](a)&=0,
\end{aligned}$$

and the separated condition at b:

$$[y,v_1](b)=0.$$

(2) For each $2 \leq r \leq k-1$, $d_a - k > r = d_b - k \geq 2$, *we let*

$$e_j = (0, 0, \cdots, 0, 1, 0, \cdots, 0)_{1 \times m_a},$$

where 1 *is in the* j*-th position.*

Set

$$A = \begin{pmatrix} \widetilde{A}_{(d_a-k)\times m_a} \\ 0_{r\times m_a} \end{pmatrix},$$

where

$$\widetilde{A} = \begin{pmatrix} e_1 + ie_2 \\ ie_1 E_{m_a} + e_2 E_{m_a} \\ e_3 \\ \vdots \\ e_{\frac{m_a}{2}} \end{pmatrix}.$$

And set

$$B = \begin{pmatrix} 0_{(d_a-k)\times 2r} \\ \widetilde{B}_{r\times 2r} \end{pmatrix},$$

where

$$\widetilde{B}_{r\times 2r} = (I_{r\times r}, \quad 0_{r\times r}).$$

Then

$$\operatorname{rank}(A : B) = d$$

and

$$\widetilde{A} E_{m_a} \widetilde{A}^* = 0, \quad \widetilde{B} E_{m_b} \widetilde{B}^* = 0.$$

Therefore

$$A E_{m_a} A^* = 0 = B E_{m_b} B^*.$$

By Theorem 7.3.1, *these separated conditions are non-real self-adjoint boundary conditions:*

$$\begin{gathered} [y, u_1](a) + i[y, u_2](a) = 0, \\ [y, u_{m_a-1}](a) - i[y, u_{m_a}](a) = 0, \\ [y, u_3](a) = 0, \\ \cdots\cdots \\ [y, u_{\frac{m_a}{2}}](a) = 0, \\ [y, v_1](b) = 0, \\ [y, v_2](b) = 0, \\ \cdots\cdots \\ [y, v_r](b) = 0. \end{gathered}$$

(ii) Let $k < d_a < n$, $k < d_b < n$ *and* $d_a = d_b$.

(1) When $d_a - k = d_b - k = 1$, *then* $d = 2$ *and* $m_a = m_b = 2$. *By Theorem* 8.2.1, *the separated self-adjoint boundary conditions, there is only one separated condition at* a *and only one separated condition at* b. *By Corollary* 8.3.5, *these conditions can always be replaced by equivalent real conditions. In this case, there are no non-real separated self-adjoint boundary conditions.*

(2) Assume that $d_a = d_b = k + r$ $(r \geq 2)$. *According to the method of (i), we can construct non-real separated self-adjoint boundary conditions.*

Here we have concentrated on the cases when $k < d_a \leq n$, $k < d_b \leq n$ and $d_a \geq d_b$. The cases when $k < d_a \leq n$, $k < d_b \leq n$ and $d_a < d_b$ are similar and hence omitted.

5. Symmetric Boundary Conditions

Theorem 6.3.3 characterizes the two-point singular boundary conditions which determine symmetric operators in terms of solutions $u_1, \cdots, u_{m_a}$ and $v_1, \cdots, v_{m_b}$ and matrices E_{m_a}, E_{m_b} which are constructed from these solutions. The proof is based on the representation of $D_{\max}$ in terms of $u_1, \cdots, u_{m_a}$ and $v_1, \cdots, v_{m_b}$ given by (4.4.7) in Theorem 4.4.4 where $u_1, \cdots, u_{m_a}$ are solutions on an interval (a, c) for some complex number λ_a and $v_1, \cdots, v_{m_b}$ are solutions on (c, b) for some complex number λ_b, with $\operatorname{Im}(\lambda_a) \neq 0 \neq \operatorname{Im}(\lambda_b)$.

Using Theorem 6.3.3 to check whether a given boundary condition is symmetric or not is not easy because the matrices E_{m_a}, E_{m_b} are not easy to compute. The next theorem is based on the real λ decomposition of $D_{\max}$ given by Theorem 7.2.4 and different matrices for E_{m_a}, E_{m_b} which are much simpler.

REMARK 8.5.1. In the next theorem we will continue to use the same notation $u_1, \cdots, u_{m_a}$ and $v_1, \cdots, v_{m_b}$ to avoid introducing more complicated notations.

THEOREM 8.5.1. *Suppose M is a symmetric expression of order $n = 2k$, $k \in \mathbb{N}$, with real coefficients. Let $a < c < b$. Assume that the deficiency indices of M on (a, c), (c, b) are d_a, d_b, respectively, and hypothesis **EH** holds. Let $u_1, u_2, \cdots, u_{m_a}$, $m_a = 2d_a - n$, and $v_1, v_2, \cdots, v_{m_b}$, $m_b = 2d_b - n$, be LC solutions constructed on (a, c), (c, b), respectively, and extended to maximal domain functions in $D_{\max} = D_{\max}(a, b)$ by Theorem 7.2.3. Using these LC solutions u_j and v_j, we define*

$$Y_{a,b} = \begin{pmatrix} Y(a) \\ Y(b) \end{pmatrix}, \; Y(a) = \begin{pmatrix} [y, u_1](a) \\ \vdots \\ [y, u_{m_a}](a) \end{pmatrix}, \quad Y(b) = \begin{pmatrix} [y, v_1](b) \\ \vdots \\ [y, v_{m_b}](b) \end{pmatrix}.$$

Assume $U \in M_{l,2d}$ has rank l, $0 \leq l \leq m_a + m_b = 2d$ and let $U = (A : B)$ with $A \in M_{l,m_a}$ consisting of the first m_a columns of U in the same order as they are in U and $B \in M_{l,m_b}$ consisting of the next m_b columns of U in the same order as they are in U. Define the operator $S(U)$ in $L^2(J, w)$ by

$$D(S(U)) = \{y \in D_{\max} : U Y_{a,b} = 0\},$$
$$S(U)y = S_{\max} y \quad for \quad y \in D(S(U)).$$

Let

$$C = C(A, B) = A E_{m_a} A^* - B E_{m_b} B^*, \text{ and let } r = \operatorname{rank} C,$$

where E_{m_a} and E_{m_b} are the simple symplectic matrices

$$E_{m_a} = ((-1)^r \delta_{r, m_a+1-s})_{r,s=1}^{m_a},$$
$$E_{m_b} = ((-1)^r \delta_{r, m_b+1-s})_{r,s=1}^{m_b}.$$

Then we have

(1) *If $l < d_a + d_b - n = d$, then $S(U)$ is not symmetric.*
(2) *If $l = d_a + d_b - n = d$, then $S(U)$ is self-adjoint if and only if $r = 0$.*
(3) *Let $l = d + s$, $0 < s \leq d$. Then $S(U)$ is symmetric if and only if $r = 2s$.*

PROOF. The proof is similar to the proof of Theorem 6.3.3 and hence omitted. □

REMARK 8.5.2. Here E_{m_a} and E_{m_b} are simple constant matrices. These are different from the matrices E_{m_a}, E_{m_b} used in Theorem 6.3.3. Note that the self-adjoint characterization of part (2) of Theorem 8.5.1 is the same as in Theorem 7.3.1.

Next we construct examples of symmetric operators based on Theorem 8.5.1. Let $l = \text{rank } U$. By (1) of Theorem 8.5.1 $S(U)$ is not symmetric when $l < d$. When $l = d$ then $S(U)$ is self-adjoint (and therefore also symmetric); when $l = 2d$, $S_{\min}$ is symmetric with no proper symmetric extension.

Here we construct examples of symmetric operators for $k > 1$ and all cases when $d < l < 2d$.

By Theorem 8.5.1 a natural question to ask is: Given $C = C(A, B) = AE_{m_a}A^* - BE_{m_b}B^*$ with $r = \text{rank } C = 0$ how can rows be added to the matrix $U = (A : B)$ such that $r = \text{rank } C = 2s$ where s is defined by $l = \text{rank } U = d + s$ for $s = 1, 2, \cdots, d-1$?

This is the approach we take here. It is based on the classification of the self-adjoint boundary conditions given in Section 8.3.

EXAMPLE 8.5.1. Let the hypotheses and notation of Theorem 8.5.1 hold and consider the symmetric expression M given by

$$My = [(p_2 y'')' + (p_1 y')]' + p_0 y = \lambda w y \ \ on\ J = (a, b),\ -\infty \leq a < b \leq \infty,$$

where

$$\frac{1}{p_2},\ p_1,\ p_0,\ w \in L_{loc}(J, \mathbb{R}),\ \ w > 0\ \ a.e.\ on\ J.$$

If a is a regular endpoint for this M then, in the discussion below, simply replace $[y, u_1](a)$, $[y, u_2](a)$, $[y, u_3](a)$, $[y, u_4](a)$ with $y(a)$, $y^{[1]}(a)$, $y^{[2]}(a)$, $y^{[3]}(a)$, respectively. Similarly, if b is regular, replace $[y, v_1](b)$, $[y, v_2](b)$, $[y, v_3](b)$, $[y, v_4](b)$ with $y(b)$, $y^{[1]}(b)$, $y^{[2]}(b)$, $y^{[3]}(b)$, respectively. If one endpoint is regular and the other singular then make this replacement only at the regular endpoint.

Suppose that $d_a = d_b = 4$, then $d = 4$, $m_a = m_b = 4$.

It follows from Theorem 8.5.1 that if $l = d + s = 4 + s$, $0 < s < 4$, then $S(U)$ is symmetric if and only if $r = 2s$. We construct examples for each $s = 1, 2, 3$.

(1) If $s = 1$, then $l = 5$ and $r = 2$.

(i) Choose

$$A_{5\times 4} = \begin{pmatrix} 1 & 0 & 0 & 0 \\ 0 & 1 & 0 & 0 \\ 0 & 0 & 0 & 0 \\ 0 & 0 & 0 & 0 \\ 0 & 0 & 0 & 0 \end{pmatrix}, \quad B_{5\times 4} = \begin{pmatrix} 0 & 0 & 0 & 0 \\ 0 & 0 & 0 & 0 \\ 1 & 0 & 0 & 0 \\ 0 & 1 & 0 & 0 \\ 0 & 0 & 1 & 0 \end{pmatrix}.$$

Clearly $\text{rank } U = \text{rank}\,(A : B) = l = 5$ and, by computation,

$$\text{rank } C = \text{rank}\,(AE_4A^* - BE_4B^*) = r = 2.$$

Therefore by Theorem 8.5.1 the operator $S(U)$ determined by the following boundary conditions is symmetric:

$$[y,u_1](a)=0,\quad [y,u_2](a)=0,$$
$$[y,v_1](b)=0,\quad [y,v_2](b)=0,\quad [y,v_3](b)=0.$$

Note that this is a symmetric operator with separated boundary conditions: there are 2 conditions at the endpoint a, 3 at b and no coupled condition i.e. no condition involving both endpoints.

(ii) Let

$$A_{5\times 4}=\begin{pmatrix}1&0&0&0\\0&1&0&0\\0&0&0&-1\\0&0&0&0\\0&0&0&0\end{pmatrix},\quad B_{5\times 4}=\begin{pmatrix}0&0&0&1\\0&0&0&0\\1&0&0&0\\0&1&0&0\\0&0&1&0\end{pmatrix}.$$

Then $\operatorname{rank} U=\operatorname{rank}(A:B)=l=5$ and $\operatorname{rank} C=r=2$. Therefore, by Theorem 8.5.1 the operator $S(U)$, determined by the following boundary conditions, is symmetric:

$$[y,u_2](a)=0,\quad [y,v_2](b)=0,\quad [y,v_3](b)=0,$$
$$[y,u_1](a)+[y,v_4](b)=0,\quad -[y,u_4](a)+[y,v_1](b)=0.$$

Note that here the symmetric operator has mixed boundary conditions: 1 separated condition at a, 2 separated conditions at b and 2 coupled conditions.

(iii) Choose

$$A_{5\times 4}=\begin{pmatrix}1&0&0&0\\0&1&0&0\\0&0&0&-1\\0&0&-1&0\\0&0&0&0\end{pmatrix},\quad B_{5\times 4}=\begin{pmatrix}0&0&0&1\\0&0&0&0\\1&0&0&0\\0&1&0&0\\0&0&1&0\end{pmatrix}.$$

Then $\operatorname{rank} U=\operatorname{rank}(A:B)=l=5$ and $\operatorname{rank} C=r=2$. Then by Theorem 8.5.1 the operator $S(U)$ determined by the following boundary conditions is symmetric:

$$\begin{aligned}&[y,u_2](a)=0,\qquad [y,v_3](b)=0,\\&[y,u_1](a)+[y,v_4](b)=0,\qquad -[y,u_4](a)+[y,v_1](b)=0,\\&-[y,u_3](a)+[y,v_2](b)=0.\end{aligned}$$

Here there is 1 separated condition at a and at b and three coupled conditions.

(2) If $s=2$, then $l=6$ and $r=4$.

(i) Choose

$$A_{6\times 4}=\begin{pmatrix}1&0&0&0\\0&1&0&0\\0&0&0&0\\0&0&0&0\\0&0&0&0\\0&0&0&0\end{pmatrix},\quad B_{6\times 4}=\begin{pmatrix}0&0&0&0\\0&0&0&0\\1&0&0&0\\0&1&0&0\\0&0&1&0\\0&0&0&1\end{pmatrix}.$$

Then $\operatorname{rank} U=\operatorname{rank}(A:B)=l=6$ and $\operatorname{rank} C=r=4$. It follows from Theorem 8.5.1 that the operator $S(U)$ determined by the following boundary conditions is symmetric:

$$[y,u_1](a)=0, \quad [y,u_2](a)=0, \quad [y,v_1](b)=0,$$
$$[y,v_2](b)=0, \quad [y,v_3](b)=0, \quad [y,v_4](b)=0.$$

(ii) Choose

$$A_{6\times 4}=\begin{pmatrix} 1 & 0 & 0 & 0\\ 0 & 1 & 0 & 0\\ 0 & 0 & 1 & 0\\ 0 & 0 & 0 & -1\\ 0 & 0 & 0 & 0\\ 0 & 0 & 0 & 0 \end{pmatrix}, \quad B_{6\times 4}=\begin{pmatrix} 0 & 0 & 0 & 0\\ 0 & 0 & 0 & 0\\ 0 & 1 & 0 & 0\\ 1 & 0 & 0 & 0\\ 0 & 0 & 1 & 0\\ 0 & 0 & 0 & 1 \end{pmatrix}.$$

Then $rankU = rank(A:B) = l = 6$ and $rankC = r = 4$. By Theorem 8.5.1, the operator $S(U)$ determined by the following boundary conditions is symmetric:

$$[y,u_1](a)=0, \quad [y,u_2](a)=0, \quad [y,v_3](b)=0, \quad [y,v_4](b)=0,$$
$$[y,u_3](a)+[y,v_2](b)=0, \quad -[y,u_4](a)+[y,v_1](b)=0.$$

Note that here there are 2 separated condition at a, 2 separated conditions at b and 2 coupled conditions.

(3) If $s=3$, then $l=7$ and $r=6$. Choose

$$A_{7\times 4}=\begin{pmatrix} 1 & 0 & 0 & 0\\ 0 & 1 & 0 & 0\\ 0 & 0 & 1 & 0\\ 0 & 0 & 0 & 0\\ 0 & 0 & 0 & 0\\ 0 & 0 & 0 & 1\\ 0 & 0 & 0 & 0 \end{pmatrix}, \quad B_{7\times 4}=\begin{pmatrix} 0 & 0 & 0 & 0\\ 0 & 0 & 0 & 0\\ 0 & 0 & 0 & 0\\ 0 & 0 & 1 & 0\\ 0 & 1 & 0 & 0\\ 0 & 0 & 0 & -i\\ 1 & 0 & 0 & 0 \end{pmatrix}.$$

It can be easily checked that rank $U = \text{rank}(A:B) = l = 7$ and rank $C = r = 6$. By Theorem 8.5.1, we see that the operator $S(U)$ determined by the following boundary conditions is symmetric:

$$[y,u_4](a) = i[y,v_4](b),$$
$$[y,u_j](a)=0, \quad [y,v_j](b)=0, \quad j=1,2,3.$$

Note the nonreal coupled condition.

Next we construct symmetric operators with separated boundary conditions for symmetric expressions M, with real coefficients, of any order $n=2k$, $k>1$. Our approach as above is to add rows to the matrices determining self-adjoint operators. Recall Corollary 8.3.1, Note that in this self-adjoint case there must be exactly $d_a - k$ separated conditions at the endpoint a and exactly $d_b - k$ separated boundary conditions at the endpoint b.

EXAMPLE 8.5.2. Let the hypotheses and notation of Theorem 8.5.1 hold. To put our construction of symmetric operators in context with the self-adjoint case, we discuss our construction for the self-adjoint case first.

Let

$$\widetilde{A}_{d\times m_a} = \begin{pmatrix} 1 & 0 & 0 & \cdots & 0 & 0 & \cdots & 0 \\ 0 & 1 & 0 & \cdots & 0 & 0 & \cdots & 0 \\ \cdots & & \cdots & \cdots & \cdots & & \cdots & \\ 0 & 0 & 0 & \cdots & 1 & 0 & \cdots & 0 \\ 0 & 0 & 0 & \cdots & 0 & 0 & \cdots & 0 \\ \cdots & & \cdots & \cdots & \cdots & \cdots & \cdots & \\ 0 & 0 & 0 & \cdots & 0 & 0 & \cdots & 0 \end{pmatrix},$$

$$\widetilde{B}_{d\times m_b} = \begin{pmatrix} 0 & 0 & 0 & \cdots & 0 & 0 & \cdots & 0 \\ \cdots & & \cdots & \cdots & \cdots & \cdots & \cdots & \\ 0 & 0 & 0 & \cdots & 0 & 0 & \cdots & 0 \\ 1 & 0 & 0 & \cdots & 0 & 0 & \cdots & 0 \\ 0 & 1 & 0 & \cdots & 0 & 0 & \cdots & 0 \\ \cdots & & \cdots & \cdots & \cdots & & \cdots & \\ 0 & 0 & 0 & \cdots & 1 & 0 & \cdots & 0 \end{pmatrix},$$

where the number of $1's$ of $\widetilde{A}$ is $d_a - k = \frac{m_a}{2}$ and the number of $1's$ of $\widetilde{B}$ is $d_b - k = \frac{m_b}{2}$.

If $d_a - k > 0$ and $d_b - k > 0$, then by Theorem 7.3.1 $U = (\widetilde{A} : \widetilde{B})$ is a self-adjoint boundary condition matrix. If $d_a - k = 0$, then there is no $\widetilde{A}$, $\widetilde{B} = (I_d, 0_d)$ and $\widetilde{B}Y(b) = 0$ determines the self-adjoint boundary conditions. If $d_b - k = 0$, then there is no $\widetilde{B}$, $\widetilde{A} = (I_d, 0_d)$ and $\widetilde{A}Y(a) = 0$ determines the self-adjoint boundary conditions.

Now we start our construction of symmetric operators with separated boundary conditions. Let rank $U = l$, $l = d + s$, $0 < s < d$. By Theorem 8.5.1 (3), for any given l, $d < l < 2d$, $S_{\min}$ has a proper symmetric extension $S(U)$ if and only if rank $C = \text{rank}\,(AE_{m_a}A^* - BE_{m_b}B^*) = r = 2s$. We construct the boundary condition matrix $U = (A : B)$ of symmetric operators $S(U)$ by adding rows to matrices $\widetilde{A}$ and $\widetilde{B}$.

(i) If $0 < s \le d_b - k < d$, let

$$e_j = (0, 0, \cdots, 0, 1, 0, \cdots, 0)_{1\times m_b}, \quad j = 1, 2, \cdots, s,$$

where 1 is located at the $(d_b - k + j)$-th column. In order insert rows $e_1, e_2, \cdots, e_s$ after the last row of matrix $\widetilde{B}$, and correspondingly insert s zero-rows after the last row of matrix $\widetilde{A}$, then we obtain the boundary condition matrix

$$U = (A : B) = \begin{pmatrix} \widetilde{A} & \widetilde{B} \\ 0_{1\times m_a} & e_1 \\ 0_{1\times m_a} & e_2 \\ \cdots & \cdots \\ 0_{1\times m_a} & e_s \end{pmatrix}.$$

By computation it follows that rank $U = l = d + s$ and rank $C = 2s$. Therefore $S(U)$ determined by $UY_{a,b} = 0$ is a symmetric operator with separated boundary conditions.

(ii) If $0 < d_b - k < s < d$, let

$$e_j = (0, 0, \cdots, 0, 1, 0, \cdots, 0)_{1\times m_b}, \quad j = 1, 2, \cdots, d_b - k,$$

where 1 is located at the $(d_b - k + j)$-th column, and let

$$h_j = (0, 0, \cdots, 0, 1, 0, \cdots, 0)_{1\times m_a}, \quad j = 1, 2, \cdots, s - (d_b - k),$$

where 1 is located at the $(d_a - k + j)$-th column. In order insert the rows e_1, $e_2, \cdots, e_{d_b-k}$ after the last row of the matrix $\widetilde{B}$, and correspondingly insert $d_b - k$ zero-rows after the last row of the matrix $\widetilde{A}$. Now continue to insert, in order, the rows h_1, h_2, $\cdots$, $h_{d_a-k+(s-d_b+k)}$ between the $(d_a - k)$-th row and $(d_a - k + 1)$-th row of $\widetilde{A}$ and then insert zero-rows into the corresponding positions of $\widetilde{B}$. Thus we obtain the boundary condition matrix $U = (A : B)$ and get the separated boundary conditions $(A : B)Y_{a,b} = 0$ of the symmetric operator $S(U)$.

(iii) If $d_b - k = d$, then $d_a - k = 0$, $m_a = 0$, $m_b = 2d$ and $\widetilde{A}$ is vacuous. In order insert rows e_1, $e_2, \cdots, e_s$ after the last row of matrix $\widetilde{B}$, then we get the matrix $B_{l\times m_b}$. The boundary conditions $BY(b) = 0$ determine symmetric operators.

(iv) If $d_b - k = 0$, then $d_a - k = d$, $m_b = 0$ and $m_a = 2d$. The construction method is similar to (iii).

6. Comments

The classification of self-adjoint regular and singular boundary conditions for problems of even order $n > 2$ into the three types, separated, coupled, and mixed is based on Wang-Sun-Zettl [**555**], [**554**] and each of these three types is realized.

The characterization, given by Theorem 8.5.1, of the boundary conditions determined by boundary matrices U which generate symmetric operators $S(U)$ is due to the authors.

CHAPTER 9

Solutions and Spectrum

1. Introduction

It is well known [**573**] that the spectrum of a self-adjoint ordinary differential operator S, $S_{\min}(Q) \subset S = S^* \subset S_{\max}(Q)$, $Q \in Z_n(J,\mathbb{R})$, $n > 1$, with Q Lagrange symmetric, constructed above in the Hilbert space $H = L^2(J,w)$, $J = (a,b)$, with w any weight function, is real, consists of eigenvalues of finite multiplicity, of essential spectrum, and of continuous spectrum. In the literature 'essential spectrum' and 'continuous spectrum' are sometimes used interchangeably, we use Weidmann's definitions from his well known book [**573**] which differentiate between these terms. There are also other parts of the spectrum we do not study here.

REMARK 9.1.1. Every eigenvalue of every operator S , $S_{\min}(Q) \subset S = S^* \subset S_{\max}(Q)$, $Q \in Z_n(J,\mathbb{R})$, $n > 1$, studied here has finite multiplicity.

PROPOSITION 9.1.1. *Any isolated point λ of the spectrum of a self-adjoint operator S, $S_{\min}(Q) \subset S = S^* \subset S_{\max}(Q)$, is an eigenvalue of S.*

Next we give Weidmann's definitions of the spectrum discussed below.

DEFINITION 9.1.1. The essential spectrum $\sigma_e(S)$ of a self-adjoint operator S in a Hilbert space H is the set of those points of $\sigma(S)$ that are either accumulation points of $\sigma(S)$ or isolated eigenvalues of infinite multiplicity. The set $\sigma_d(S) = \sigma(S)\backslash\sigma_e(S)$ is called the discrete spectrum of S and consists of the isolated eigenvalues of finite multiplicity for the operators S studied here. Below, by the multiplicity of an eigenvalue we mean its geometric multiplicity. We say that the spectrum of S is discrete if $\sigma_e(S)$ is empty.

DEFINITION 9.1.2. Let S be a self-adjoint operator on an abstract Hilbert space H. Let H_p denote the closed linear hull of all eigenfunctions of S, we call $H_p = H_p(S)$ the discontinuous subspace of H with respect to S. The orthogonal complement of H_p is called the continuous subspace of H with respect to S. This is denoted by $H_c = H_c(S)$. We denote by S_p, S_c the restrictions of S to H_p, H_c, respectively. These operators are called the (spectral) discontinuous, and continuous parts of S, respectively.

DEFINITION 9.1.3. The continuous spectrum $\sigma_c(S)$ of S is defined as the spectrum of S_c. The point spectrum $\sigma_p(S)$ is the set of eigenvalues of S.

REMARK 9.1.2. The point spectrum $\sigma_p(S)$ is also the set of eigenvalues of S_p, however, in general, we only have that $\sigma(S_p) = \overline{\sigma_p(S)}$. The set $\sigma_c(S)$ is closed and $\sigma(S) = \overline{\sigma_p(S)} \cup \sigma_c(S)$. We say that S has pure point spectrum if $H_p = H$, i.e. $\sigma(S) = \overline{\sigma_p(S)}$, see page 209 in [**573**].

A number λ is an eigenvalue of a self-adjoint operator S if the corresponding differential equation $My = \lambda wy$ on J has a nontrivial solution in H which satisfies the boundary condition which determines S. On the other hand, the essential spectrum is independent of the boundary conditions and thus depends only on the coefficients, including the weight function w, of the equation. This dependence is implicit and highly complicated. The coefficients and the weight function also determine the deficiency index d of the minimal operator $S_{\min}$ determined by the equation. This is the number of linearly independent solutions in H for nonreal values of the spectral parameter λ and this number is independent of λ provided $\text{Im}(\lambda) \neq 0$.

REMARK 9.1.3. For real values of λ the number of linearly independent solutions $r(\lambda)$ which lie in H varies with λ. The spectrum of every self-adjoint operator S is real. Hence the function $r(\lambda)$ contains the spectral information of every self-adjoint operator S. Using the construction of LC and LP solutions in Section 7.2 we extract some of this information. Most of the results in this chapter are surprisingly recent.

The contrasting behavior of $r(\lambda)$ for the two singular endpoint case from the one singular endpoint case has some interesting consequences. In the case of only one singular endpoint we have $r(\lambda) \leq d$ for every real λ, whereas in the two singular endpoint case $r(\lambda)$ may assume values less than d, equal to d, or greater than d. The case $r(\lambda) > d$ leads to the surprising and counter-intuitive result that this value of λ is an eigenvalue of *every* self-adjoint extension, i.e. for any given self-adjoint boundary condition, there are eigenfunctions of this λ which satisfy the given boundary condition. The one singular endpoint case is discussed in Section 9.2, the two singular endpoint case in Section 9.3.

2. Only One Singular Endpoint

We study spectral properties of the self-adjoint realizations S of the equation

$$My = \lambda wy \text{ on } J = (a, b), \quad -\infty \leq a < b \leq \infty \tag{9.2.1}$$

in the Hilbert space $H = L^2(J, w)$ where $M = M_Q$, $Q \in Z_n(J, \mathbb{R})$, $n = 2k$, $k \geq 1$, $Q = Q^+$, w is a weight function on J, $d(M) = d$. In this section, we assume that the endpoint a is regular. Our emphasis in this chapter is on the higher dimensional case when $k > 1$, for a more comprehensive discussion of the second order case, e.g. dependence of the eigenvalues on the boundary conditions, see the book [**620**].

Recall from Chapter 4 that

$$k \leq d \leq 2k = n,$$

and that for every $\lambda \in \mathbb{R}$ we have $r(\lambda) \leq d$ by Lemma 7.2.1.

THEOREM 9.2.1. *The following results hold:*

1. *If $r(\lambda) < d$ for some $\lambda \in \mathbb{R}$, then λ is in the essential spectrum of every self-adjoint extension S of $S_{\min}$. In particular, if $r(\lambda) < d$ for every $\lambda \in \mathbb{R}$, then $\sigma_e(S) = (-\infty, \infty)$ for every self-adjoint extension S.*
2. *If, for some $\lambda \in \mathbb{R}$, $r(\lambda) = d$, then λ is an eigenvalue of geometric multiplicity d for some self-adjoint realization S.*

PROOF. Part (1) is well known [**574**]. See also the proof given in [**440**] for a special case, it extends readily to our hypotheses. For part (2), see [**531**], let

$\chi_1, \cdots, \chi_d$ be linearly independent real solutions in H for some $\lambda_1 \in \mathbb{R}$. From the proof of Lemma 7.2.1 it follows that the operator S_u with domain D_u given by

$$D_u = D_{\min} \dot{+} \operatorname{span}\{\chi_1, \cdots, \chi_d\}$$

is a self-adjoint extension of $S_{\min}$. Hence each χ_j, $j = 1, \cdots, d$ is an eigenfunction of this λ_1. □

Next we explore the relationship between $r(\lambda)$ and the continuous spectrum for arbitrary deficiency index d.

THEOREM 9.2.2. *Assume there exists an open interval* $I = (\mu_1, \mu_2)$, $-\infty \leq \mu_1 < \mu_2 \leq \infty$, *of the real line such that the equation* (9.2.1) *has d linearly independent solutions which lie in H for every* $\lambda \in I$. *Then*

(1) for any self-adjoint realization S of (9.2.1), *the intersection* $\sigma_c(S) \cap I$ *is empty.*

(2) for any self-adjoint realization S of (9.2.1), *the point spectrum* $\sigma_p(S)$ *is nowhere dense in I.*

PROOF. See [**256**] for part (1) and [**531**] for part (2). The proof of each part uses the construction of LC and LP solutions given in Section 7.2. Also, see the next Remark. □

REMARK 9.2.1. We comment on this theorem. When $d = n$ the conclusions follow from the well known fact that the spectrum of every self-adjoint extension is discrete. The special case when $d = k$ and $w = 1$ is proved in Weidmann [**574**]. The extension to general w is routine. The extension to intermediate deficiency indices d, $k < d < n$ is not routine. There are two major obstacles: (i) When $d = k$ there is no boundary condition at the singular endpoint. When $k < d < n$ there are exactly d self-adjoint boundary conditions including some at the singular endpoint. What are they? This answer is given in Section 7.3 in terms of LC solutions. Obstacle (ii) involves an approximation method which depends on the LC and LP solutions constructed in Section 7.2. It is somewhat similar in spirit to the approximations based on 'inherited' boundary conditions used in an algorithm of the Bailey-Everitt-Zettl code SLEIGN2 for the computation of eigenvalues of singular Sturm-Liouville problems.

The continuous spectrum is contained in the essential spectrum. Can the conclusion of part (1) of this theorem be strengthened to 'the intersection $\sigma_e(S) \cap I$ is empty'? This is conjectured in [**256**] but the answer is no even in the second order case where the question dates back to Hartman and Wintner [**284**].

THEOREM 9.2.3. *Assume that* $My = -y'' + qy = \lambda y$ *has an* L^2 *solution for all* λ *in some interval* $I = (\mu_1, \mu_2)$. *Then for every self-adjoint extension of* $S_{\min}$ *:*

(a) There is no continuous spectrum in I.

(b) The point spectrum σ_p *is nowhere dense in I, i.e. its closure does not contain a nonempty open set.*

Hartman and Wintner [**284**] as well as others conjectured that part (b) could be improved to: σ_p has no accumulation points of eigenvalues in I. But this is false. In fact, the next theorem not only disproves this conjecture, but also shows that (a) is sharp!

REMARK 9.2.2. Remling [**498**] has clarified the complicated relationship between the essential spectrum and the real numbers λ for which $r(\lambda) = d$. In particular he showed that, in general, $r(\lambda) = d$ for all λ in some open interval I, does not imply that there is no essential spectrum in I. In 1996 Remling [**498**] proved that, in the second order case, "continuous spectrum is empty in I" cannot be strengthened to "essential spectrum is empty in I" And the continuous spectrum result of this theorem is 'best possible'.

THEOREM 9.2.4 (Remling). *Let I be a finite, open interval, and let $I_1 \subset I$ be a closed, nowhere dense set. Then there exists a potential q such that:*
1) $\sigma_e = I_1$,
2) $My = -y'' + qy = \lambda y$ has an L^2-solution for all $\lambda \in I$.

Note that I_1 can be an uncountable set.

Thus a natural question is: under what additional condition is the essential spectrum empty in an interval I? This question is answered by Hao et. al. in [**256**] for the general even order case $n = 2k$, $k \geq 1$, using Assumption (A) stated next, and the following theorem.

Assumption (A): There exists a domain G in the complex plane containing $I = (\alpha, \beta)$ with α,β on the boundary of G, and for equation (9.2.1) and for $\lambda \in I$ there exist real solutions $\varphi_1(t, \lambda), \cdots, \varphi_d(t, \lambda)$ which lie in H, and are analytic on G for each fixed $t \in J$.

THEOREM 9.2.5. *Consider the equation* (9.2.1). *If $r(\lambda) = d$ for all λ in some open interval $I = (\alpha, \beta)$, and* ***Assumption (A)*** *holds, then the eigenvalues of every self-adjoint realization have no accumulation point in I.*

PROOF. Let S be an arbitrary self-adjoint extension of $S_{\min}$. By the characterization of self-adjoint domains given in Theorem 7.2.2, there exist complex matrices $A_{d\times n}$, $B_{d\times m}$ such that $\operatorname{rank}(A : B) = d$, $AE_nA^* = BE_mB^*$, and the domain of S is given by:

$D(S) = \{y \in D_{\max}$:

$$A\begin{pmatrix} y(a) \\ \vdots \\ y^{[n-1]}(a) \end{pmatrix} + B\begin{pmatrix} [y, u_1](b) \\ \vdots \\ [y, u_m](b) \end{pmatrix} = \begin{pmatrix} 0 \\ \vdots \\ 0 \end{pmatrix}\} \tag{9.2.2}$$

where u_j, $j = 1, 2, ..., m$ ($m = 2d - 2k$), are LC solutions of differential equation

$$My = \lambda_0 w\, y \quad on \quad J = (a, b)$$

for some fixed $\lambda_0 \in (\alpha, \beta)$. The self-adjoint boundary conditions (9.2.2) consist of the system of d equations

$$U_i(y) = \sum_{j=1}^{n} a_{ij} y^{[j-1]}(a) + \sum_{j=1}^{m} b_{ij}[y, u_j](b) = 0, \quad i = 1, \cdots, d. \tag{9.2.3}$$

Let $y(\cdot, \lambda)$ be an eigenfunction of S for some $\lambda \in I$, and let $\varphi_j(\cdot, \lambda)$, $j = 1, \cdots, d$, denote d linearly independent solutions which satisfy **Assumption** (A). Then y is a nontrivial linear combination of the solutions $\varphi_j(\cdot, \lambda)$, $j = 1, \cdots, d$, i.e.

$$y(\cdot, \lambda) = \sum_{j=1}^{d} c_j \varphi_j(\cdot, \lambda), \quad c_j \in \mathbb{C},$$

and $y(\cdot,\lambda)$ satisfies the boundary conditions (9.2.3). Substituting $y(\cdot,\lambda)$ into these boundary conditions, we have

$$U_i(y) = U_i(\sum_{j=1}^{d} c_j\varphi_j(\cdot,\lambda)) = \sum_{j=1}^{d} c_j U_i(\varphi_j(\cdot,\lambda)) = 0, \quad i = 1,\cdots,d. \tag{9.2.4}$$

Let

$$\Delta(\lambda) = \det\left(U_i(\varphi_j(\cdot,\lambda))\right)_{1\le i,j\le d}$$

denote the determinant of the matrix of coefficients of the system (9.2.4). Therefore (9.2.4) has a nontrivial solutions $c_1,\cdots,c_d$ if and only if $\Delta(\lambda) = 0$. This shows that $\lambda \in I$ is in $\sigma_p(S)$ if and only if $\Delta(\lambda) = 0$.

Since

$$U_i(\varphi_r(\cdot,\lambda)) = \sum_{j=1}^{n} a_{ij}\varphi_r^{[j-1]}(a,\lambda) + \sum_{j=1}^{m} b_{ij}[\varphi_r,u_j](b,\lambda), \quad i,r = 1,\cdots,d,$$

we have

$$\begin{aligned}
\Delta(\lambda) &= \det(U_i(\varphi_r(\cdot,\lambda))) \\
&= \det\left[\begin{bmatrix} \sum_{j=1}^{n} a_{1j}\varphi_1^{[j-1]}(a,\lambda) & \cdots & \sum_{j=1}^{n} a_{1j}\varphi_d^{[j-1]}(a,\lambda) \\ \cdots & \cdots & \cdots \\ \sum_{j=1}^{n} a_{dj}\varphi_1^{[j-1]}(a,\lambda) & \cdots & \sum_{j=1}^{n} a_{dj}\varphi_d^{[j-1]}(a,\lambda) \end{bmatrix} + \begin{bmatrix} \sum_{j=1}^{m} b_{1j}[\varphi_1,u_j](b,\lambda) & \cdots & \sum_{j=1}^{m} b_{1j}[\varphi_d,u_j](b,\lambda) \\ \cdots & \cdots & \cdots \\ \sum_{j=1}^{m} b_{dj}[\varphi_1,u_j](b,\lambda) & \cdots & \sum_{j=1}^{m} b_{dj}[\varphi_d,u_j](b,\lambda) \end{bmatrix}\right] \\
&= \det\left[A\begin{bmatrix} \varphi_1(a,\lambda) & \cdots & \varphi_d(a,\lambda) \\ \cdots & \cdots & \cdots \\ \varphi_1^{[n-1]}(a,\lambda) & \cdots & \varphi_d^{[n-1]}(a,\lambda) \end{bmatrix} + B\begin{bmatrix} [\varphi_1,u_1](b,\lambda) & \cdots & [\varphi_d,u_1](b,\lambda) \\ \cdots & \cdots & \cdots \\ [\varphi_1,u_m](b,\lambda) & \cdots & [\varphi_d,u_m](b,\lambda) \end{bmatrix}\right].
\end{aligned}$$

By the assumption that, for fixed $t \in J$, $\varphi_1(t,\lambda),\cdots,\varphi_d(t,\lambda)$ are analytic on G, we conclude that $\Delta(\lambda)$ is an analytic functions of λ in the domain G of the complex plane which contains the real interval (α,β). Note that $\Delta(\lambda)$ is not identically zero on G since all eigenvalues of S are real. Therefore, from the well known distribution of the zeros of analytic functions, we obtain that the eigenvalues of S have no accumulation point in the interval $I = (\alpha,\beta)$. This completes the proof. □

REMARK 9.2.3. This theorem is proven by constructing, for any given self-adjoint realization S, a characteristic function $\Delta(\lambda)$ whose zeros in the interval (α,β) are precisely the eigenvalues of S in this interval. The construction of $\Delta(\lambda)$ uses LC solutions. Thus the proof of this theorem provides a very good illustration of the use of the LC and LP solutions constructed in Chapter 7 to get information about the spectrum.

THEOREM 9.2.6. *Let the notation and hypotheses of Theorem 9.2.5 hold. Then there is no essential spectrum in I for any self-adjoint realization S of* (9.2.1).

PROOF. Let S be a self-adjoint realization of (9.2.1). By Theorem 9.2.2, $\sigma_c(S)\cap I$ is empty and from Theorem 9.2.5 we have that $[\overline{\sigma_p(S)} \setminus \sigma_d(S)]$ is empty. Thus we have

$$\sigma_e(S) = \sigma_c(S) \cup [(\overline{\sigma_p(S)}) \setminus \sigma_d(S)]$$

and the conclusion follows. □

3. Two Singular Endpoints

In this section both endpoints may be singular. As mentioned in Section 9.1 above the behavior of $r(\lambda)$ in the two singular endpoint case is dramatically different from the one singular endpoint case. Each endpoint has an influence which is independent of the other endpoint. So we study the equation (9.2.1) on the intervals

$$J = (a,b), \quad J_a = (a,c), \quad \textit{and} \quad J_b = (c,b), \quad -\infty \le a < c < b \le \infty$$

in the Hilbert spaces $H = L^2(J,w)$, $H_a = L^2(J_a,w)$, $H_b = L^2(J_b,w)$. Here c is an arbitrarily chosen point in J. Note that the results of Section 9.2 apply to both intervals (a,c) and (c,b) since c is a regular endpoint for both.

Recall that $Q \in Z_n(J,\mathbb{R})$ implies $Q \in Z_n(J_a,\mathbb{R})$, $Q \in Z_n(J_b,\mathbb{R})$ and recall the notation: d_a for the deficiency index of $S_{\min}(a,c)$ in H_a, d_b denotes the deficiency index of $S_{\min}(c,b)$ in H_b, and d denotes the deficiency index of $S_{\min}(a,b)$ in H.

The next lemma summarizes some basic facts for the one regular endpoint case. It is stated here for convenience to make the comparison with the two singular endpoint case easier.

LEMMA 9.3.1. *Let $Q \in Z_n(J,\mathbb{R})$, $n = 2k$, $k \ge 1$, assume that Q is Lagrange symmetric and w is a weight function on J.*

(1) *Then $M = M_Q$ is a symmetric differential expression on J, J_a and on J_b.*
(2) *The deficiency indices d_a and d_b are independent of the choice of $c \in (a,b)$.*
(3) *For any $\lambda \in \mathbb{R}$, $r_a(\lambda) \le d_a$, $r_b(\lambda) \le d_b$, and strict inequality can occur.*
(4) *We have*
$$k \le d_a,\ d_b \le 2k = n$$
and all values of d_a, d_b in this range are realized.
(5)
$$d = d_a + d_b - n.$$
(6) *We have*
$$0 \le d \le 2k = n$$
and all values in this range occur.
(7) *If one endpoint is regular and $r(\lambda) = d$, then λ is an eigenvalue of multiplicity d for some self-adjoint extension.*
(8) *If one endpoint is regular and $r(\lambda) = d$ for all λ in some open interval I, then there is no continuous spectrum in I for any self-adjoint extension S. Moreover, the eigenvalues of any self-adjoint extension are nowhere dense in I.*

The next theorem shows that the behavior of $r_a(\lambda)$ on (a,c) and the behavior of $r_b(\lambda)$ on (c,b) affect the spectrum on the whole interval (a,b).

THEOREM 9.3.1. *Let the hypothesis and notation of Lemma 9.3.1 hold. Then*

(1) *We have*

$$\sigma_e(a,b) = \sigma_e(a,c) \cup \sigma_e(c,b). \tag{9.3.1}$$

(2) *If* $r_a(\lambda) < d_a$ *or* $r_b(\lambda) < d_b$, *then* $\lambda \in \sigma_e(a,b)$.
(3) *If* $r_a(\lambda) + r_b(\lambda) < d_a + d_b$, *then* $\lambda \in \sigma_e(a,b)$.
(4) *If* $\lambda \notin \sigma_e(a,b)$, *then*

$$r_a(\lambda) + r_b(\lambda) = d_a + d_b.$$

(5)

$$r_a(\lambda) + r_b(\lambda) - n \le r(\lambda) \le \min\{r_a(\lambda), r_b(\lambda)\}.$$

(6) *If* $\lambda \notin \sigma_e(a,b)$, *then*

$$d = d_a + d_b - n \le r(\lambda) \le \min\{r_a(\lambda), r_b(\lambda)\}.$$

PROOF. Parts (2), (3), (4) follow from (1) and Theorem 9.2.1, (6) follows from (4) and (5). For details of the proof of (5) see Section 4 of [**257**]. For smooth coefficients (1) follows from Theorem 4, p.1438 of Dunford and Schwartz [**111**]. This proof can readily be adapted to our hypotheses on the coefficients, see Hao, et. al. [**257**].

An alternative proof of (1) can be constructed from the 'two-interval' theory developed by Everitt and Zettl in [**205**], [**199**] and discussed in Chapters 12 and 13. Consider the 'two-interval' minimal operator $S_{2\min}$ given by

$$S_{2\min} = S_{\min}(a,c) \dotplus S_{\min}(c,b)$$

in the direct sum space $L^2((a,c),w) \dotplus L^2((c,b),w)$ which can be identified with $H = L^2((a,b),w)$. Let S_a, S_b be self-adjoint extensions of $S_{\min}(a,c)$ and $S_{\min}(c,b)$, respectively and let $S = S_a \dotplus S_b$. Then S is a self-adjoint extension in H. It is well known that the essential spectrum of the direct sum of two self-adjoint operators in Hilbert space is the union of their essential spectra and (9.3.1) follows. □

REMARK 9.3.1. For problems with only one singular endpoint the 'Decomposition Method of Glazman' [**240**] shows that the essential spectrum depends only on the coefficients near the singular endpoint. In Theorem 9.3.1 both endpoints may be singular. The 'two-interval' proof of part (1) of Theorem 9.3.1 essentially consists in showing that the two one interval results for the intervals (a,c) and (c,b) can be combined to prove (9.3.1). Although this is conceptually simple the technical details involve the 'two-interval' theory of Everitt and Zettl [**205**], [**199**] as described above and the 'Naimark Patching Lemma' which 'connects' these two intervals through the interior to obtain this result for the whole interval (a,b).

THEOREM 9.3.2. *Assume that* $Q \in Z_n(J,\mathbb{R})$, $n = 2k$, $k \ge 1$, *is Lagrange symmetric and* w *is a weight function on* J. *Suppose* a *and* b *are singular. If* $r(\lambda) < d$, *then* $\lambda \in \sigma_e(a,b)$.

PROOF. Suppose $\lambda \notin \sigma_e(a,b)$. Then $r_a(\lambda) = d_a$ and $r_b(\lambda) = d_b$ since otherwise $r_a(\lambda) < d_a$ or $r_b(\lambda) < d_b$ would imply that $\lambda \in \sigma_e(a,b)$ by (2) of Theorem 9.3.1. But $r_a(\lambda) = d_a$ and $r_b(\lambda) = d_b$ implies, by (5) of Theorem 9.3.1, that

$$r(\lambda) \ge r_a(\lambda) + r_b(\lambda) - n = d_a + d_b - n = d$$

which contradicts the hypothesis $r(\lambda) < d$. □

REMARK 9.3.2. We comment on Theorem 9.3.2. Theorem 11.1 in Weidmann [**574**] proves this result under the additional assumption that there exists a self-adjoint extension for which λ is not an eigenvalue and he comments on pages 162, 163 that (1) this theorem is the basis for all other results in Chapter 11 of [**574**] and

(2) that he does not know if the additional assumption is really necessary. Theorem 9.3.2 has eliminated this additional condition used in [**574**].

The contrasting behavior of $r(\lambda)$ for the two singular endpoint case from the one singular endpoint case has some interesting consequences. In the case of only one singular endpoint we have $r(\lambda) \leq d$ whereas in the two singular endpoint case $r(\lambda)$ may assume values less than d, equal to d, or greater than d. The case $r(\lambda) > d$ leads to a surprising and counter intuitive result that this value of λ is an eigenvalue of *every* self-adjoint extension, i.e. for any given self-adjoint boundary condition, there are eigenfunctions of λ which satisfy this boundary condition (see an example in [**257**]).

THEOREM 9.3.3. *Assume that* $Q \in Z_n(J, \mathbb{R})$, $n = 2k$, $k \geq 1$, *is Lagrange symmetric and* w *is a weight function on* J. *Suppose* a *and* b *are singular. If, for some* $\lambda \in \mathbb{R}$, $d < r(\lambda) < \min\{r_a(\lambda), r_b(\lambda)\}$, *then* λ *is an eigenvalue of every self-adjoint realization of* (9.2.1).

PROOF. See [**257**]. □

Next we give a specific example where $r(\lambda) > d$.

EXAMPLE 9.3.1. Consider the Hermite differential expression M given by

$$My = -y'' + t^2 y = \lambda y \quad on \quad J = (-\infty, \infty).$$

It is well known that M is LP at both endpoints. Hence $d = 0$, $S_{\min}$ is self-adjoint with no proper self-adjoint extension and its spectrum is discrete and given by

$$\sigma(S_{\min}) = \{\lambda_n = 2n + 1, \ n = 0, 1, 2, ...\}.$$

Hence

$$r(\lambda_n) = 1 > 0 = d.$$

On the relationship between the essential spectrum and the numbers of real parameter square-integrable solutions of differential equations, in his book Weidmann conjectured if there exist 'sufficiently many' L^2-solutions of $(M - \lambda)u = 0$, for $\lambda \in I = (\mu_1, \mu_2)$, then I contains no points of the essential spectrum [**574**]. But as mentioned in Section 9.2 C. Remling [**498**] has clarified the complicated relationship between the essential spectrum and the real numbers λ for which $r(\lambda) = d$. In fact knowing only the number of real parameter solutions is not sufficient to guarantee the discreteness of the spectrum of ordinary differential operators.

For the two singular endpoints case, using the 'two-interval' theory given in Chapters 12 and 13 and the above theorem we have the following result. Recall

Assumption (A): There exists a domain G in the complex plane containing $I = (\alpha, \beta)$ with α, β on the boundary of G, and for equation (9.2.1) there exist solutions $u_1(t, \lambda), \cdots, u_{d_a}(t, \lambda)$ on (a, c) and $v_1(t, \lambda), \cdots, v_{d_b}(t, \lambda)$ on (c, b) which lie in the appropriate Hilbert space H, are real valued for $\lambda \in I$, and are analytic on G for each fixed $t \in J$.

THEOREM 9.3.4. *Assume that* $Q \in Z_n(J, \mathbb{R})$, $n = 2k$, $k \geq 1$, *is Lagrange symmetric and* w *is a weight function on* J. *Each endpoint may be regular or singular. If* $r_a(\lambda) = d_a$ *on* (a, c), $r_b(\lambda) = d_b$ *on* (c, b) *for all* λ *in some open interval* $I = (\alpha, \beta)$ *and Assumption (A) holds, then there is no essential spectrum in* I *for any self-adjoint extension* $S(a, c)$, $S(c, b)$, $S(a, b)$. *In particular, the eigenvalues of none of these extensions* $S(a, c)$, $S(c, b)$, $S(a, b)$ *can have an accumulation point in* I.

PROOF. From the above theorems we have $\sigma_e(a,c)\cap I=\emptyset$ and $\sigma_e(c,b)\cap I=\emptyset$. Therefore by (1) of Theorem 9.3.1 we conclude that $\sigma_e(a,b)\cap I=\emptyset$. Since such an accumulation point of eigenvalues would be in the essential spectrum, we may conclude that the eigenvalues of none of these extensions $S(a,c)$, $S(c,b)$, $S(a,b)$ can have an accumulation point in I. □

4. Comments

To a large extent this Chapter is based on the 2011 and 2012 papers of Hao et. al. [**256**], [**257**]. The LC solutions play a critical role in Sections 9.2 and 9.3. These were first constructed by Wang et. al. in [**553**] for the one singular endpoint case and then extended to two singular endpoints in [**253**]. As mentioned in Remark 9.3.2 an open problem stated in Weidmann's 1987 book [**574**] was solved using these LC solutions. Although the LC solutions are so called in analogy to the Weyl limit-circle solutions in the second order case $n=2$; when $n>2$ not all solutions in the Hilbert space $L^2(J,w)$ are LC solutions in general.

CHAPTER 10

Coefficients, the Deficiency Index, Spectrum

1. Introduction

There is a vast literature studying the determination of the deficiency index $d(M)$ from the coefficients of M including the weight function. Giving a comprehensive account of these results is well beyond the scope of this book.

Most of the results are for the constant weight function $w = 1$ and the interval $J = (0, \infty)$ with 0 a regular endpoint. See the books and monographs by Coddington and Levinson [**102**], Dunford and Schwartz [**111**], Naimark [**440**], Atkinson [**23**], Rofe-Beketov and Kholkin [**502**] (with its 941 references), Weidmann [**574**], Kauffman, Read and Zettl [**326**], Kwong and Zettl [**380**].

And the papers by Ahlbrandt, Hinton and Lewis [**4**], Atkinson, Eastham and McLeod [**24**], Behncke and Focke [**49**], Bennewitz [**56**], Bradley [**78**], Eastham [**118**],[**119**], [**120**], Eastham and Thompson [**125**], Eastham and Zettl [**126**], Evans, Kwong and Zettl [**138**], Evans [**132**], [**133**], Everitt [**150**], [**154**], [**155**], [**157**], [**163**], Everitt, Giertz and McLeod [**169**], Everitt, Giertz and Weidmann [**170**], Everitt, Knowles and Read [**175**], Evans and Zettl [**140**], [**141**], Harris [**268**], [**280**], Hinton [**291**], Kalf [**313**], Knowles [**333**], [**334**], [**335**], Kurss [**367**], Kwong [**368**], [**369**], [**370**], Kwong and Zettl [**381**], Levinson [**398**], Niessen [**448**], [**449**], Niessen and Zettl [**454**], Ong [**455**], Patula and Waltman [**458**], Patula and Wong [**459**], Pleijel [**465**], [**466**], [**467**], Read [**486**], [**487**], [**488**], [**489**], Walter [**546**], Weidmann [**569**], Wong and Zettl [**585**], Zettl [**611**], [**618**], [**610**], [**613**], [**598**], [**614**], Boruvka [**76**], Eastham [**121**] , Hartman [**283**], Hille [**288**], Hinton and Schafer [**298**], Jörgens and Rellich [**311**], Kaper, Kwong and Zettl [**322**], Müller-Pfeiffer [**437**], and [**252**], [**240**], [**48**], [**47**], [**51**], [**179**], [**144**], [**427**], [**384**].

This list of books and papers is not up to date and not intended to be comprehensive. Despite a voluminous literature the case $n = 2$ is still unsolved in general, see [**620**]. Much less is known for $n > 2$.

As discussed in Section 4.2 the dependence of the deficiency index on the coefficients is very delicate. For some expressions M of order $n = 2k$, if $d(M) > k$ it is possible to change the coefficients in a sequence of intervals of arbitrary lengths going to infinity such that for the changed expression M, $d(M) = k$. On the other hand if the deficiency index is maximal, i.e. $d(M) = 2k$, there seem to be no such 'interval' type conditions on the coefficients known.

In most of this massive collection of papers the authors found conditions on the coefficients for the limit-point case $d(M) = k$ when $J = (a, \infty)$, $w = 1$, and a is a regular endpoint. Using asymptotic methods some conditions are known for $d = 2k = n$ and $d = k + 1$ and very few for $k + 1 < d < 2k$.

In Section 10.2 we illustrate an approach used by Zettl [**616**] for the second order expression

$$My = -(py')' + qy.$$

Instead of starting with p and q and searching for conditions to determine d, start with p and search for q such that $d(M) = 2$. Starting with a bounded q, $d(M) = 1$ for all p. So this method is limited. On the other hand it indicates that the leading coefficient plays a prominent role. When $p > 0$ some authors have a tendency to prove a theorem for $p = 1$ and then claim that this theorem holds for the transformed equation with $p > 0$. Such a claim may be misleading.

Although the result in Section 10.2 is very special it does suggest the question : Given a subset of the coefficients of M is there an algorithm - perhaps with parameter functions - to find the remaining coefficients for $d(M) = d$, for a given d, $k \leq d \leq 2k$. In this search the leading and trailing coefficients may be dominant in the sense that starting with them intermediate coefficients can be found.

There is also a voluminous literature on finding conditions for the spectrum to be bounded below and discrete. In Section 10.3 we discuss an approach of Kwong and Zettl in [**384**]. This approach also seems not to have been widely used in the literature. It's main features are that it finds conditions on the coefficients *and on the weight function* w and uses a method based on norm inequalities in weighted Hilbert spaces from their monograph [**385**].

2. An Algorithm for the Construction of the Maximal Deficiency Index

Consider the equation

$$\begin{aligned} &My = -(py')' + qy = 0 \quad on\ J = (a,b), \ -\infty \leq a < b \leq \infty, \\ &\frac{1}{p}, q \in L_{loc}(J,\mathbb{R}), \quad p > 0. \end{aligned} \tag{10.2.1}$$

THEOREM 10.2.1. *Given any p satisfying* (10.2.1) *there exists a q satisfying* (10.2.1) *such that $d(M) = 2$.*

PROOF. Let u be any positive $L^2(J,\mathbb{R})$ function such that $(pu')' \in L_{loc}(J,\mathbb{R})$. There exist such functions u since the minimal operator of the nonoscillatory equation $(pu')' - u = 0$ has closed range, i.e. $\lambda = 0$ is a point of regular type for the minimal operator and hence this equation has a positive L^2 solution. Let $a < c < b$ and define

$$y(t) = u(t) \sin \int_c^t \frac{1}{pu^2}, \quad z(t) = u(t) \cos \int_c^t \frac{1}{pu^2}.$$

Now we show that y and z are linearly independent solutions of equation (10.2.1) for

$$q = [\frac{1}{pu^4} + \frac{(pu')'}{u}].$$

Note that

$$y'(t) = u'(t) \sin \int_c^t \frac{1}{pu^2} + \frac{u(t)}{p(t)u^2(t)} \cos \int_c^t \frac{1}{pu^2},$$

$$(py')'(t) = [\frac{1}{pu^4} + \frac{(pu')'}{u}](t)u(t) \sin \int_c^t \frac{1}{pu^2} = q(t)y(t).$$

Hence y is an $L^2(J)$ solution of (10.2.1). Similarly, z is an $L^2(J)$ solution of (10.2.1) and the conclusion follows. □

REMARK 10.2.1. Note that for any leading coefficient p satisfying (10.2.1) we constructed a potential function q in terms of a parameter function u such that $d(M) = 2$. For a given q there does not, in general, exist a p such that $d(M) = 2$. For example if $J = (0, \infty)$ and q is bounded on J then $d(M) = 1$. This shows that the influence of p and q on the determination of $d(M)$ is quite different.

THEOREM 10.2.2. *Let M be given by* (10.2.1) *and define M^s, $s = 2, 3, 4, \cdots$ as in Section 4.3. Then*

$$d(M^s) = 2s, \quad s = 2, 3, 4, \cdots$$

PROOF. This follows from Corollary 4.3.2. □

REMARK 10.2.2. Theorem 10.2.2 was proven by Zettl in [**616**] for smooth coefficients p, q using classical derivatives $y^{(i)}$ and the classical construction of the powers M^s, $s = 2, 3, 4, \cdots$. Here using quasi-derivatives only the local integrability conditions on the coefficients are used. From Corollary 4.3.2 it also follows that each M^s is partially separated.

REMARK 10.2.3. Very few conditions on the coefficients are known for $d(M)$ to be maximal. In [**616**] there are explicit examples given for the coefficients p and q as powers of t. There are many papers giving conditions on the coefficients for $d(M) = k$ when the order of M is $2k$. There are a few papers with asymptotic conditions on the coefficients for $d(M) = k + 1$ but very few - other than the examples of Glazman, Orlov and Read - for $k+1 < d < 2k$. We know of no general construction of coefficients to determine $d(M) = d$ for $k + 1 < d < 2k$.

3. Discreteness Conditions

In this section we discuss discreteness conditions on the coefficients for the spectrum to be bounded below and discrete for the two term equation:

$$My = (-1)^n (py^{(n)})^{(n)} + qy = \lambda wy \quad on \ J = (0, \infty), \tag{10.3.1}$$

where $n \geq 1$ is a positive integer, λ is the spectral parameter, 0 is a regular endpoint, p, q, w are real valued, and satisfy

$$p > 0, \ w > 0 \ \ a.e.; \quad \frac{1}{p}, w, q \in L_{loc}(0, \infty). \tag{10.3.2}$$

Let

$$q = q^+ - q^- \ \ where \ \ q^+(t) = \max\{0, q(t)\}, \ \ for \ \ t \in J. \tag{10.3.3}$$

DEFINITION 10.3.1. Following Hinton and Lewis [**297**] we say that M has property BD if the spectrum of every self-adjoint extension of the minimal operator $S_{\min}(M)$ is bounded below and discrete. In this case for each self-adjoint extension S we have:

$$\sigma(S) = \{\lambda_n : n = 0, 1, 2, 3, \cdots\}$$

and the eigenvalues can be ordered to satisfy the inequalities:

$$-\infty < \lambda_0 \leq \lambda_1 \leq \lambda_2 \leq \lambda_3 \leq \cdots .$$

Only results are stated here. For proofs and extensions of these results to equations with middle terms see the 1981 paper of Kwong and Zettl [**384**]. These proofs are based on certain norm inequalities in weighted L^p spaces and are different from most (possibly all) earlier proofs of similar results given in the above mentioned literature. As mentioned above most of the earlier results were for the constant weight function $w = 1$.

DEFINITION 10.3.2. Given a positive number a and a positive function $f(t)$ let

$$Q(t, a, f(t)) = \inf \int_I q^+(x)dx, \tag{10.3.4}$$

where the infimum is taken over all intervals $I \subset ([t, t + af(t)] \cap J)$ of length $3^{1-n}af(t)$.

THEOREM 10.3.1. *Let M be given by (10.3.1). Then M has property BD if there exists a positive function f on J such that*

$$\lim_{a\to 0}\{\limsup_{t\to\infty}(af(t))^{2(n-1)}\int_t^{t+af(t)}(w+q^-)\int_t^{t+af(t)}\frac{1}{p}\} = 0 \tag{10.3.5}$$

and for each $a > 0$

$$\lim_{t\to\infty}\int_t^{t+af(t)}(w+q^-)/Q(t, a, f(t)) = 0. \tag{10.3.6}$$

The next theorem shows that the complicated condition (10.3.6) can be simplified if the function $w + q^-$ satisfies a mild restriction.

THEOREM 10.3.2. *Let M be given by (10.3.1). Then M has property BD if, in addition to (10.3.5) the function $w + q^-$ satisfies*

$$\int_t^{t+af(t)}(w+q^-) \le K\int_s^{s+3^{1-n}af(s)}(w+q^-) \tag{10.3.7}$$

for some fixed $K > 0$, all sufficiently large t, all sufficiently small $a > 0$, and for all s such that $[s, s + 3^{1-n}af(s)] \subset [t, t + af(t)]$ and for each $a > 0$

$$\lim_{t\to\infty}\int_t^{t+af(t)}(w+q^-)/\int_t^{t+af(t)}q^+ = 0. \tag{10.3.8}$$

Next we make some remarks and illustrate these conditions.

REMARK 10.3.1. Condition (10.3.8) can be viewed as an extension of the well known Molchanov criterion. The role of the function f in conditions (10.3.5) to (10.3.8) is to allow the intervals of integration to have varying lengths in contrast with the Molchanov condition. The presence of f enlarges the class of functions p, q, w which satisfy these conditions. For some classes of functions condition (10.3.8) is necessary and sufficient for the spectrum to be bounded below and discrete even when $f = 1$.

The next theorems are special cases of Theorems 10.3.1 and 10.3.2. They are stated as theorems to make it easier to make comparisons with previous results and their statements are less technical and therefore more transparent than these two theorems.

THEOREM 10.3.3. *For $t \geq 0$. Assume $q(t) \geq -c\,w(t)$ for some $c > 0$. Suppose $w(t) \leq t^{\alpha}u(t)$ and $p(t) \geq t^{\alpha}v(t)$ for some $\alpha \geq 0$ where u and v are locally integrable functions.*

(1) *If, for some positive numbers d and D, $0 < d \leq u(t) \leq D < \infty$, and $v(t) \geq d$, then (10.3.8) with $f(t) = 1$ is sufficient for BD to hold.*
(2) *If, for some positive numbers d and D, $0 < d \leq u(t) \leq D < \infty$ and $v(t) \leq D$, then (10.3.8) with $f(t) = 1$ is necessary for BD to hold.*

REMARK 10.3.2. The special case $p = w = f = 1$ of Theorem 10.3.3 is the well known Molchanov criterion.

In the second order case condition (10.3.7) is not needed, so for the convenience of the reader we state this result in the next theorem.

THEOREM 10.3.4. *Consider*

$$My = \frac{1}{w}[-(py')' + qy] \quad on\ J = (0, \infty),$$

where p, q, w satisfy (10.3.2) and 0 is a regular endpoint. Then M has property BD if there exists some positive function f such that

$$\lim_{a\to 0}\{\limsup_{t\to\infty} \int_t^{t+af(t)} (w + q^-) \int_t^{t+af(t)} \frac{1}{p}\} = 0$$

and for each $a > 0$

$$\lim_{t\to\infty} \int_t^{t+af(t)} (w + q^-) / \int_t^{t+af(t)} q^+ = 0.$$

By Theorem 10.3.3 for some classes of coefficients (10.3.8) is necessary and sufficient for property BD to hold. Nevertheless, (10.3.8) can be weakened if (10.3.5) is appropriately strengthened as in the next theorem.

THEOREM 10.3.5. *Let M be given by (10.3.1) to (10.3.3). Then M has property BD if there exists a function f such that for some positive number a the following conditions hold:*

$$\lim_{t\to\infty} f(t)^{2(n-1)} \int_t^{t+af(t)} (w + q^-) \int_t^{t+af(t)} \frac{1}{p} = 0 \tag{10.3.9}$$

and

$$\lim_{t\to\infty} \int_t^{t+af(t)} (w + q^-)/Q(t, a, f(t)) = 0.$$

Next we mention some Corollaries followed by Examples.

COROLLARY 10.3.1. *Assume (10.3.7) holds. Then M has property BD if (10.3.8) and (10.3.9) hold for some positive function f and some positive number a.*

COROLLARY 10.3.2. *Let $n = 1$. Then M has property BD if*

$$\lim_{t\to\infty} \int_t^{t+af(t)} (w + q^-) \int_t^{t+af(t)} \frac{1}{p} = 0 \tag{10.3.10}$$

and

$$\lim_{t\to\infty} \int_t^{t+af(t)} (w + q^-) / \int_t^{t+af(t)} q^+ = 0 \tag{10.3.11}$$

hold for some positive function f and some positive number a.

COROLLARY 10.3.3. *Let $n = 1$. Suppose for some positive numbers c, b and B*

$$q(t) \geq -c, \quad w(t) \leq B, \quad t \in J, \quad \lim_{t\to\infty} p(t) = \infty$$

or

$$q(t) \geq -c, \quad 0 < b \leq p(t), \quad \lim_{t\to\infty} w(t) = 0.$$

Then M has property BD if

$$\lim_{t\to\infty} \int_t^{t+a} q^+ = \infty \tag{10.3.12}$$

for some $a > 0$.

REMARK 10.3.3. Note that (10.3.12) allows q to be identically constant on a sequence of intervals I_n of length less than a. T. T. Read has shown that property BD can hold even for coefficients q with the property that $q(t) \to -\infty$ as $t \to \infty$ through some sequence of intervals.

COROLLARY 10.3.4. *Let $n = 1$ or $n > 1$ and (10.3.7) hold. Then M has property BD if*

$$q^- \in L^s(J), \quad \frac{1}{p} \in L^u(J), \quad w \in L^v(J), \quad 1 \leq s, u, v \leq \infty$$

and

$$\liminf_{t\to\infty} \int_t^{t+a} q \geq c > 0, \quad \textit{for some } a > 0.$$

Next we give some examples to illustrate some of the above conditions.

EXAMPLE 10.3.1. Let $p(t) = t$, $w(t) = 1$ and $q(t) \geq 0$. Then (10.3.5) holds with $f(t) = \sqrt{t}$. Hence property BD holds by Theorem 10.3.2 if, in addition, q satisfies

$$\lim_{t\to\infty} \sqrt{t} / \int_t^{t+a\sqrt{t}} q = 0$$

for each fixed $a > 0$.

EXAMPLE 10.3.2. Let $w(t) = 1$, $q(t) \geq 0$, $p(t) = t^{-1}$, $f(t) = t^{-1/2}$. Then (10.3.5) holds. Hence property BD holds by Theorem 10.3.2 if, in addition, for each fixed $a > 0$

$$\lim_{t\to\infty} \sqrt{t} \int_t^{t+a/\sqrt{t}} q = \infty.$$

EXAMPLE 10.3.3. Let $n = 1$, $p(t) = t^{-1}$, $w(t) = 1$, $q^-(t) = 0$, $a = 1$, $f(t) = t^{-1/2-\varepsilon}$, $\varepsilon > 0$. Then (10.3.10) is satisfied. Therefore (10.3.11) is satisfied and hence property BD holds.

EXAMPLE 10.3.4. Let $n = 1$, $p(t) = t$, $w(t) = 1$, $q^-(t) = 0$, $a = 1$, $f(t) = t^{1/2+\varepsilon}$, $\varepsilon > 0$. Then (10.3.10) is satisfied. Therefore (10.3.11) holds implies that the property BD holds.

4. Comments

Section 10.2 is based on the paper [**616**]. The method used in the proof may be more interesting than the result. Here the idea is: Given some of the coefficients, find the remaining ones so that the deficiency index $d = k$ or $k + s$, for $s = 1, \cdots, k$. Of particular interest are the cases $d = k$ and $d = 2k$. It seems that the leading coefficient may be dominant in this search.

Section 10.3 is based on the Kwong-Zettl paper [**384**] which was strongly influenced by the work of T.T. Read [**489**], [**488**]. The conditions given in this paper are of a new type and significantly improved the then existing results. Of particular interest is the approach; it is based on some norm inequalities in weighted L^p spaces from the Kwong-Zettl monograph [**385**]. In most of the papers in the previous literature the weight function is the constant $w = 1$ even in the second order case, here we include w with the other coefficients p, q.

Another feature of the paper [**384**], not discussed here, is that results are proven for the two term expression (10.3.1), then extended to a general expression with middle terms. Note that this idea is similar to the idea used in Section 10.2. Here results are proven for the two 'outermost coefficients' q_k and q_0 and then extended to include the intermediate coefficients $q_j, j = 1, \cdots, k-1$. So, in this sense, the intermediate terms are 'perturbations' of the outer ones.

CHAPTER 11

Dissipative Operators

1. Introduction

Everitt and Markus (EM) [**184**], [**185**] characterized the self-adjoint domains of differential expressions $M = M_Q$, $Q \in Z_n(J, \mathbb{C})$, $J = (a, b)$, $n \in \mathbb{N}_2$, $Q = Q^+$, in terms of subspaces of symplectic geometry spaces. They proved that, given such an expression M whose minimal operator $S_{\min}$ has equal deficiency indices d, there exists a one-to-one correspondence between the set of all self-adjoint operators S satisfying $S_{\min} \subset S = S^* \subset S_{\max}$ in the Hilbert space $H = L^2(J, w)$ and the set of all complete Lagrangian subspaces of the symplectic geometry space $G = D_{\max}/D_{\min}$ with symplectic product $[\cdot : \cdot]$ given by

$$[\widetilde{f} : \widetilde{g}] = [f + D_{\min} : g + D_{\min}] := [f : g] = [f, g]_a^b, \quad \text{for all } \widetilde{f}, \widetilde{g} \in G.$$

where $[f, g]$ is the Lagrange bracket of $f, g \in D_{\max}$.

In this chapter, for the minimal operator $S_{\min}$ generated by M, we establish:

(1) A one to one correspondence between the set $\{S\}$ of all symmetric extensions of $S_{\min}$ in H and the set $\{L\}$ of all Lagrangian subspaces of the complex symplectic space G.

(2) A one to one correspondence between the set $\{T_D\}$ of all dissipative extensions of $S_{\min}$ in H and the set $\{\mathfrak{D}\}$ of all dissipative subspaces of G.

(3) A one-to-one correspondence between the set $\{T_{sD}\}$ of all strictly dissipative extensions of $S_{\min}$ in H and the set $\{\mathfrak{D}_s\}$ of all strictly dissipative subspaces in the complex symplectic space G.

In Section 11.2 we extend the EM symplectic geometry space theory by introducing the concepts of 'dissipative' ('strictly dissipative') subspaces of the symplectic space G; this is based on the Yao et.al. paper [**591**]. In Sections 11.3 and 11.4 we develop some algebraic properties of these dissipative and strictly dissipative subspaces and establish a 1-1 correspondence between the subspaces $\{L\}$, $\{\mathfrak{D}\}$, $\{\mathfrak{D}_s\}$ of G and the corresponding sets $\{S\}, \{T_D\}, \{T_{sD}\}$ of differential operators in H. In Sections 11.5 and 11.6 we specialize to the case when $M = M_Q$, $Q \in Z_n(J, \mathbb{R})$, $n = 2k$, $k \geq 1$, $Q = Q^+$, characterize the dissipative extensions of $S_{\min}$ in H in terms of LC solutions, and give the symplectic geometry characterization of symmetric operators.

2. Concepts for Complex Symplectic Geometry Spaces

For the convenience of the reader we start with some general definitions from [**184**], [**185**].

DEFINITION 11.2.1. A complex symplectic space S is a complex vector space, together with a prescribed symplectic form $[\cdot : \cdot] : S \times S \to \mathbb{C}$, i.e.

$$\text{a complex-valued function} \quad u, v \to [u : v] \quad \text{for } S \times S \to \mathbb{C}$$

satisfying the following properties, for all vectors $u, v, w \in S$ and $c_1, c_2 \in \mathbb{C}$,

(i) Linearity property in its first entry:

$$[c_1 u + c_2 v : w] = c_1[u : w] + c_2[v : w].$$

(ii) Skew-Hermitian property:

$$[u : v] = -\overline{[v : u]}.$$

(iii) Non-degeneracy property:

$$[u : S] = 0 \ \ \textit{implies} \ u = 0.$$

Note that properties (i) and (ii) of Definition 11.2.1 imply that

$$[u : c_1 v + c_2 w] = \bar{c}_1[u : v] + \bar{c}_2[u : w],$$

for all vectors $u, v, w \in S$ and $c_1, c_2 \in \mathbb{C}$.

Linear submanifolds of a complex symplectic space S need not be complex symplectic spaces, since the induced symplectic form can be degenerate on them. In this chapter we only consider complex symplectic spaces S of finite dimension $D \geq 0$. The case $D = 0$ consists of a single point and is not further discussed here. In the finite dimensional cases considered here each linear submanifold is a linear subspace of S, it is closed in the usual topology of S, as in $\mathbb{C}^D$.

It is these finite dimensional symplectic subspaces which are related to self-adjoint, symmetric, and dissipative operators in the Hilbert space $H = L^2(J, w)$ as we will see below.

DEFINITION 11.2.2. A linear subspace L in the complex symplectic space S is called Lagrangian in case $[L : L] = 0$, that is,

$$[u : v] = 0 \quad \textit{for all } u, v \in L.$$

Furthermore, a Lagrangian subspace $L \subset S$ is said to be complete in case

$$u \in S \textit{ and } [u : L] = 0 \ \text{ imply } \ u \in L.$$

DEFINITION 11.2.3. Let S be a complex symplectic space with symplectic form $[\cdot : \cdot]$. Then linear subspaces S_- and S_+ are symplectic ortho-complements in S, written as

$$S = S_- \oplus S_+,$$

in case

$$(i)\ S = \operatorname{span}\{S_-, S_+\}, \qquad (ii)\ [S_- : S_+] = 0.$$

In this case $S_- \cap S_+ = 0$, so S is the direct sum of S_- and S_+, i.e. every $u \in S$ has a unique decomposition $u = u_- + u_+$ with $u_- \in S_-$ and $u_+ \in S_+$. For a symplectic orthogonal direct sum decomposition $S = S_- \oplus S_+$ each of S_- and S_+ is itself a complex symplectic space, since the symplectic form induced by $[\cdot : \cdot]$ is non-degenerate on S_- and on S_+.

Next we define dissipative, strictly dissipative, maximal dissipative; accretive, strictly accretive and maximal accretive subspaces of complex symplectic spaces. These definitions from Yao et. al. [**591**] seem to be new but are minor variants of known definitions.

DEFINITION 11.2.4. A linear subspace $\mathfrak{D}$ of a complex symplectic space S is called dissipative in case

$$\operatorname{Im}[u : u] \geq 0 \quad \textit{for all vectors } u \in \mathfrak{D}.$$

A linear subspace $\mathfrak{A}$ of S is called accretive in case

$$\operatorname{Im}[u : u] \leq 0 \quad \textit{for all vectors } u \in \mathfrak{A}.$$

DEFINITION 11.2.5. A dissipative subspace $\mathfrak{D} \subset S$ is said to be maximal dissipative, if for any dissipative subspace $\widetilde{\mathfrak{D}}$ such that $\mathfrak{D} \subseteq \widetilde{\mathfrak{D}}$, we have $\mathfrak{D} = \widetilde{\mathfrak{D}}$.

An accretive subspace $\mathfrak{A} \subset S$ is said to be maximal accretive, if for any accretive subspace $\widetilde{\mathfrak{A}}$ such that $\mathfrak{A} \subseteq \widetilde{\mathfrak{A}}$, we have $\mathfrak{A} = \widetilde{\mathfrak{A}}$.

DEFINITION 11.2.6. A dissipative subspace $\mathfrak{D}$ of S is called strictly dissipative in case

$$\operatorname{Im}[u : u] > 0 \quad \textit{for} \text{ all } u \in \mathfrak{D}, \ u \neq 0.$$

An accretive subspace $\mathfrak{A}$ of S is called strictly accretive in case

$$\operatorname{Im}[v : v] < 0 \quad \textit{for} \text{ all } v \in \mathfrak{A}, \ v \neq 0.$$

We denote a strictly dissipative (accretive) subspaces by $\mathfrak{D}_s$ ($\mathfrak{A}_s$).

Note that a Lagrangian subspace is both dissipative and accretive.

Example 1. Consider the complex linear space $S = \mathbb{C}^3$ with the prescribed symplectic products $[e_1 : e_1] = i$, $[e_2 : e_2] = i$, $[e_3 : e_3] = -i$, and $[e_i : e_j] = 0$ for $i \neq j$, $i, j = 1, 2, 3$, for the customary basis vectors:

$$e_1 = (1, 0, 0), \quad e_2 = (0, 1, 0), \quad e_3 = (0, 0, 1).$$

That is, we use the skew-Hermitian matrix $H = \operatorname{diag}\{i, i, -i\}$ to define the symplectic structure on $\mathbb{C}^3$.

Define $L = \operatorname{span}\{e_2 + e_3\} = \{(0, c, c)\}$ for all $c \in \mathbb{C}$, then L is a Lagrangian subspace, but not a complete Lagrangian subspace, since $[e_1 : L] = 0$ but $e_1 \notin L$.

Define $\mathfrak{D} = \operatorname{span}\{e_1, e_2 + e_3\}$, then $\mathfrak{D}$ is a dissipative subspace since for any $\alpha, \beta \in \mathbb{C}$,

$$\operatorname{Im}[\alpha e_1 + \beta(e_2 + e_3) : \alpha e_1 + \beta(e_2 + e_3)] = \operatorname{Im}[\alpha e_1 : \alpha e_1] \geq 0.$$

Furthermore, $\mathfrak{D}$ is a maximal dissipative subspace. Assume that there exists a dissipative subspace $\mathfrak{D}^*$ such that $\mathfrak{D} \subset \mathfrak{D}^* \subseteq S$. Then there is an $\widehat{e} \in \mathfrak{D}^*$, with $\widehat{e} \notin \mathfrak{D}$ such that $\{e_1,\ e_2 + e_3,\ \widehat{e}\} \in S$ are linearly independent. Hence $\mathfrak{D}^* = S$. Note that S is not a dissipative subspace, since $\operatorname{Im}[e_3 : e_3] < 0$. This contradiction shows that $\mathfrak{D}$ is maximal.

3. Finite Dimensional Complex Symplectic Spaces and Their Dissipative Subspaces

In this section we develop some algebraic properties of the dissipative and strictly dissipative subspaces defined in Section 11.2 which will be applied in the following sections to differential operators in H.

The next definition defines symplectic invariants of S.

DEFINITION 11.3.1. In a complex symplectic space S, with symplectic form $[\cdot:\cdot]$, and finite dimension $D \geq 1$, define the following symplectic invariants of S:

$$p = \max\{dimension\ of\ complex\ linear\ subspaces\ satisfying\ \operatorname{Im}[v:v] \geq 0\},$$
$$q = \max\{dimension\ of\ complex\ linear\ subspaces\ satisfying\ \operatorname{Im}[v:v] \leq 0\}.$$

The pair of non-negative integers (p, q) is called the signature of S and p is called the positivity index, q the negativity index.

In addition, we define the Lagrangian index $\triangle$ and the excess Ex of S:

$$\triangle = \max\{dimension\ of\ complex\ Lagrangian\ subspace\ of\ S\},$$

$$Ex = p - q, \quad excess\ of\ positivity\ over\ negativity\ index\ of\ S.$$

LEMMA 11.3.1. *Consider a complex symplectic space S, with symplectic form $[\cdot:\cdot]$ and finite dimension $D \geq 1$. Let $\mathfrak{D} \subseteq S$ be a dissipative subspace. Then*

(1) $\dim \mathfrak{D} \leq p$.

(2) There are non-zero dissipative subspaces if and only if $p \neq 0$.

(3) A dissipative subspace $\mathfrak{D}$ is maximal if and only if $\dim \mathfrak{D} = p$.

PROOF. This follows immediately from the above definitions. Note that when $p = 0$, the trivial subspace consisting of the single zero vector is the unique dissipative subspace of S. □

THEOREM 11.3.1. *Consider a complex symplectic space S with symplectic form $[\cdot:\cdot]$ and finite dimension $D \geq 1$. Choose any basis of S, with corresponding coordinates and skew-Hermitian nonsingular matrix H determined by $[\cdot:\cdot]$, and note that H is congruent to a diagonal matrix of the form* $\operatorname{diag}\{i, i, \cdots, i, -i, -i, \cdots, -i\}$. *Then conclude that the symplectic invariants of S are related to H by:*

$$p = \ number\ of\ (+i)\ terms\ on\ the\ diagonal,$$
$$q = \ number\ of\ (-i)\ terms\ on\ the\ diagonal,$$

with $0 \leq p, q \leq D$, hence the diagonal form for H is unique. Also we have

$$D = p + q, \quad Ex = p - q,$$

$$\triangle = \min\{p, q\} = \frac{1}{2}(D - |Ex|) \leq \frac{1}{2}D, \quad with\ equality\ if\ and\ only\ if\ \ Ex = 0.$$

PROOF. See [**184**], [**185**] for a proof. Also see the next Remark. □

REMARK 11.3.1. See Example 1 in [**185**], we note that each complex symplectic space S with finite dimension $D \geq 1$ is isomorphic to the complex vector space $\mathbb{C}^D$ with a suitable complex symplectic form. Let $\{e_1, e_2, \cdots, e_p, a_1, a_2, \cdots, a_q\}$ denote the basis of S such that the corresponding skew-Hermitian matrix of the symplectic form for S becomes

$$H = \begin{pmatrix} iI_p & 0 \\ 0 & -iI_q \end{pmatrix},$$

where I_s is the identity matrix of order $s \geq 0$ for $s = p, q$. Then the symplectic product of vectors $u, v \in S$ is given by

$$[u:v] = (u_1 \cdots u_D)H(v_1 \cdots v_D)^*.$$

THEOREM 11.3.2. *Consider a complex symplectic space S with symplectic form $[\cdot:\cdot]$ and finite dimension $D \geq 1$, and let p, q denote the positivity and the negativity indices of S. Let*

$$S_+ = \operatorname{span}\{e_1, \cdots, e_p\},$$
$$S_- = \operatorname{span}\{a_1, \cdots, a_q\}.$$

Then

(1) $S = S_+ \oplus S_-$ *and* $[S_+ : S_-] = 0$.
(2) S_+ *is a maximal dissipative subspace, also a strictly dissipative subspace.*
(3) S_- *is a maximal accretive subspace, also a strictly accretive subspace.*

PROOF. It is evident that $S = \operatorname{span}\{S_+, S_-\}$ and $[S_+ : S_-] = 0$. So item (1) holds.

For each $f = \sum_{j=1}^{p} \alpha_j e_j \in S_+$, $\alpha_1, \cdots, \alpha_p \in \mathbb{C}$, we have

$$[f : f] = i(|\alpha_1|^2 + \cdots + |\alpha_p|^2).$$

Therefore S_+ is a dissipative subspace, in fact a strictly dissipative subspace.

Now we prove S_+ is a maximal dissipative subspace. Assume that there exists a dissipative subspace S^* such that $S_+ \subset S^* \subseteq S$. Then there exists $\widehat{u} \in S^*$, but $\widehat{u} \notin S_+$. So $\widehat{u} = \widehat{u}_+ + \widehat{u}_-$, $\widehat{u}_+ \in S_+$, $\widehat{u}_- \in S_-$ and $\widehat{u}_- \neq 0$. Note that $\widehat{u}_- \in S^*$ but

$$\operatorname{Im}[\widehat{u}_- : \widehat{u}_-] < 0.$$

This contradicts the assumption that S^* is a dissipative subspace. Therefore S_+ is a maximal dissipative subspace. Similarly S_- is a maximal accretive subspace, also a strictly accretive subspace. □

DEFINITION 11.3.2. Consider a complex symplectic space S, with symplectic form $[\cdot:\cdot]$ and finite dimension $D \geq 1$. Let $\mathfrak{D} \subseteq S$ be a dissipative subspace.

(1) We call an element $u \in \mathfrak{D}$ a Lagrangian element if $[u : u] = 0$.
(2) We call an element $u \in \mathfrak{D}$ a dissipative element if $\operatorname{Im}[u : u] > 0$.

THEOREM 11.3.3. *Consider a complex symplectic space S with symplectic form $[\cdot:\cdot]$ and finite dimension $D \geq 1$. Let $\mathfrak{D} \subseteq S$ be a dissipative subspace with $\dim \mathfrak{D} = r \leq p$. Then the set*

$$\mathfrak{D}_L = \{f : f \in \mathfrak{D}, [f : f] = 0\} \subseteq \mathfrak{D}$$

of all the Lagrangian elements of $\mathfrak{D}$ is a Lagrangian subspace of S.

PROOF. For all $f, g \in \mathfrak{D}_L$, $\alpha, \beta \in \mathbb{C}$, we have $\alpha f + \beta g \in \mathfrak{D}$, and then $\operatorname{Im}[\alpha f + \beta g : \alpha f + \beta g] \geq 0$. Since

$$[\alpha f + \beta g : \alpha f + \beta g] = |\alpha|^2 [f : f] + |\beta|^2 [g : g] + 2i \operatorname{Im} \alpha\overline{\beta}[f : g] = 2i \operatorname{Im} \alpha\overline{\beta}[f : g],$$

we have

$$\operatorname{Im} \alpha\overline{\beta}[f : g] \geq 0.$$

Since $\alpha, \beta \in \mathbb{C}$ are arbitrary, we obtain $[f : g] = 0$ and then $[\alpha f + \beta g : \alpha f + \beta g] = 0$. Therefore $\mathfrak{D}_L$ is a Lagrangian subspace of S. □

THEOREM 11.3.4. *Let the notations of Theorem 11.3.3 hold, and let $\mathfrak{D}$ be a dissipative subspace of S with $\dim \mathfrak{D} = r$. Let $\dim \mathfrak{D}_L = r_L$, and let $\{a_1, \cdots, a_{r_L}\}$ be a basis for $\mathfrak{D}_L$. Then we can select a basis for $\mathfrak{D}$:*

$$\{a_1, \cdots, a_{r_L}, e_1, \cdots, e_{r_s}\}, \quad r_L + r_s = r$$

such that

(1) $\mathfrak{D}_s = \text{span}\{e_1, \cdots, e_{r_s}\}$ *is a strictly dissipative subspace, and*

$$[\mathfrak{D}_s : \mathfrak{D}_L] = 0.$$

(2) *The matrix of the symplectic form for* $\mathfrak{D}$ *induced from* S *becomes*

$$H_{\mathfrak{D}} = \begin{pmatrix} iI_{r_s} & 0 \\ 0 & 0_{r_L \times r_L} \end{pmatrix}.$$

PROOF. Assume that $\{a_1, \cdots, a_{r_L}, e_1, \cdots, e_{r_s}\}$, $r_s + r_L = r$, is a basis for $\mathfrak{D}$. Consider the definition of $\mathfrak{D}_L$ given in Theorem 11.3.3. Here $e_j \notin \mathfrak{D}_L$ and $\text{Im}[e_j : e_j] > 0$, $j = 1, \cdots, r_s$. Without loss of generality, we may choose $e_1, \cdots, e_{r_s}$ satisfying

$$[e_j : e_j] = i, \quad [e_j : e_k] = 0, \quad j \neq k, \qquad j, k = 1, \cdots, r_s.$$

(1) For each $\phi = \sum_{j=1}^{r_s} \beta_j e_j \in \mathfrak{D}_s$, $\beta_1, \cdots, \beta_{r_s} \in \mathbb{C}$, we have

$$[\phi : \phi] = i(|\beta_1|^2 + \cdots + |\beta_{r_s}|^2).$$

Therefore $\text{Im}[\phi : \phi] \geq 0$, and $[\phi : \phi] = 0$ if and only if $\phi = 0$. This shows $\mathfrak{D}_s = \text{span}\{e_1, \cdots, e_{r_s}\}$ is a strictly dissipative subspace.

For all $f \in \mathfrak{D}_L$, $\phi \in \mathfrak{D}_s$, *and* $\alpha, \beta \in \mathbb{C}$, since $\alpha f + \beta\phi \in \mathfrak{D}$, we have

$$\text{Im}[\alpha f + \beta\phi : \alpha f + \beta\phi] \geq 0.$$

Note that

$$\begin{aligned}[\alpha f + \beta\phi : \alpha f + \beta\phi] &= |\alpha|^2[f : f] + |\beta|^2[\phi : \phi] + 2\,i\,\text{Im}\,\alpha\overline{\beta}[f : \phi] \\ &= |\beta|^2[\phi : \phi] + 2\,i\,\text{Im}\,\alpha\overline{\beta}[f : \phi].\end{aligned}$$

By the arbitrariness of α, $\beta \in \mathbb{C}$, we have $[f : \phi] = 0$, i.e. $[\mathfrak{D}_L : \mathfrak{D}_s] = 0$.

Item (2) follows immediately from item (1). □

THEOREM 11.3.5. *Consider a complex symplectic space* S, *with symplectic form* $[\cdot : \cdot]$, *and finite dimension* $D \geq 1$. *Then* $\mathfrak{D} \subseteq S$ *is a dissipative subspace if and only if there exist a Lagrangian subspace* $\mathfrak{D}_L \subseteq \mathfrak{D}$ *and a strictly dissipative subspace* $\mathfrak{D}_s \subseteq \mathfrak{D}$ *such that*

$$\mathfrak{D} = \mathfrak{D}_L \oplus \mathfrak{D}_s,$$

where ' $\oplus$ *' means* $\mathfrak{D}_L$ *is symplectic orthogonal to* $\mathfrak{D}_s$.

PROOF. The necessity is obtained directly from the previous theorem. To prove the sufficiency, take any $u \in \mathfrak{D}$.

If $u \in \mathfrak{D}_L$, $[u : u] = 0$, i.e. $\text{Im}[u : u] = 0$.

If $u \in \mathfrak{D}_s$, $\text{Im}[u : u] > 0$.

If $u \notin \mathfrak{D}_L$ and $u \notin \mathfrak{D}_s$ then $u = f + g$, $f \in \mathfrak{D}_L$, $g \in \mathfrak{D}_s$, $f \neq 0$, $g \neq 0$, and then

$$\text{Im}[u : u] = \text{Im}[f + g : f + g] = \text{Im}[g : g] > 0.$$

This completes the proof. □

Example 2. Consider the complex symplectic space $S = \mathbb{C}^2$ with standard basis vectors $e_1 = (1, 0)$, $e_2 = (0, 1)$ satisfying $[e_1 : e_1] = i$, $[e_2 : e_2] = -i$, and other symplectic products are zero. That is, we use the skew-Hermitian matrix $H = \text{diag}\{i, -i\}$ to define the symplectic structure on $S = \mathbb{C}^2$.

(1) Define $\mathfrak{D}_1 = \text{span}\{e_1\}$. Then $\mathfrak{D}_1$ is a maximal and strictly dissipative subspace.

(2) Define $\mathfrak{D}_2 = \operatorname{span}\{2e_1 + e_2\}$. Then $\mathfrak{D}_2$ is a maximal and strictly dissipative subspace.

4. Applications of Symplectic Geometry to Ordinary Differential Operators

In this section we study the relationship between complex symplectic space and differential operators. In particular we find a natural one to one correspondence between the set of all (strictly) dissipative extensions of the minimal operator $S_{\min}$ in H and the set of all (strictly) dissipative subspaces in the complex symplectic space $G = D_{\max}/D_{\min}$.

THEOREM 11.4.1. *Let $M = M_Q$, $Q \in Z_n(J, \mathbb{C})$, $Q = Q^+$, and let w be a weight function. Let $\mathcal{N}(S)$ denote the null space of an operator S. Let (d^-, d^+) denote the deficiency indices of $S_{\min}(M)$ and consider the equation*

$$My = \lambda wy.$$

It follows from the von Neumann formula applied to ordinary differential operators (4.4.2) that:

(1) *There exist d^+ linearly independent solutions u_j, $j = 1, \cdots, d^+$ in $L^2(J, w)$ which lie in $\mathcal{N}(S_{\max} - iI)$ and satisfy*

$$(u_k, u_j) = \frac{1}{2}\delta_{kj}, \quad k, j = 1, \cdots, d^+.$$

(2) *There exist d^- linearly independent solutions $v_1, v_2, \cdots, v_{d^-}$ in $L^2(J, w)$ which lie in $\mathcal{N}(S_{\max} + iI)$ and satisfy*

$$(v_k, v_j) = \frac{1}{2}\delta_{kj}, \quad k, j = 1, \cdots, d^-.$$

(3)

$$(u_k, v_j) = 0, \quad k = 1, \cdots, d^+; \ j = 1, \cdots, d^-.$$

Here $(\cdot, \cdot)$ denotes the inner product in the Hilbert space $H = L^2(J, w)$.

PROOF. Since the space $\mathcal{N}(S_{\max} - iI) \dotplus \mathcal{N}(S_{\max} + iI)$ is a $(d^+ + d^-)$ dimensional subspace of the Hilbert space H, there exist orthonormal bases

$$\widehat{u}_1, \widehat{u}_2, \cdots, \widehat{u}_{d^+}, \widehat{v}_1, \widehat{v}_2, \cdots, \widehat{v}_{d^-}$$

such that $\widehat{u}_1, \widehat{u}_2, \cdots, \widehat{u}_{d^+}$ are in $\mathcal{N}(S_{\max} - iI)$, and $\widehat{v}_1, \widehat{v}_2, \cdots, \widehat{v}_{d^-}$ are in $\mathcal{N}(S_{\max} + iI)$. Let

$$u_k = \frac{1}{\sqrt{2}}\widehat{u}_k, \quad v_j = \frac{1}{\sqrt{2}}\widehat{v}_j, \quad k = 1, 2, \cdots, d^+; \ j = 1, 2, \cdots, d^-,$$

then it follows that the bases $u_1, u_2, \cdots, u_{d^+}, v_1, v_2, \cdots, v_{d^-}$ satisfy the properties (1), (2), (3) of Theorem 11.4.1. □

COROLLARY 11.4.1. *Let the notations and hypotheses of Theorem 11.4.1 hold. Then*

$$D_{\max} = D_{\min} \dotplus \{u_1, u_2, \cdots, u_{d^+}\} \dotplus \{v_1, v_2, \cdots, v_{d^-}\}.$$

PROOF. This follows immediately from Theorem 4.4.2. □

4.1. Dissipative and Accretive Extensions of the Minimal Operator.

DEFINITION 11.4.1. A linear operator T_D on H with dense domain $D(T_D)$ is called dissipative if

$$\operatorname{Im}(T_D f, f) \geq 0 \quad \textit{for all } f \in D(T_D).$$

A linear operator T_A on H with dense domain $D(T_A)$ is called accretive if

$$\operatorname{Im}(T_A f, f) \leq 0 \quad \textit{for all } f \in D(T_A).$$

DEFINITION 11.4.2. A dissipative (accretive) operator T_D (T_A) is called maximal dissipative (maximal accretive) if it does not have any proper dissipative (accretive) extension in H.

Next we define strictly dissipative (strictly accretive) extensions of symmetric operators.

DEFINITION 11.4.3. A dissipative extension T_D of a symmetric operator T on H is called strictly dissipative if

$$\operatorname{Im}(T_D f,\, f) > 0 \quad \textit{for every } f \in D(T_D) \setminus D(T).$$

An accretive extension T_A of a symmetric operator T is called strictly accretive if

$$\operatorname{Im}(T_A f,\, f) < 0 \quad \textit{for every } f \in D(T_A) \setminus D(T).$$

We denote the strictly dissipative (accretive) extension as T_{sD} (T_{sA}).

REMARK 11.4.1. We note that

(1) Since a linear operator T is accretive if and only if $-T$ is dissipative, all results concerning dissipative operators can be immediately transferred to accretive operators.
(2) A symmetric operator is both dissipative and accretive.

LEMMA 11.4.1. *Let the notations and hypotheses of Theorem* 11.4.1 *hold. Then there are proper dissipative extensions of* $S_{\min}$ *if and only if* $d^+ \neq 0$.

PROOF. *Sufficiency.* Let r be an integer with $0 < r \leq d^+$, and define

$$T = S_{\max}|_D, \quad D = D_{\min} \dotplus \operatorname{span}\{u_1, \cdots, u_r\},$$

where $u_1, u_2, \cdots, u_r \in \mathcal{N}(S_{\max} - iI)$. Then T is a dissipative extension of $S_{\min}$. In fact, for each $v \in D$, $v = f + u$ where $f \in D_{\min}$, $u \in \operatorname{span}\{u_1, \cdots, u_r\}$,

$$\begin{aligned}(Tv, v) &= (S_{\max} f, f) + (S_{\max} u, u) + (S_{\max} f, u) + (S_{\max} u, f) \\ &= (S_{\min} f, f) + i\|u\|^2 + (f, S_{\max} u) + (iu, f) \\ &= (S_{\min} f, f) + i\|u\|^2 - i(f, u) + i(u, f).\end{aligned}$$

Since $-i(f,u) + i(u,f)$ is real $\operatorname{Im}(T(f+u),\, f+u) = \|u\|^2 \geq 0$. So T is a proper dissipative extension of $S_{\min}$.

Necessity. If there exist proper dissipative extensions of $S_{\min}$, then $d^+ \neq 0$. If not, let $d^+ = 0$, by Corollary 11.4.1 we have

$$D_{\max} = D_{\min} \dotplus \mathcal{N}(S_{\max} + iI).$$

For any $f \in D_{\max}$, let $f = f_1 + f_2$, $f_1 \in D_{\min}$, $f_2 \in \mathcal{N}(S_{\max} + iI)$, then

$$\begin{aligned}(S_{\max} f, f) &= (S_{\max} f_1, f_1) + (S_{\max} f_1, f_2) + (S_{\max} f_2, f_1) + (S_{\max} f_2, f_2) \\ &= (S_{\min} f_1, f_1) + (f_1, S_{\max} f_2) - i(f_2, f_1) - i\|f_2\|^2 \\ &= (S_{\min} f_1, f_1) + i(f_1, f_2) - i(f_2, f_1) - i\|f_2\|^2.\end{aligned}$$

This shows that for all $f \in D_{\max}$, $\operatorname{Im}(S_{\max} f, f) = -\|f_2\|^2 \le 0$, with equality if and only if $f_2 = 0$. Therefore $S_{\min}$ has no proper dissipative extension. □

By Lemma 11.4.1, we have the following result.

COROLLARY 11.4.2. *Let the notations and hypotheses of Theorem* 11.4.1 *hold. Then there are proper accretive extensions of* $S_{\min}$ *if and only if* $d^- \neq 0$.

4.2. A Relationship between G and M. Next we define and investigate the structures of a complex symplectic space G, with its Lagrangian subspaces L, dissipative subspaces $\mathfrak{D}$ and accretive subspaces $\mathfrak{A}$, which arise in connection with boundary value problems associated with a symmetric differential expression $M = M_Q$, $Q \in Z_n(J, \mathbb{C})$, $Q = Q^+$.

Define the endpoint complex vector space for M as the quotient space:

$$G = D_{\max}/D_{\min},$$

where each element $\widetilde{f}$ is a coset $\widetilde{f} = \{f + D_{\min}\}$ of some function $f \in D_{\max}$. So there is a natural projection Ψ of $D_{\max}$ onto G by

$$\Psi : D_{\max} \to G, \quad f \to \widetilde{f} = \Psi f = \{f + D_{\min}\}. \tag{11.4.1}$$

In the following, for simplicity, sometimes we use $\widetilde{f} = f + D_{\min}$.

The endpoint space G becomes a complex symplectic space with the symplectic product $[\cdot : \cdot]$ inherited from $D_{\max}$ as follows:

$$[\widetilde{f} : \widetilde{g}] = [f + D_{\min} : g + D_{\min}] := [f : g],$$

where the skew-Hermitian form $[f : g]$ is defined by

$$[f : g] = \int_a^b \{\overline{g} M f - f \overline{Mg}\} = [f,\, g]_a^b, \quad \text{for all } f, g \in D_{\max}.$$

Hence

$$D_{\min} = \{f \in D_{\max} \;:\; [f : D_{\max}] = 0\}.$$

LEMMA 11.4.2. *Let the notation and hypotheses of Theorem* 11.4.1 *hold, then*

$$([u_k : u_j])_{d^+ \times d^+} = iI_{d^+}, \quad ([v_k : v_j])_{d^- \times d^-} = -iI_{d^-}.$$

Here I_{d^+}, I_{d^-} *are the identity matrices of order* d^+ *and* d^-, *respectively.*

PROOF. From

$$[u_k : u_j] = \int_a^b \{\overline{u}_j M(u_k) - u_k \overline{M u_j}\} = \int_a^b \{\overline{u}_j i w u_k - u_k \overline{i w u_j}\} = 2i(u_k, u_j),$$

$$[v_k : v_j] = \int_a^b \{\overline{v}_j M(v_k) - v_k \overline{M v_j}\} = \int_a^b \{\overline{v}_j(-i w v_k) - v_k \overline{(-i w v_j)}\} = -2i(v_k, v_j),$$

and from Theorem 11.4.1, we have

$$\begin{aligned}([u_k : u_j])_{d^+ \times d^+} &= iI_{d^+}, \\ ([v_k : v_j])_{d^- \times d^-} &= -iI_{d^-}.\end{aligned}$$

□

LEMMA 11.4.3. *The symplectic invariants of the complex symplectic space G with the skew-Hermitian form $[\cdot:\cdot]$ defined above satisfy*

$$p = d^+, q = d^-, \ \dim G = d^+ + d^-, \ \textit{and } Ex = d^+ - d^-.$$

PROOF. From Proposition 1 in [**184**], we know that the symplectic invariants of G are related to the deficiency indices $d^{\pm}$ of a symmetric differential expression M by $p = d^+$, $q = d^-$. □

THEOREM 11.4.2. *Let the notation and hypotheses of Theorem 11.4.1 hold, and let $G = D_{\max}/D_{\min}$ be the complex symplectic space as defined above. Then*

(1) $G = \text{span}\{\widetilde{u}_1, \cdots, \widetilde{u}_{d^+}, \ \widetilde{v}_1, \cdots, \widetilde{v}_{d^-}\}$, *where* $\widetilde{u}_j = u_j + D_{\min}$, $j = 1, \cdots, d^+$; $\widetilde{v}_k = v_k + D_{\min}$, $k = 1, \cdots, d^-$.
(2) *G is symplectic isomorphic to a complex symplectic space* $\mathbb{C}^{d^+ + d^-}$.
(3) *For the basis $\{\widetilde{u}_1, \cdots, \widetilde{u}_{d^+}, \ \widetilde{v}_1, \cdots, \widetilde{v}_{d^-}\}$, the associated skew-Hermitian matrix of the symplectic form for G is as follows*

$$H = \begin{pmatrix} iI_{d^+} & 0 \\ 0 & -iI_{d^-} \end{pmatrix}.$$

(The notation H for this matrix should not be confused with our Hilbert space notation $H = L^2(J, w)$.)

PROOF. By Corollary 11.4.1, item (1) can be obtained directly; from Example 1 in [**185**], item (2) follows. Note that $\forall u_j \in \mathcal{N}(S_{max} - iI)$, $v_k \in \mathcal{N}(S_{\max} + iI)$,

$$[\widetilde{u}_j : \widetilde{v}_k] = [u_j : v_k] = \int_a^b \overline{v}_k M u_j - u_j \overline{M v_k} = \int_a^b \overline{v}_k (iwu_j) - u_j \overline{(-iwv_k)} = 0.$$

Connecting this with Lemma 11.4.2, the associated skew-Hermitian matrix is

$$H = \begin{pmatrix} iI_{d^+} & 0 \\ 0 & -iI_{d^-} \end{pmatrix}.$$ □

Using Theorem 11.4.2 we next give a symplectic orthogonal direct sum decomposition of the complex symplectic space G in terms of dissipative and accretive subspaces.

THEOREM 11.4.3. *Let $G = D_{\max}/D_{\min}$ be the complex symplectic space associated with M, and let the notation and hypotheses of Theorem 11.4.2 hold. Let*

$$\widetilde{S}_+ = \text{span}\{\widetilde{u}_1, \widetilde{u}_2, \cdots, \widetilde{u}_{d^+}\},$$
$$\widetilde{S}_- = \text{span}\{\widetilde{v}_1, \widetilde{v}_2, \cdots, \widetilde{v}_{d^-}\}.$$

Then

(1) $G = \widetilde{S}_+ \oplus \widetilde{S}_-$ *with* $[\widetilde{S}_+ : \widetilde{S}_-] = 0$.
(2) *$\widetilde{S}_+$ is a maximal dissipative subspace, also a strictly dissipative subspace.*
(3) *$\widetilde{S}_-$ is a maximal accretive subspace, also a strictly accretive subspace.*

PROOF. It follows from Theorem 11.4.2 that $G = \text{span}\{\widetilde{S}_+, \widetilde{S}_-\}$ and $[\widetilde{S}_+ : \widetilde{S}_-] = 0$. From Lemma 11.4.2

$$[\widetilde{u}_k : \widetilde{u}_j] = [u_k : u_j] = 2i(u_k, u_j) = i\delta_{kj}, \quad k, j = 1, \cdots, d^+. \tag{11.4.2}$$

For any $\widetilde{u} = \sum_{j=1}^{d^+} c_j\widetilde{u}_j \in \widetilde{S}_+$, $c_1, \cdots, c_{d^+} \in \mathbb{C}$, from (11.4.2), we have

$$\begin{aligned}[\widetilde{u} : \widetilde{u}] &= [\sum_{j=1}^{d^+} c_j\widetilde{u}_j : \sum_{j=1}^{d^+} c_j\widetilde{u}_j] \\ &= |c_1|^2[\widetilde{u}_1 : \widetilde{u}_1] + |c_2|^2[\widetilde{u}_2 : \widetilde{u}_2] + \cdots + |c_{d^+}|^2[\widetilde{u}_{d^+} : \widetilde{u}_{d^+}] \\ &= i(|c_1|^2 + |c_2|^2 + \cdots + |c_{d^+}|^2).\end{aligned}$$

Hence $\operatorname{Im}[\widetilde{u} : \widetilde{u}] \geq 0$, and $\operatorname{Im}[\widetilde{u} : \widetilde{u}] = 0$ if and only if $\widetilde{u} = 0$. Therefore $\widetilde{S}_+$ is a strictly dissipative subspace. Similarly, $\widetilde{S}_-$ is a strictly accretive subspace.

Now to prove that $\widetilde{S}_+$ is a maximal dissipative subspace. Assume that there exists a dissipative subspace $\widetilde{S}^*$ of G such that $\widetilde{S}_+ \subset \widetilde{S}^*$. Then there exists $\widetilde{\chi} \in \widetilde{S}^*$, but $\widetilde{\chi} \notin \widetilde{S}_+$. By (1), let $\widetilde{\chi} = \widetilde{\chi}_+ + \widetilde{\chi}_-$, where $\widetilde{\chi}_+ \in \widetilde{S}_+, \widetilde{\chi}_- \in \widetilde{S}_-$ and $\widetilde{\chi}_- \neq 0$. Note that $\widetilde{\chi}_- \in \widetilde{S}^*$ but

$$\operatorname{Im}[\widetilde{\chi}_- : \widetilde{\chi}_-] < 0.$$

This contradicts the assumption that $\widetilde{S}^*$ is a dissipative subspace. Therefore, $\widetilde{S}_+$ is a maximal dissipative subspace. Similarly, $\widetilde{S}_-$ is a maximal accretive subspace. □

4.3. Symmetric and Dissipative Operators, and Symplectic Geometry. In this subsection we investigate the 1-1 correspondences from Hilbert space H to the complex symplectic space G as defined above for the three classes of operators: symmetric, dissipative and strictly dissipative.

THEOREM 11.4.4. *Let the hypotheses and notation of Theorem* 11.4.2 *hold. Then*

(1) $S_{\min}$ *has symmetric extensions in* H *if and only if* G *has Lagrangian subspaces.*
(2) *There is a one to one correspondence between the set* $\{S\}$ *of all symmetric extensions of* $S_{\min}$ *and the set* $\{L\}$ *of all Lagrangian subspaces of* $G = D_{\max}/D_{\min}$.

PROOF. (1) Let S be any symmetric extension of the minimal operator $S_{\min}$ with domain $D(S)$, i.e.

$$S_{\min} \subseteq S \subset S^* \subseteq S_{\max}$$

and therefore

$$D_{\min} \subseteq D(S) \subset D(S^*) \subseteq D_{\max}.$$

Define the subset L of the complex symplectic space G:

$$L = D(S)/D_{\min}$$

which consists of all $\widetilde{f} = \{f + D_{\min}\}$ with $f \in D(S)$.

Let Ψ be the natural projection map:

$$\Psi : D_{\max} \longrightarrow G, \quad f \longrightarrow \widetilde{f} = \{f + D_{\min}\}. \tag{11.4.3}$$

Note that L is the image of the linear manifold $D(S) \subseteq D_{\max}$ under the projection map Ψ. Hence L is a linear subspace of G. For all $\widetilde{f} = \{f + D_{\min}\}$, $\widetilde{g} = \{g + D_{\min}\}$, where f, $g \in D(S)$, since S is symmetric, we have

$$[\widetilde{f} : \widetilde{g}] = [f : g] = (S_{\max}f, g) - (f, S_{\max}g) = (Sf, g) - (f, Sg) = 0.$$

This shows that L is a Lagrangian subspace of G.

On the other hand, let L be any Lagrangian subspace of G. We need to find a corresponding symmetric extension S of $S_{\min}$ such that $D(S)$ satisfies $\Psi D(S) = L$. Now define

$$D(S) = \Psi^{-1}L = \{h \in D_{\max} : \widetilde{h} \in L\}.$$

So $D(S)$ is a submanifold of H and

$$D_{\min} \subseteq D(S) \subseteq D_{\max}.$$

Define S as the restriction of S_{max} on $D(S)$. In the following, we prove that S is symmetric.

Since L is a Lagrangian subspace of G, we have for all $f,\, g \in D(S)$,

$$0 = [\widetilde{f} : \widetilde{g}] = [f : g] = (S_{\max}f, g) - (f, S_{\max}g) = (Sf, g) - (f, Sg).$$

Therefore S is symmetric and $\Psi D(S) = L = D(S)/D_{\min}$.

(2) Let the map induced by Ψ defined in (11.4.3) be

$$\mathbf{\Psi} : \{S\} \longrightarrow \{L\}, \quad S \longrightarrow L = D(S)/D_{\min}.$$

By (1), clearly $\mathbf{\Psi}$ is surjective. Next we prove that $\mathbf{\Psi}$ is injective. Let S_1 and S_2 be two different symmetric extensions of $S_{\min}$, and therefore they are restrictions of $S_{\min}^* = S_{\max}$. Let $D(S_1)$ and $D(S_2)$ be the domains of S_1 and S_2 respectively. So $D(S_1) \neq D(S_2)$. Hence there exists a $u \in D(S_1)$, but $u \notin D(S_2)$ (or vice visa). Note that $u \notin D_{\min}$. Then $\widetilde{u} = \{u + D_{\min}\}$ satisfies $\widetilde{u} \in L_1 = D(S_1)/D_{\min}$ but $\widetilde{u} \notin L_2 = D(S_2)/D_{\min}$. This shows that $L_1 \neq L_2$, and thus $\mathbf{\Psi}$ is injective. Therefore the map $\mathbf{\Psi}$ is bijective, i.e. there is a one to one correspondence between the set $\{S\}$ of all symmetric extensions of $S_{\min}$ and the set $\{L\}$ of all Lagrangian subspaces of G. □

THEOREM 11.4.5. *Let the hypotheses and notation of Theorem* 11.4.2 *hold. Then there exists a natural one-to-one correspondence between the set* $\{T_{sD}\}$ *of all strictly dissipative extensions of* $S_{\min}$ *in* H *and the set* $\{\widetilde{\mathfrak{D}}_s\}$ *of all strictly dissipative subspaces of* G.

PROOF. Let T_{sD} , with domain $D(T_{sD})$, be any strictly dissipative extension of $S_{\min}$, $D_{\min} \subset D(T_{sD}) \subseteq D_{\max}$. By the natural map defined in (11.4.1), we define the subset $\mathfrak{D}_s$ of G:

$$\mathfrak{D}_s = \Psi D(T_{sD}) = D(T_{sD})/D_{\min} = \{f + D_{\min} | f \in D(T_{sD})\}.$$

Clearly $\mathfrak{D}_s$ is a linear subspace of G. Since T_{sD} is a strictly dissipative operator, we have

$$\operatorname{Im}(T_{sD}f, f) > 0 \quad \text{for every } f \in D(T_{sD})\backslash D_{\min}.$$

For any $\widetilde{f} = \{f + D_{\min}\} \in \mathfrak{D}_s$ and $\widetilde{f} \neq 0$, i.e. $f \in D(T_{sD})\backslash D_{\min}$,

$$\begin{aligned}[\widetilde{f} : \widetilde{f}] &= [f : f] = (S_{\max}f, f) - (f, S_{\max}f) \\ &= (T_{sD}f, f) - (f, T_{sD}f) \\ &= 2i \operatorname{Im}(T_{sD}f, f).\end{aligned}$$

Therefore $\operatorname{Im}[\widetilde{f} : \widetilde{f}] > 0$. This shows that $\mathfrak{D}_s$ is a strictly dissipative subspace.

On the other hand, let $\mathfrak{D}_s$ be any strictly dissipative space of G. By the natural map Ψ, we define

$$D(T_{sD}) = \Psi^{-1}\mathfrak{D}_s = \{f \in D_{\max} | f + D_{\min} \in \mathfrak{D}_s\}.$$

Then $D(T_{sD})$ is a linear submanifold of H and $D_{\min} \subset D(T_{sD}) \subseteq D_{\max}$.

Define $T_{sD} = S_{\max}|_{D(T_{sD})}$. For every $f \in D(T_{sD})$, but $f \notin D_{\min}$, then $\widetilde{f} = \{f + D_{\min}\} \in \mathfrak{D}_s$ and $\widetilde{f} \neq 0$,

$$[\widetilde{f} : \widetilde{f}] = [f : f] = (T_{sD}f, f) - (f, T_{sD}f) = 2i\,\mathrm{Im}(T_{sD}f, f).$$

Since $\mathfrak{D}_s$ is strictly dissipative, we have $\mathrm{Im}[\widetilde{f} : \widetilde{f}]>0$, $\widetilde{f}\neq 0$, and then $\mathrm{Im}(T_{sD}f, f) > 0$ for all $f \in D(T_{sD}) \setminus D_{\min}$. Hence T_{sD} is a strictly dissipative operator and $\mathfrak{D}_s = D(T_{sD})/D_{\min}$.

Now define the map induced by Ψ:

$$\mathbf{\Psi} : \{T_{sD}\} \longrightarrow \{\mathfrak{D}_s\}, \quad T_{sD} \longrightarrow \mathfrak{D}_s = D(T_{sD})/D_{\min},$$

which is surjective. In the following, we prove that $\mathbf{\Psi}$ is injective. Let T_{s1} and T_{s2} be any two different strictly dissipative extensions of $S_{\min}$, whose domains are $D(T_{s1})$ and $D(T_{s2})$ respectively. So $D(T_{s1}) \neq D(T_{s2})$. Therefore there exists $u \in D(T_{s1})$, but $u \notin D(T_{s2})$ (or vice visa). Note that $u \notin D_{\min}$. Then $\widetilde{u} = \{u + D_{\min}\}$ satisfies $\widetilde{u} \in \mathfrak{D}_{s1} = D(T_{s1})/D_{\min}$, but $\widetilde{u} \notin \mathfrak{D}_{s2} = D(T_{s2})/D_{\min}$. This shows $\mathfrak{D}_{s1} \neq \mathfrak{D}_{s2}$. Therefore $\mathbf{\Psi}$ is injective. Thus $\mathbf{\Psi}$ defines a one-to-one correspondence between $\{T_{sD}\}$ and $\{\mathfrak{D}_s\}$. □

THEOREM 11.4.6. *Let the hypotheses and notation of Theorem* 11.4.2 *hold. Then there exists a natural one-to-one correspondence between the set* $\{T_D\}$ *of all dissipative extensions of* $S_{\min}$ *in* H *and the set* $\{\mathfrak{D}\}$ *of all dissipative subspaces of* G. *Namely, for each such dissipative extension* T_D *with domain* $D(T_D) \subseteq D_{\max}$, *the corresponding dissipative subspace* $\mathfrak{D}$ *is defined by*

$$\mathfrak{D} = D(T_D)/D_{\min}.$$

PROOF. This proof is similar to the proof of Theorem 11.4.5 and hence omitted. □

5. LC Representation of Dissipative Operators

In this section we specialize to the case when $M = M_Q$, $Q \in Z_n(J, \mathbb{R})$, $n = 2k$, $k \geq 1$, $Q = Q^+$. For this case we characterize the dissipative extensions of $S_{\min}$ and the dissipative subspaces of G in terms of LC solutions given in Section 7.2.

The next theorem establishes the connection between the structure of the complex symplectic space $G = D_{\max}/D_{\min}$ with the symplectic product $[\cdot : \cdot]$ defined in Section 11.4 and LC solutions.

REMARK 11.5.1. Note that for this even order expression M with real coefficients the deficiency indices $d^+ = d^- = d$ and, from Lemma 11.4.3, the symplectic invariants of G satisfy $p = q = d$, $\dim G = 2d$, and $Ex = 0$.

THEOREM 11.5.1. *Let the notation and hypotheses of Theorem* 7.2.3 *hold and let* $G = D_{\max}/D_{\min}$. *Then*

(1) $G = \mathrm{span}\{\widetilde{u}_1, \cdots, \widetilde{u}_{m_a}\} \oplus \mathrm{span}\{\widetilde{v}_1, \cdots, \widetilde{v}_{m_b}\}$.
(2) *G is the complexification of the unique real symplectic space* $\mathbb{R}^{2d}$.
(3) *G is symplectic isomorphic to a complex symplectic space* $\mathbb{C}^{2d}$.
(4) *For the basis* $\{\widetilde{u}_1, \cdots, \widetilde{u}_{m_a}, \widetilde{v}_1, \cdots, \widetilde{v}_{m_b}\}$, *let*

$$H = \begin{pmatrix} -U_{m_a \times m_a} & 0 \\ 0 & V_{m_b \times m_b} \end{pmatrix},$$

then this matrix H (there should be no confusion with the Hilbert space H) is a skew-Hermitian matrix, and for all $\widetilde{f} = (f_1, \cdots, f_{m_a}, \widehat{f}_1, \cdots, \widehat{f}_{m_b})$, $\widetilde{g} = (g_1, \cdots, g_{m_a}, \widehat{g}_1, \cdots, \widehat{g}_{m_b})$ in G,

$$[\widetilde{f} : \widetilde{g}] = (f_1, \cdots, f_{m_a}, \widehat{f}_1, \cdots, \widehat{f}_{m_b}) H (g_1, \cdots, g_{m_a}, \widehat{g}_1, \cdots, \widehat{g}_{m_b})^*.$$

Here $\widetilde{u}_i = \{u_i + D_{\min}\}$, $\widetilde{v}_j = \{v_j + D_{\min}\}$, $i = 1, \cdots, m_a$; $j = 1, 2, \cdots, m_b$; u_i and v_j are (real valued) LC solutions of the endpoints a and b respectively. The matrices $U = (-1)^k E_{m_a}$ and $V = (-1)^k E_{m_b}$ are defined in Theorem 7.2.3.

PROOF. By Theorem 7.2.3 and the decomposition of the maximal domain given in Theorem 7.2.4, we obtain (1); from Theorem 1 in [**185**] and Remark 11.5.1, we obtain (2); from Example 1 in [**185**], (3) follows. Now, to prove (4): for all $\widetilde{f}$, $\widetilde{g} \in G$,

$$\begin{aligned}
&[\widetilde{f} : \widetilde{g}] \\
&= [f_1 \widetilde{u}_1 + \cdots + f_{m_a} \widetilde{u}_{m_a} + \widehat{f}_1 \widetilde{v}_1 + \cdots + \widehat{f}_{m_b} \widetilde{v}_{m_b} \\
&\qquad : g_1 \widetilde{u}_1 + \cdots + g_{m_a} \widetilde{u}_{m_a} + \widehat{g}_1 \widetilde{v}_1 + \cdots + \widehat{g}_{m_b} \widetilde{v}_{m_b}] \\
&= (f_1, \cdots, f_{m_a}, \widehat{f}_1, \cdots, \widehat{f}_{m_b}) \\
&\qquad \cdot \begin{pmatrix} [u_1 : u_1] & \cdots & [u_1 : u_{m_a}] & [u_1 : v_1] & \cdots & [u_1 : v_{m_b}] \\ \cdots & \cdots & \cdots & \cdots & \cdots & \cdots \\ [u_{m_a} : u_1] & \cdots & [u_{m_a} : u_{m_a}] & [u_{m_a} : v_1] & \cdots & [u_{m_a} : v_{m_b}] \\ [v_1 : u_1] & \cdots & [v_1 : u_{m_a}] & [v_1 : v_1] & \cdots & [v_1 : v_{m_b}] \\ \cdots & \cdots & \cdots & \cdots & \cdots & \cdots \\ [v_{m_b} : u_1] & \cdots & [v_{m_b} : u_{m_a}] & [v_{m_b} : v_1] & \cdots & [v_{m_b} : v_{m_b}] \end{pmatrix} \\
&\qquad \cdot (g_1, \cdots, g_{m_a}, \widehat{g}_1, \cdots, \widehat{g}_{m_b})^* \\
&= (f_1, \cdots, f_{m_a}, \widehat{f}_1, \cdots, \widehat{f}_{m_b}) \\
&\qquad \cdot \begin{pmatrix} -([u_i, u_j](a)) & 0 \\ 0 & ([v_i, v_j](b)) \end{pmatrix} (g_1, \cdots, g_{m_a}, \widehat{g}_1, \cdots, \widehat{g}_{m_b})^* \\
&= (f_1, \cdots, f_{m_a}, \widehat{f}_1, \cdots, \widehat{f}_{m_b}) \\
&\qquad \cdot \begin{pmatrix} -U_{m_a \times m_a} & 0 \\ 0 & V_{m_b \times m_b} \end{pmatrix} (g_1, \cdots, g_{m_a}, \widehat{g}_1, \cdots, \widehat{g}_{m_b})^* \\
&= (f_1, \cdots, f_{m_a}, \widehat{f}_1, \cdots, \widehat{f}_{m_b}) H (g_1, \cdots, g_{m_a}, \widehat{g}_1, \cdots, \widehat{g}_{m_b})^*.
\end{aligned}$$

Clearly H is a skew-Hermitian matrix. □

REMARK 11.5.2. In Theorem 11.5.1, since G is symplectic isomorphic to $\mathbb{C}^{2d}$, we don't distinguish it from $\mathbb{C}^{2d}$. The next theorem characterizes the domains of dissipative extensions of $S_{\min}$ in terms of LC solutions.

THEOREM 11.5.2. *Let the hypotheses and notation of Theorem 11.5.1 hold. A linear submanifold $D(T_D)$ of $D_{\max}$ is the domain of a dissipative extension T_D of $S_{\min}$ in H if and only if there exist*

(11.5.1) *linearly independent vectors* $\gamma_i, \alpha_j \in \mathbb{C}^{2d}$, $i = 1, \cdots, r_L$, $j = 1, \cdots, r_s$, *satisfying*

$$[\gamma_i : \gamma_j] = 0, \quad i, j = 1, \cdots, r_L, \tag{11.5.2}$$

$$[\gamma_i : \alpha_j] = 0, \quad i = 1, \cdots, r_L, \quad j = 1, \cdots, r_s, \tag{11.5.3}$$

(11.5.4) $$\mathrm{Im}[\sum_{j=1}^{r_s} c_j\alpha_j : \sum_{j=1}^{r_s} c_j\alpha_j] > 0, \quad c_1, \cdots, c_{r_s} \in \mathbb{C} \text{ are not all zero,}$$

such that

$$D(T_D) = D_{\min} \dot{+} \mathrm{span}\{w_1, w_2, \cdots w_{r_L}\} \dot{+} \mathrm{span}\{\chi_1, \chi_2, \cdots, \chi_{r_s}\},$$

where

$$w_i = \gamma_i W, \quad \chi_j = \alpha_j W, \qquad i = 1, \cdots, r_L, \ \ j = 1, \cdots, r_s,$$
$$W = (u_1, \cdots, u_{m_a}, v_1, \cdots, v_{m_b})^T.$$

PROOF. *Necessity.* Let $D(T_D)$ be the domain of a dissipative extension T_D. By Theorem 11.4.6, the corresponding $\mathfrak{D} = D(T_D)/D_{\min}$ is a dissipative subspace of G, then from Theorem 11.3.5 there exists a Lagrangian subspace $\mathfrak{D}_L \subseteq \mathfrak{D}$ and a strictly dissipative subspace $\mathfrak{D}_s \subseteq \mathfrak{D}$ such that

(11.5.5) $$\mathfrak{D} = \mathfrak{D}_L \oplus \mathfrak{D}_s.$$

By Theorem 11.5.1, G is symplectic isomorphic to the complex symplectic space $\mathbb{C}^{2d}$, and we let $\widetilde{u}_1, \cdots, \widetilde{u}_{m_a}, \widetilde{v}_1, \cdots, \widetilde{v}_{m_b}$ be a basis of G and let

$$\widetilde{W} = (\widetilde{u}_1, \cdots, \widetilde{u}_{m_a}, \widetilde{v}_1, \cdots, \widetilde{v}_{m_b})^T.$$

Let $r_L = \dim \mathfrak{D}_L$, $r_s = \dim \mathfrak{D}_s$. Then there exist linearly independent complex vectors

$$\gamma_i, \ \alpha_j \in \mathbb{C}^{2d}, \quad i = 1, \cdots, r_L, \ \ j = 1, \cdots, r_s$$

such that

$$\widetilde{w}_i = \gamma_i \widetilde{W}, \quad i = 1, \cdots, r_L, \quad \text{and} \quad \widetilde{\chi}_j = \alpha_j \widetilde{W}, \quad j = 1, \cdots, r_s$$

are bases for $\mathfrak{D}_L$ and $\mathfrak{D}_s$ respectively. Since $\mathfrak{D}_L$ is a Lagrangian subspace, we have

$$[\gamma_i : \gamma_j] = [\widetilde{w}_i : \widetilde{w}_j] = 0, \quad i, j = 1, \cdots, r_L.$$

By (11.5.5), we have

$$[\gamma_i : \alpha_j] = [\widetilde{w}_i : \widetilde{\chi}_j] = 0, \quad i = 1, \cdots, r_L, \ \ j = 1, \cdots, r_s.$$

Hence (11.5.2) and (11.5.3) hold. From the definition of strictly dissipative subspaces we obtain α_j, $j = 1, \cdots, r_s$ satisfy (11.5.4). Let

$$w_i = \gamma_i W, \quad \chi_j = \alpha_j W, \quad i = 1, \cdots, r_L, \ \ j = 1, \cdots, r_s,$$
$$W = (u_1, \cdots, u_{m_a}, v_1, \cdots, v_{m_b})^T,$$

then $\widetilde{w}_i = \{w_i + D_{\min}\}$, $\widetilde{\chi}_j = \{\chi_j + D_{\min}\}$. From

$$\mathfrak{D} = \mathrm{span}\{\widetilde{w}_1, \widetilde{w}_2, \cdots \widetilde{w}_{r_L}\} \oplus \mathrm{span}\{\widetilde{\chi}_1, \widetilde{\chi}_2, \cdots, \widetilde{\chi}_{r_s}\},$$

we obtain the domain of the dissipative extension T_D:

$$D(T_D) = D_{\min} \dot{+} \mathrm{span}\{w_1, w_2, \cdots w_{r_L}\} \dot{+} \mathrm{span}\{\chi_1, \chi_2, \cdots, \chi_{r_s}\}.$$

Sufficiency. Let $\widetilde{w}_i = \{w_i + D_{\min}\}$, $i = 1, \cdots, r_L$, i.e. $\widetilde{w}_i = \gamma_i \widetilde{W}$, where $\widetilde{W} = (\widetilde{u}_1, \cdots, \widetilde{u}_{m_a}, \widetilde{v}_1, \cdots, \widetilde{v}_{m_b})^T$. It follows from (11.5.1) and (11.5.2) that $\mathfrak{D}_L = \mathrm{span}\{\widetilde{w}_1, \widetilde{w}_2, \cdots \widetilde{w}_{r_L}\}$ is an r_L-dimensional Lagrangian subspace of G.

Let

$$\widetilde{\chi}_j = \alpha_j \widetilde{W}, \quad j = 1, \cdots, r_s,$$

where $\widetilde{\chi}_j = \{\chi_j + D_{\min}\}$. From (11.5.1) and (11.5.4), we can obtain that $\mathfrak{D}_s = \mathrm{span}\{\widetilde{\chi}_1, \widetilde{\chi}_2, \cdots, \widetilde{\chi}_{r_s}\}$ is an r_s-dimensional strictly dissipative subspace of G. By (11.5.3), the subspace $\mathfrak{D}_L$ is symplectic orthogonal to $\mathfrak{D}_s$. It follows from Theorem

11.3.5 that $\mathfrak{D}_L \oplus \mathfrak{D}_s$ is a dissipative subspace of G. Therefore, by Theorem 11.4.6, we obtain

$$D(T_D) = D_{\min} \dot{+} \operatorname{span}\{w_1, w_2, \cdots, w_{r_L}\} \dot{+} \operatorname{span}\{\chi_1, \chi_2, \cdots, \chi_{r_s}\}$$

is the domain of a dissipative extension T_D of $S_{\min}$ in H. □

REMARK 11.5.3. Note that $[\gamma_i : \gamma_j]$, $[\alpha_i : \alpha_j]$ etc. which appear in Theorem 11.5.2 and the next Corollary as well as Theorem 11.5.3 below, the symplectic product in $\mathbb{C}^{2d}$, $[\gamma_i : \gamma_j] = \gamma_i H \gamma_j^*$, where H is the skew-Hermitian matrix defined in Theorem 11.5.1. As mentioned in Remark 11.5.2 since G is symplectic isomorphic to $\mathbb{C}^{2d}$, we don't distinguish it from its corresponding element in G.

COROLLARY 11.5.1. *Let the hypotheses and notation of Theorem* 11.5.1 *hold. A linear submanifold $D(T_{sD})$ of $D_{\max}$ is the domain of an r_s-dimensional strictly dissipative extension T_{sD} of $S_{\min}$ if and only if there exist linearly independent vectors $\alpha_j \in \mathbb{C}^{2d}$, $j = 1, \cdots, r_s$ satisfying*

$$\operatorname{Im}[\sum_{j=1}^{r_s} c_j \alpha_j : \sum_{j=1}^{r_s} c_j \alpha_j] > 0, \tag{11.5.6}$$

where the complex numbers $c_1, \cdots, c_{r_s}$ are not all zero, such that

$$D(T_{sD}) = D_{\min} \dot{+} \operatorname{span}\{\chi_1, \chi_2, \cdots, \chi_{r_s}\},$$

where $\chi_j = \alpha_j W$, $j = 1, \cdots, r_s$, and $W = (u_1, \cdots, u_{m_a}, v_1, \cdots, v_{m_b})^T$.

PROOF. Since $D(T_{sD})$ is the domain of a strictly dissipative extension T_{sD} of $S_{\min}$, there are no nontrivial Lagrangian elements in the strictly dissipative subspace $D(T_{sD})/D_{\min}$ and thus this Corollary follows from Theorem 11.5.2. □

REMARK 11.5.4. Note that conditions (11.5.4) and (11.5.6) are not easy to check. Below we will give alternative conditions which are easier to check.

Now we extend condition (11.5.4) to get a sufficient condition for $D(T_D)$ to be the domain of a dissipative extension.

THEOREM 11.5.3. *Let the hypotheses and notation of Theorem* 11.5.1 *hold. Assume that there exist linearly independent vectors*

$$\gamma_i, \alpha_j \in \mathbb{C}^{2d}, \quad i = 1, \cdots, r_L, \quad j = 1, \cdots, r_s$$

satisfying

$$[\gamma_i : \gamma_j] = 0, \quad i, j = 1, \cdots, r_L.$$

$$\operatorname{Im}[\alpha_j : \alpha_j] > 0, \quad j = 1, \cdots, r_s. \tag{11.5.7}$$

$$[\gamma_i : \alpha_j] = 0, \qquad i = 1, \cdots, r_L, \ j = 1, \cdots, r_s.$$

$$[\alpha_i : \alpha_j] = 0, \quad i \neq j, \ i, j = 1, \cdots, r_s. \tag{11.5.8}$$

Then

$$D(T_D) = D_{\min} \dot{+} \operatorname{span}\{w_1, w_2, \cdots w_{r_L}\} \dot{+} \operatorname{span}\{\chi_1, \chi_2, \cdots, \chi_{r_s}\}.$$

is the domain of a dissipative extension T_D. Here $w_i = \gamma_i W$, $\chi_j = \alpha_j W$, $i = 1, \cdots, r_L$, $j = 1, \cdots, r_s$, $W = (u_1, \cdots, u_{m_a}, v_1, \cdots, v_{m_b})^T$.

PROOF. We know that (11.5.7) and (11.5.8) are sufficient but not necessary conditions for

$$\operatorname{Im}[\sum_{j=1}^{r_s} c_j\alpha_j : \sum_{j=1}^{r_s} c_j\alpha_j] > 0,$$

where $c_1, \cdots, c_{r_s}$ are not all zero. From Theorem 11.5.2, the result follows. □

REMARK 11.5.5. Although we gave the characterization of the domains of dissipative extensions in terms of LC solutions in Theorem 11.5.2, Corollary 11.5.1 and Theorem 11.5.3, it may be complicated to check a specific dissipative extension, since the characterization of a dissipative subspace with the basis $\widetilde{u}_1, \cdots, \widetilde{u}_{m_a}, \widetilde{v}_1, \cdots, \widetilde{v}_{m_b}$ is complicated. Next a new basis for G in terms of LC solutions is given which makes this easier.

LEMMA 11.5.1. *Let the notation and hypotheses of Theorem 7.2.3 hold. Then there exists a nonsingular $m_a \times m_a$ matrix Q_a and a nonsingular $m_b \times m_b$ matrix Q_b such that*

$$Q_a(-U_{m_a\times m_a})Q_a^* = \operatorname{diag}\{\underbrace{i, \cdots, i}_{\frac{m_a}{2}}, \underbrace{-i, \cdots, -i}_{\frac{m_a}{2}}\},$$

$$Q_b(V_{m_b\times m_b})Q_b^* = \operatorname{diag}\{\underbrace{i, \cdots, i}_{\frac{m_b}{2}}, \underbrace{-i, \cdots, -i}_{\frac{m_b}{2}}\}.$$

PROOF. (1) When k is even and $\frac{m_a}{2}$ is even, let

$$Q_a = \frac{1}{\sqrt{2}}\left(\begin{array}{cccccc|cccccc}
i & & & & & & & & & & & 1\\
 & 1 & & & & & & & & & i & \\
 & & i & & & & & & & 1 & & \\
 & & & 1 & & & & & i & & & \\
 & & & & \ddots & & & \iddots & & & & \\
 & & & & & 1 & i & & & & & \\
\hline
1 & & & & & & & & & & & i\\
 & i & & & & & & & & & 1 & \\
 & & 1 & & & & & & & i & & \\
 & & & i & & & & & 1 & & & \\
 & & & & \ddots & & & \iddots & & & & \\
 & & & & & i & 1 & & & & &
\end{array}\right).$$

(2) When k is even and $\frac{m_a}{2}$ is odd, let

$$Q_a = \frac{1}{\sqrt{2}}\left(\begin{array}{cccccc|cccccc}
i & & & & & & & & & & & 1\\
 & 1 & & & & & & & & & i & \\
 & & i & & & & & & & 1 & & \\
 & & & 1 & & & & & i & & & \\
 & & & & \ddots & & & \iddots & & & & \\
 & & & & & i & 1 & & & & & \\
\hline
1 & & & & & & & & & & & i\\
 & i & & & & & & & & & 1 & \\
 & & 1 & & & & & & & i & & \\
 & & & i & & & & & 1 & & & \\
 & & & & \ddots & & & \iddots & & & & \\
 & & & & & 1 & i & & & & &
\end{array}\right).$$

(3) When k is odd and $\frac{m_a}{2}$ is even, let

$$Q_a = \frac{1}{\sqrt{2}} \left(\begin{array}{cccccc|cccccc}
1 & & & & & & & & & & & i \\
& i & & & & & & & & & 1 & \\
& & 1 & & & & & & & i & & \\
& & & i & & & & & 1 & & & \\
& & & & \ddots & & & \iddots & & & & \\
& & & & & i & 1 & & & & & \\
\hline
i & & & & & & & & & & & 1 \\
& 1 & & & & & & & & & i & \\
& & i & & & & & & & 1 & & \\
& & & 1 & & & & & i & & & \\
& & & & \ddots & & & \iddots & & & & \\
& & & & & 1 & i & & & & &
\end{array} \right).$$

(4) When k is odd and $\frac{m_a}{2}$ is odd, let

$$Q_a = \frac{1}{\sqrt{2}} \left(\begin{array}{cccccc|cccccc}
1 & & & & & & & & & & & i \\
& i & & & & & & & & & 1 & \\
& & 1 & & & & & & & i & & \\
& & & i & & & & & 1 & & & \\
& & & & \ddots & & & \iddots & & & & \\
& & & & & 1 & i & & & & & \\
\hline
i & & & & & & & & & & & 1 \\
& 1 & & & & & & & & & i & \\
& & i & & & & & & & 1 & & \\
& & & 1 & & & & & i & & & \\
& & & & \ddots & & & \iddots & & & & \\
& & & & & i & 1 & & & & &
\end{array} \right).$$

For the above four cases, it is easy to verify that

$$Q_a(-U_{m_a \times m_a})Q_a^* = \operatorname{diag}\{\underbrace{i, \cdots, i}_{\frac{m_a}{2}}, \underbrace{-i, \cdots, -i}_{\frac{m_a}{2}}\}.$$

Similarly Q_b can be constructed. □

THEOREM 11.5.4. *Let the notation and hypotheses of Lemma* 11.5.1 *and Theorem* 11.5.1 *hold, and let*

$$(\widetilde{x}_1, \cdots, \widetilde{x}_{\frac{m_a}{2}}, \widetilde{x}_{\frac{m_a}{2}+1}, \cdots, \widetilde{x}_{m_a})^T = Q_a(\widetilde{u}_1, \cdots, \widetilde{u}_{m_a})^T,$$

$$(\widetilde{z}_1, \cdots, \widetilde{z}_{\frac{m_b}{2}}, \widetilde{z}_{\frac{m_b}{2}+1}, \cdots, \widetilde{z}_{m_b})^T = Q_b(\widetilde{v}_1, \cdots, \widetilde{v}_{m_b})^T.$$

Then

(1) *Each of* $\operatorname{span}\{\widetilde{x}_1, \cdots, \widetilde{x}_{\frac{m_a}{2}}\}$ *and* $\operatorname{span}\{\widetilde{z}_1, \cdots, \widetilde{z}_{\frac{m_b}{2}}\}$ *is a strictly dissipative subspaces of* G.

(2) *Each of* $\operatorname{span}\{\widetilde{x}_{\frac{m_a}{2}+1}, \cdots, \widetilde{x}_{m_a}\}$ *and* $\operatorname{span}\{\widetilde{z}_{\frac{m_b}{2}+1}, \cdots, \widetilde{z}_{m_b}\}$ *is a strictly accretive subspaces of* G.

PROOF. From (4) of Theorem 11.5.1, $([\widetilde{u}_i : \widetilde{u}_j])_{m_a \times m_a} = -U_{m_a \times m_a}$, $([\widetilde{v}_i : \widetilde{v}_j])_{m_b \times m_b} = V_{m_b \times m_b}$, and from Lemma 11.5.1,

$$Q_a(-U_{m_a \times m_a})Q_a^* = \operatorname{diag}\{\underbrace{i, \cdots, i}_{\frac{m_a}{2}}, \underbrace{-i, \cdots, -i}_{\frac{m_a}{2}}\},$$

$$Q_b(V_{m_b \times m_b})Q_b^* = \operatorname{diag}\{\underbrace{i, \cdots, i}_{\frac{m_b}{2}}, \underbrace{-i, \cdots, -i}_{\frac{mb}{2}}\}.$$

Therefore we have

$$([\widetilde{x}_i : \widetilde{x}_j])_{m_a \times m_a} = Q_a([\widetilde{u}_i : \widetilde{u}_j])Q_a^* = \operatorname{diag}\{\underbrace{i, \cdots, i}_{\frac{m_a}{2}}, \underbrace{-i, \cdots, -i}_{\frac{m_a}{2}}\}. \tag{11.5.9}$$

By the definition of strictly dissipative subspaces, it is easy to see that item (1) holds. Similarly

$$([\widetilde{z}_i : \widetilde{z}_j])_{m_b \times m_b} = Q_b([\widetilde{v}_i : \widetilde{v}_j])Q_b^* = \operatorname{diag}\{\underbrace{i, \cdots, i}_{\frac{m_b}{2}}, \underbrace{-i, \cdots, -i}_{\frac{m_b}{2}}\} \tag{11.5.10}$$

and item (2) can be proven. □

THEOREM 11.5.5. *Let the notation and hypotheses of Theorem* 11.5.4 *hold. Let*

$$\widetilde{D} = \operatorname{span}\{\widetilde{x}_1, \cdots, \widetilde{x}_{\frac{m_a}{2}}, \widetilde{z}_1, \cdots, \widetilde{z}_{\frac{m_b}{2}}\},$$

$$\widetilde{A} = \operatorname{span}\{\widetilde{x}_{\frac{m_a}{2}+1}, \cdots, \widetilde{x}_{m_a}, \widetilde{z}_{\frac{m_b}{2}+1}, \cdots, \widetilde{z}_{m_b}\}.$$

Then

$$G = \widetilde{D} \oplus \widetilde{A} \quad \text{with} \quad [\widetilde{D} : \widetilde{A}] = 0.$$

$\widetilde{D}$ is a maximal dissipative subspace of G.
$\widetilde{A}$ is a maximal accretive subspace of G.

PROOF. From the construction of the new basis of G: $\widetilde{x}_1, \cdots, \widetilde{x}_{\frac{m_a}{2}}, \widetilde{z}_1, \cdots, \widetilde{z}_{\frac{m_b}{2}}$, $\widetilde{x}_{\frac{m_a}{2}+1}, \cdots, \widetilde{x}_{m_a}, \widetilde{z}_{\frac{m_b}{2}+1}, \cdots, \widetilde{z}_{m_b}$, and the properties of $\widetilde{x}_j$, $\widetilde{z}_i$ given in (11.5.9) and (11.5.10), we obtain that $G = \widetilde{D} \oplus \widetilde{A}$ and the skew-Hermitian matrix associated with the new basis is

$$\widetilde{H} = \begin{pmatrix} iI_d & 0 \\ 0 & -iI_d \end{pmatrix}.$$

Using the method provided in the proof of Theorem 11.4.3, we may conclude that $\widetilde{D}$ is a maximal dissipative subspace of G, and $\widetilde{A}$ is a maximal accretive subspace of G. □

6. Symplectic Geometry Characterization of Symmetric Operators

In this section we give the symplectic geometry characterization of symmetric operators generated by an even order symmetric differential expression $M = M_Q$, $Q \in Z_n(J, \mathbb{R})$, $n = 2k, k \geq 1$, $Q = Q^+$.

THEOREM 11.6.1. *Let the notation and hypotheses of Theorem* 7.2.3, *Theorem* 8.5.1 *and Theorem* 11.5.1 *hold. Then a linear submanifold $D(S)$ of $D_{\max}$ is the domain of symmetric extension S of $S_{\min}$ if and only if the following conditions hold:*

(1) *there exist l $(d \leq l \leq 2d)$ linearly independent vectors*

$$\gamma_j = (a_{j1}, \cdots, a_{jm_a}, b_{j1}, \cdots, b_{jm_b}) \in \mathbb{C}^{2d}, \quad j = 1, 2, \cdots, l;$$

(2)

$$\operatorname{rank} \begin{pmatrix} [\gamma_1 : \gamma_1] & \cdots & [\gamma_l : \gamma_1] \\ \cdots & \cdots & \cdots \\ [\gamma_1 : \gamma_l] & \cdots & [\gamma_l : \gamma_l] \end{pmatrix} = 2(l - d); \tag{11.6.1}$$

(3)

$$D(S) = \{y \in D_{\max} : [y : w_j] = 0,\ j = 1, 2, \cdots, l\}, \tag{11.6.2}$$

where

$$w_j = \left(-\bar{a}_{j1}, \cdots, -\bar{a}_{jm_a}, \bar{b}_{j1}, \cdots, \bar{b}_{jm_b}\right) W,$$
$$W = (u_1, \cdots, u_{m_a}, v_1, \cdots, v_{m_b})^T,$$

and $u_1, \cdots, u_{m_a}, v_1, \cdots, v_{m_b}$ *are LC solutions constructed in Theorem* 7.2.3.

PROOF. By Theorem 8.5.1, $D(S)$ is the domain of symmetric extension S of $S_{\min}$ if and only if there exist complex matrices $A = (a_{ij})_{l\times m_a}$ and $B = (b_{ij})_{l\times m_b}$ such that $\text{rank}(A : B) = l$, $d \le l \le 2d$, $\text{rank}(AE_{m_a}A^* - BE_{m_b}B^*) = 2(l-d)$, and then $D(S)$ is characterized by $(A : B)Y_{a,b} = 0$, where

$$Y_{a,b} = ([y, u_1](a), \cdots, [y, u_{m_a}](a),\ [y, v_1](b), \cdots, [y, v_{m_b}](b))^T.$$

Let

$$\gamma_j = (a_{j1}, \cdots, a_{jm_a}, b_{j1}, \cdots, b_{jm_b}) \in \mathbb{C}^{2d}, \quad j = 1, 2, \cdots, l.$$

Then $\text{rank}(A : B) = l$ is equivalent to the fact that $\gamma_1, \cdots, \gamma_l$ are linearly independent. By Theorem 11.5.1, $G = D_{\max}/D_{\min}$ is symplectic isomorphic to a complex symplectic space $\mathbb{C}^{2d}$ and therefore

$$\begin{aligned}\begin{pmatrix} [\gamma_1 : \gamma_1] & \cdots & [\gamma_1 : \gamma_l] \\ \cdots & \cdots & \cdots \\ [\gamma_l : \gamma_1] & \cdots & [\gamma_l : \gamma_l] \end{pmatrix} &= (A \ \ B)\begin{pmatrix} -U_{m_a\times m_a} & 0 \\ 0 & V_{m_b\times m_b} \end{pmatrix}\begin{pmatrix} A^* \\ B^* \end{pmatrix} \\ &= -AUA^* + BVB^* \\ &= (-1)^k(BE_{m_b}B^* - AE_{m_a}A^*)\end{aligned}$$

Hence $\text{rank}(AE_{m_a}A^* - BE_{m_b}B^*) = 2(l-d)$ is equivalent to (11.6.1). Note that $[y : w_j] = [y, w_j](b) - [y, w_j](a)$. It is easy to check that (11.6.2) is equivalent to $(A : B)Y_{a,b} = 0$. This completes the proof. □

7. Comments

This chapter was influenced by the methods used by Yao, et. al. in [**590**], [**591**], but we have made a number of changes. In particular, the symplectic geometry characterization of symmetric operators and its 1-1 correspondence with their Hilbert space characterization are published here for the first time.

Part 3

Two-Interval Problems

CHAPTER 12

Two-Interval Symmetric Domains

1. Introduction

Motivated by applications, in particular the paper of Boyd [**77**] and its references, in 1986 Everitt and Zettl [**205**] introduced a framework for the rigorous study of Sturm-Liouville problems which have a singularity in the interior of the domain interval since the existing theory did not cover such cases. The Boyd paper, which was based on several previous papers by Atmospheric Scientists, studies eddies in the atmosphere using a mathematical model based on the Sturm-Liouville problem

$$-y'' + \frac{1}{x}\, y = \lambda y, \quad y(-1) = 0 = y(1), \quad -1 < x < 1.$$

Note that 0 is a singular point in the interior of the underlying interval $(-1, 1)$ and therefore the 1-interval theory discussed above in Chapters 5 and 6 does not hold.

The framework introduced in [**205**] is the direct sum of Hilbert spaces, one for each interval $(-1, 0)$ and $(0, 1)$. The primary goal of this study is the characterization of all self-adjoint realizations from the two intervals. A simple way of getting self-adjoint operators in the direct sum space is to take the direct sum of self-adjoint operators from the separate spaces. However, there are many self-adjoint operators in the direct sum space which are not obtained this way. These 'new' self-adjoint operators involve interactions between the two intervals.

Mukhtarov and Yakubov [**436**] observed that the set of self-adjoint operator realizations developed in [**205**] can be further enlarged by using different multiples of the usual inner products associated with each of the intervals. In [**532**], [**556**], Sun, Wang and Zettl use the Mukhtarov and Yakubov modification of the Everitt-Zettl theory in [**205**] to obtain more general self-adjoint two-interval boundary conditions.

In this chapter we extend the Sun-Wang-Zettl method for self-adjoint boundary conditions from $n = 2$ to symmetric boundary conditions for general $n \geq 2$ with the self-adjoint boundary conditions as a special case. Here we consider the equations

$$\begin{aligned} &M_r y = \lambda w_r y \;\; on \;\; J_r = (a_r, b_r), \;\; -\infty \leq a_r < b_r \leq \infty, \;\; \lambda \in \mathbb{C}, \\ &w_r \in L_{loc}(J_r), \;\; w_r > 0, \; a.e. \; on \; J_r, \;\; r = 1, 2, \end{aligned}$$

where $M_r = M_{Q_r} \in Z_n(J_r)$, $Q_r = Q_r^+$ and w_r is a weight function in J_r. In the general theory discussed in Sections 12.2 and 12.3 the right endpoint b_1 of J_1 need not be equal to the left endpoint a_2 of J_2. The intervals J_r are arbitrary non-degenerate intervals: they may be disjoint, overlap, or be identical with the same or different differential expressions on each interval. In Section 12.4 we apply the general 2-interval theory to the case when the right endpoint b_1 of J_1 is equal to the left endpoint a_2 of J_2 and show that this generates regular and singular discontinuous boundary conditions. A number of regular and singular examples of symmetric and self-adjoint discontinuous boundary conditions are given in Section 12.5.

Following Mukhtarov and Yakubov [**436**], we use the inner product

$$\langle \mathbf{f},\mathbf{g}\rangle = h\ (f_1,g_1)_1 + k\ (f_2,g_2)_2, \quad h>0,\ k>0, \tag{12.1.1}$$

where $(\cdot,\cdot)_r$ denotes the usual inner product in $L^2(J_r,w_r)$ and study operator theory in the framework of the direct sum space

$$H = (L^2(J_1,w_1)\dotplus L^2(J_2,w_2),\ \langle\cdot,\cdot\rangle). \tag{12.1.2}$$

In particular we characterize the boundary conditions which determine symmetric operators S in H which satisfy

$$S_{\min}\subset S\subset S^*\subset S_{\max}$$

where the operators $S_{\min}$ and $S_{\max}$ with domains $D_{\min}$ and $D_{\max}$ are given by

$$D_{\max} = D_{1\max} + D_{2\max}, \quad D_{\min} = D_{1\min} + D_{2\min};$$

$$S_{\max} = S_{1\max} + S_{2\max}, \quad S_{\min} = S_{1\min} + S_{2\min}.$$

There should be no confusion between this notation here in Part III for the 2-interval theory and the similar notation used in Parts I and II for the 1-interval theory.

Elements of H will be denoted in bold face type: $\mathbf{f} = \{f_1,f_2\}$ with $f_1\in L^2(J_1,w_1)$ and $f_2\in L^2(J_2,w_2)$.

As in the one interval case the 2-interval Lagrange sesquilinear form is fundamental to the study of boundary value problems.

DEFINITION 12.1.1. Let

$$[\mathbf{f},\mathbf{g}] = h\,[f_1,g_1]_1(b_1) - h\,[f_1,g_1]_1(a_1) + k\,[f_2,g_2]_2(b_2) - k\,[f_2,g_2]_2(a_2),$$

where

$$[f_r,g_r]_r = i^n\sum_{s=0}^{n-1}(-1)^{n+1-s}\overline{g}_r^{[n-s-1]}f_r^{[s]}, \quad r=1,2,\quad i=\sqrt{-1}.$$

Here we use a subscript r to denote the r-th interval. Below, for simplicity, the subscript r is omitted when it is clear from the context.

We comment on the constants h,k.

REMARK 12.1.1. Note that this 'new' inner product with the Mukhtarov-Yakubov constants h,k is an inner product in H for any positive numbers h and k. The elements of this Hilbert space H are the same as those of the usual direct sum Hilbert space when $h=1=k$, thus these spaces are differentiated from each other only by their inner product. The set of square-integrable functions in these different Hilbert spaces H is the same but the inner product in H changes when h and k change. As we will see below the parameters h,k extend the set of boundary conditions which generate symmetric operators in H.

From another perspective, the Hilbert space H can be viewd as a 'usual' direct sum space when $h=1=k$ but with w_1 replaced by $h\,w_1$ and w_2 replaced by $k\,w_2$.

Note that $w>0$ ensures that $L^2(J,w)$ is a Hilbert space. However, if $w<0$ on J we can multiply the equation by -1 to obtain

$$-My = \lambda(-w)\,y \quad on\ J,$$

and observe that the 1-interval theory developed above applies since there is no sign restriction on the leading coefficient of M . Also the boundary conditions are homogeneous and thus invariant with respect to multiplication by -1.

2. Two-Interval Minimal and Maximal Operators

Our proof of the 2-interval symmetric domain characterization in the next section is long and complicated. So for the benefit of the reader we discuss the next two lemmas and recall a theorem from Section 4.4. These are used in the proof.

Let $a_r < c_r < b_r$, $r = 1, 2$. For the rest of this chapter we assume that the deficiency indices on each interval (a_r, c_r) and (c_r, b_r) are equal and denote them by d_{a_r}, d_{b_r}, respectively. Let d_r^+, d_r^- denote the deficiency indices of $S_{\min}(a_r, b_r)$. Then $d_r^+ = d_r^- = d_{a_r} + d_{b_r} - n$, and the common value is denoted by d_r called the deficiency index of $S_{\min}(a_r, b_r)$, i.e. $d_r = d_{a_r} + d_{b_r} - n$. By Lemma 4.4.1, $[\frac{n+1}{2}] \leq d_{a_r}, d_{b_r} \leq n$. By [**199**], the deficiency index $d = d^+ = d^-$ of $S_{\min}$ is given by $d = d_1 + d_2$. Let $m_{a_r} = 2d_{a_r} - n$, $m_{b_r} = 2d_{b_r} - n$.

The next two lemmas will be used in Section 12.3 and are stated here for the convenience of the reader. They are routine extensions from the 1-interval theory.

LEMMA 12.2.1. *We have*

(1)
$$\begin{aligned} S_{\min}^* &= S_{1\min}^* + S_{2\min}^* = S_{1\max} + S_{2\max} = S_{\max}; \\ S_{\max}^* &= S_{1\max}^* + S_{2\max}^* = S_{1\min} + S_{2\min} = S_{\min}. \\ D_{\max} &= D(S_{\max}) = D(S_{1\max}) + D(S_{2\max}); \\ D_{\min} &= D(S_{\min}) = D(S_{1\min}) + D(S_{2\min}). \end{aligned}$$

(2) *The minimal operator $S_{\min}$ is a closed, symmetric, densely defined operator in the Hilbert space H.*

LEMMA 12.2.2. *The two-interval minimal domain $D_{\min}$ can be characterized as*

$$D_{\min} = \{\mathbf{f} \in D_{\max} : [\mathbf{f}, \mathbf{g}] = 0, \ \ \mathbf{g} \in D_{\max}\}$$

and recall the Green's formula for each interval

$$(S_{r\max} f_r, g_r)_r - (f_r, S_{r\max} g_r)_r = [f_r, g_r]_r(b_r) - [f_r, g_r]_r(a_r), \quad f_r, g_r \in D(S_{r\max}).$$

The generalized Green's formula holds:

$$\langle S_{\max}\mathbf{f}, \mathbf{g}\rangle - \langle \mathbf{f}, S_{\max}\mathbf{g}\rangle = [\mathbf{f}, \mathbf{g}], \quad \mathbf{f}, \mathbf{g} \in D_{\max}. \tag{12.2.1}$$

Next we recall a result from Section 4.4.

THEOREM 12.2.1. *Let $Q_r \in Z_n(J_r)$, $J_r = (a_r, b_r)$, $-\infty \leq a_r < b_r \leq \infty$, satisfy $Q_r = Q_r^+$ let $M_r y = M_{Q_r} y = \lambda w_r y$ be the corresponding symmetric differential equation. Let $a_r < c_r < b_r$. Fix $\lambda_{a_r} \in \mathbb{C}$, $\lambda_{b_r} \in \mathbb{C}$ with $\operatorname{Im}(\lambda_{a_r}) \neq 0 \neq \operatorname{Im}(\lambda_{b_r})$. For $r = 1, 2$, we have*

a:

(1) *there exist linearly independent solutions u_{r1} $u_{r2}, \cdots, u_{rm_{a_r}}$ of $M_r y = \lambda_{a_r} w_r y$ on (a_r, c_r) such that the $m_{a_r} \times m_{a_r}$ matrix*

$$E_{m_{a_r}} = \begin{pmatrix} [u_{r1}, u_{r1}] & \cdots & [u_{rm_{a_r}}, u_{r1}] \\ \vdots & \vdots & \vdots \\ [u_{r1}, u_{rm_{a_r}}] & \cdots & [u_{rm_{a_r}}, u_{rm_{a_r}}] \end{pmatrix} (a_r)$$

is nonsingular;

(2) $u_{r1}\ u_{r2}, \cdots, u_{rm_{a_r}}$ *can be extended to* (a_r, b_r) *such that the extended functions, still denoted by* $u_{r1}\ u_{r2}, \cdots, u_{rm_{a_r}}$ *are in* $D_{r\,\max}$ *and are identically* 0 *near* b_r;

(3) $u_{r1}\ u_{r2}, \cdots, u_{rm_{a_r}}$ *are linearly independent modulo* $D_{r\,\min}$;

b:

(1) *there exist linearly independent solutions* $v_{r1},\ v_{r2},\ \cdots,\ v_{rm_{b_r}}$ *of* $M_r y = \lambda_{b_r} w_r y$ *on* (c_r, b_r) *such that the* $m_{b_r} \times m_{b_r}$ *matrix*

$$E_{m_{b_r}} = \begin{pmatrix} [v_{r1}, v_{r1}] & \cdots & [v_{rm_{b_r}}, v_{r1}] \\ \vdots & \vdots & \vdots \\ [v_{r1}, v_{rm_{b_r}}] & \cdots & [v_{rm_{b_r}}, v_{rm_{b_r}}] \end{pmatrix} (b_r)$$

is nonsingular;

(2) $v_{r1},\ v_{r2},\ \cdots,\ v_{rm_{b_r}}$ *can be extended to* (a_r, b_r) *such that the extended functions, still denoted by* $v_{r1},\ v_{r2},\ \cdots,\ v_{rm_{b_r}}$ *are in* $D_{r\,\max}$ *and are identically* 0 *near* a_r;

(3) $v_{r,1},\ v_{r2},\ \cdots,\ v_{rm_{b_r}}$ *are linearly independent modulo* $D_{r\,\min}$;

c:

The maximal domain for each interval has the following representation:

(12.2.2)
$$D_{\max}(a_r, b_r) = D_{\min}(a_r, b_r) \dotplus \text{span}\{u_{r1}, \cdots, u_{rm_{a_r}}\} \dotplus \text{span}\{v_{r1}, \cdots, v_{rm_{b_r}}\}, \quad r = 1, 2.$$

PROOF. This is given by Theorem 4.4.4. □

3. Two-Interval Symmetric Domains

We start with some preliminary definitions and results.

LEMMA 12.3.1. *For any complex numbers* $\alpha_{r1},\ \alpha_{r2}, \cdots, \alpha_{rm_{a_r}},\ \beta_{r1}, \beta_{r2}, \cdots, \beta_{rm_{b_r}}, r = 1, 2,$ *there exists* $\mathbf{y} = \{y_1, y_2\} \in D_{\max}$ *such that*

$$[y_r, u_{r1}](a_r) = \alpha_{r1}, \quad [y_r, u_{r2}](a_r) = \alpha_{r2}, \quad \cdots, \quad [y_r, u_{rm_{a_r}}](a_r) = \alpha_{rm_{a_r}},$$
$$[y_r, v_{r1}](b_r) = \beta_{r1}, \quad [y_r, v_{r2}](b_r) = \beta_{r2}, \quad \cdots, \quad [y_r, v_{rm_{b_r}}](b_r) = \beta_{rm_{b_r}}.$$

PROOF. This is obtained by Theorem 4.4.6. □

DEFINITION 12.3.1. For any $\mathbf{y} = \{y_1, y_2\} \in D_{\max}$ define

$$Y(a_r) = \begin{pmatrix} [y_r, u_{r1}]_r(a_r) \\ \vdots \\ [y_r, u_{rm_{a_r}}]_r(a_r) \end{pmatrix}, \quad Y(b_r) = \begin{pmatrix} [y_r, v_{r1}]_r(b_r) \\ \vdots \\ [y_r, v_{rm_{b_r}}]_r(b_r) \end{pmatrix},$$

$$Y_{a_r, b_r} = \begin{pmatrix} Y(a_r) \\ Y(b_r) \end{pmatrix}, \quad r = 1, 2, \quad \mathbf{Y} = \begin{pmatrix} Y_{a_1, b_1} \\ Y_{a_2, b_2} \end{pmatrix} \tag{12.3.1}$$

and recall that the Lagrange brackets $[y_r, u_{rj}]_r(a_r)$ and $[y_r, v_{rj}]_r(b_r)$ exist as finite limits.

DEFINITION 12.3.2. A matrix $U \in M_{l,2d}$ with rank l, $0 \leq l \leq 2d$, is called a boundary condition matrix and for $\mathbf{y} \in D_{\max}$ and for $\mathbf{Y}$ given by (12.3.1)

$$U\mathbf{Y} = 0$$

is called a boundary condition. The null space of U is denoted by $\mathcal{N}(U)$ and $\mathcal{R}(U)$ denotes its range, U^* is the conjugate transpose of U.

DEFINITION 12.3.3. Suppose $U \in M_{l,2d}$ is a boundary condition matrix. Define an operator $S(U)$ in H by

$$D(S(U)) = \{\mathbf{y} \in D_{\max} : U\mathbf{Y} = 0\},$$
$$S(U)\mathbf{y} = S_{\max}\mathbf{y} \quad \text{for} \quad \mathbf{y} \in D(S(U)). \tag{12.3.2}$$

REMARK 12.3.1. If $l = 0$, then $U = 0$ and $S(U) = S_{\max}$. If $l = 2d$ and I_{2d} denotes the identity matrix then $S(I_{2d}) = S_{\min}$. For any boundary condition matrix U, $D(S(U))$ is a linear submanifold of $D_{\max}$. Note that

$$S_{\min} \subset S(U) \subset S_{\max}$$

and consequently, since $S_{\max}$ is a closed finite dimensional extension of $S_{\min}$, it follows that every operator $S(U)$ is a closed finite dimensional extension of $S_{\min}$.

THEOREM 12.3.1. *Let the notation and hypotheses of Theorem* 12.2.1 *hold. Then*

$$D_{\min} = \{\mathbf{y} \in D_{\max} : [y_r, u_{rj}](a_r) = 0, \quad \text{for } j = 1, \cdots, m_{a_r};$$
$$[y_r, v_{rj}](b_r) = 0, \quad \text{for } j = 1, \cdots, m_{b_r}, \quad r = 1, 2\}.$$

PROOF. This can be obtained by Theorem 4.4.5. □

THEOREM 12.3.2. *Assume that* $U \in M_{l,2d}$, $\operatorname{rank} U = l$, $d \leq l \leq 2d$. *Let* $\mathbf{y}, \mathbf{z} \in D_{\max}$ *and define* $\mathbf{Y}$, $\mathbf{Z}$ *by* (12.3.1). *Let*

$$P_r = \begin{pmatrix} E_{m_{a_r}}^{-1} & 0 \\ 0 & -E_{m_{b_r}}^{-1} \end{pmatrix}, \quad r = 1, 2; \quad P = \begin{pmatrix} hP_1 & 0 \\ 0 & kP_2 \end{pmatrix}, \tag{12.3.3}$$

and note that $P_r^* = -P_r$, $(P_r^{-1})^* = -P_r^{-1}$, *and* $P^* = -P$, $(P^{-1})^* = -P^{-1}$. *Then* $S(U)$ *is symmetric if and only if*

$$\mathbf{Z}^* P \mathbf{Y} = 0, \quad \text{for all } \mathbf{y}, \mathbf{z} \in D(S(U)). \tag{12.3.4}$$

PROOF. It follows from (12.2.1) and the definition of $S(U)$ given in (12.3.2) that $S(U)$ is symmetric if and only if for all $\mathbf{y}, \mathbf{z} \in D(S(U))$,

$$\langle S_{\max}\mathbf{y}, \mathbf{z}\rangle - \langle \mathbf{y}, S_{\max}\mathbf{z}\rangle = \langle S(U)\mathbf{y}, \mathbf{z}\rangle - \langle \mathbf{y}, S(U)\mathbf{z}\rangle = [\mathbf{y}, \mathbf{z}] = 0.$$

From the above two-interval maximal domain definition and characterization we have $\mathbf{y} = \{y_1, y_2\}, \mathbf{z} = \{z_1, z_2\} \in D_{\max}$ can be represented by the following form: for $r = 1, 2$,

$$y_r = y_{r0} + c_{r1}u_{r1} + c_{r2}u_{r2} + \cdots + c_{rm_{a_r}}u_{rm_{a_r}} + h_{r1}v_{r1} + h_{r2}v_{r2} + \cdots + h_{rm_{b_r}}v_{rm_{b_r}},$$

$$z_r = z_{r0} + \widehat{c}_{r1}u_{r1} + \widehat{c}_{r2}u_{r2} + \cdots + \widehat{c}_{rm_{a_r}}u_{rm_{a_r}} + \widehat{h}_{r1}v_{r1} + \widehat{h}_{r2}v_{r2} + \cdots + \widehat{h}_{rm_{b_r}}v_{rm_{b_r}},$$

where $y_{r0}, z_{r0} \in D_{r\min}$ and $c_{rj}, \widehat{c}_{rj} \in \mathbb{C}$, $j = 1, \cdots, m_{a_r}$; $h_{rj}, \widehat{h}_{rj} \in \mathbb{C}$, $j = 1, \cdots, m_{b_r}$.

By calculation, it follows that

$$\begin{pmatrix} [y_r, v_{r1}](b_r) \\ \vdots \\ [y_r, v_{rm_{b_r}}](b_r) \end{pmatrix} = E_{m_{b_r}} \begin{pmatrix} h_{r1} \\ \vdots \\ h_{rm_{b_r}} \end{pmatrix},$$

$$\begin{pmatrix} [z_r, v_{r1}](b_r) \\ \vdots \\ [z_r, v_{rm_{b_r}}](b_r) \end{pmatrix} = E_{m_{b_r}} \begin{pmatrix} \widehat{h}_{r1} \\ \vdots \\ \widehat{h}_{rm_{b_r}} \end{pmatrix}.$$

Therefore

$$\begin{aligned}
&[y_r, z_r]_r(b_r)\\
&= (\overline{\overline{h}}_{r1},\ \overline{\overline{h}}_{r2},\ \cdots,\overline{\overline{h}}_{rm_{b_r}})E_{m_{b_r}}\begin{pmatrix} h_{r1}\\ h_{r2}\\ \vdots\\ h_{rm_{b_r}}\end{pmatrix}\\
&= (\overline{[z_r, v_{r1}](b_r)},\cdots,\overline{[z_r, v_{rm_{b_r}}](b_r)})(E_{m_{b_r}}^{-1})^* E_{m_{b_r}} E_{m_{b_r}}^{-1}\begin{pmatrix} [y_r, v_{r1}](b_r)\\ \vdots\\ [y_r, v_{rm_{b_r}}](b_r)\end{pmatrix}\\
&= -(\overline{[z_r, v_{r1}](b_r)},\cdots,\overline{[z_r, v_{rm_{b_r}}](b_r)})E_{m_{b_r}}^{-1}\begin{pmatrix} [y_r, v_{r1}](b_r)\\ \vdots\\ [y_r, v_{rm_{b_r}}](b_r)\end{pmatrix}.
\end{aligned}$$

Similarly, we have

$$\begin{aligned}
[y_r, z_r]_r(a_r) =&(\overline{\overline{c}}_{r1},\ \overline{\overline{c}}_{r2},\ \cdots,\overline{\overline{c}}_{rm_{a_r}})E_{m_{a_r}}\begin{pmatrix} c_{r1}\\ c_{r2}\\ \vdots\\ c_{rm_{a_r}}\end{pmatrix}\\
=&-(\overline{[z_r, u_{r1}](a_r)},\cdots,\overline{[z_r, u_{rm_{a_r}}](a_r)})E_{m_{a_r}}^{-1}\begin{pmatrix} [y_r, u_{r1}](a_r)\\ \vdots\\ [y_r, u_{rm_{a_r}}](a_r)\end{pmatrix}.
\end{aligned}$$

Hence

$$\begin{aligned}
&[y_r,z_r](b_r) - [y_r, z_r](a_r)\\
&\quad= (\overline{[z_r, u_{r1}](a_r)},\cdots,\overline{[z_r, u_{rm_{a_r}}](a_r)},\overline{[z_r, v_{r1}](b_r)},\cdots,\overline{[z_r, v_{rm_{b_r}}](b_r)})P_r\\
&\quad\quad\cdot\begin{pmatrix} [y_r, u_{r1}](a_r)\\ \vdots\\ [y_r, u_{rm_{a_r}}](a_r)\\ [y_r, v_{r1}](b_r)\\ \vdots\\ [y_r, v_{rm_{b_r}}](b_r)\end{pmatrix}\\
&\quad= Z_{a_r,b_r}^* P_r Y_{a_r,b_r}.
\end{aligned}$$

Then

$$\begin{aligned}
[\mathbf{y},\mathbf{z}] &= h[y_1, z_1]_1(b_1) - h[y_1, z_1]_1(a_1) + k[y_2, z_2]_2(b_2) - k[y_2, z_2]_2(a_2)\\
&= hZ_{a_1,b_1}^* P_1 Y_{a_1,b_1} + kZ_{a_2,b_2}^* P_2 Y_{a_2,b_2}\\
&= \begin{pmatrix} Z_{a_1,b_1}^* & Z_{a_2,b_2}^*\end{pmatrix}\begin{pmatrix} hP_1 & 0\\ 0 & kP_2\end{pmatrix}\begin{pmatrix} Y_{a_1,b_1}\\ Y_{a_2,b_2}\end{pmatrix}\\
&= \mathbf{Z}^* P\mathbf{Y}.
\end{aligned}$$

Therefore the operator $S(U)$ is symmetric if and only if

$$[\mathbf{y},\mathbf{z}] = 0 \quad \textit{for all } \mathbf{y},\mathbf{z} \in D(S(U)),$$

i.e.

$$\mathbf{Z}^* P \mathbf{Y} = 0 \quad \textit{for all } \mathbf{y}, \mathbf{z} \in D(S(U)). \qquad \square$$

LEMMA 12.3.2. *Each of the following statements is equivalent to* (12.3.4):

(1) *For all* $Y, Z \in \mathcal{N}(U)$, $Z^* P Y = 0$;
(2) $\mathcal{N}(U) \perp P(\mathcal{N}(U))$;
(3) $P(\mathcal{N}(U)) \subset \mathcal{N}(U)^{\perp} = \mathcal{R}(U^*)$;
(4) $\mathcal{N}(U) \subset \mathcal{R}(P^{-1}U^*)$.

Here P *is defined by* (12.3.3).

PROOF. In the following, we prove the equivalence of (12.3.4) and (1). For all $Y, Z \in \mathcal{N}(U)$, let

$$Y = \begin{pmatrix} y_1 \\ y_2 \\ \vdots \\ y_{2d} \end{pmatrix}, \quad Z = \begin{pmatrix} z_1 \\ z_2 \\ \vdots \\ z_{2d} \end{pmatrix}.$$

Then $UY = 0$ and $UZ = 0$. By Lemma 12.3.1, there exist $\mathbf{y} = \{y_1, y_2\}, \mathbf{z} = \{z_1, z_2\} \in D_{\max}$ such that

$$\mathbf{Y} = \begin{pmatrix} [y_1, u_{11}](a_1) \\ \vdots \\ [y_1, u_{1m_{a_1}}](a_1) \\ [y_1, v_{11}](b_1) \\ \vdots \\ [y_1, v_{1m_{b_1}}](b_1) \\ [y_2, u_{21}](a_2) \\ \vdots \\ [y_2, u_{2m_{a_2}}](a_2) \\ [y_2, v_{21}](b_2) \\ \vdots \\ [y_2, v_{2m_{b_2}}](b_2) \end{pmatrix} = Y, \quad \mathbf{Z} = \begin{pmatrix} [z_1, u_{11}](a_1) \\ \vdots \\ [z_1, u_{1m_{a_1}}](a_1) \\ [z_1, v_{11}](b_1) \\ \vdots \\ [z_1, v_{1m_{b_1}}](b_1) \\ [z_2, u_{21}](a_2) \\ \vdots \\ [z_2, u_{2m_{a_2}}](a_2) \\ [z_2, v_{21}](b_2) \\ \vdots \\ [z_2, v_{2m_{b_2}}](b_2) \end{pmatrix} = Z,$$

and $U\mathbf{Y} = U\mathbf{Z} = 0$. Therefore $\mathbf{y}, \mathbf{z} \in D(S(U))$, and by (12.3.4) we have

$$Z^* P Y = \mathbf{Z}^* P \mathbf{Y} = 0.$$

Hence, for all $Y, Z \in \mathcal{N}(U)$, $Z^* P Y = 0$.

On the other hand, for all $\mathbf{y}, \mathbf{z} \in D(S(U))$, $U\mathbf{Y} = U\mathbf{Z} = 0$. This shows that $\mathbf{Y}, \mathbf{Z} \in \mathcal{N}(U)$. If (1) holds, then $\mathbf{Z}^* P \mathbf{Y} = 0$. Therefore, when (1) holds, then for any $\mathbf{y}, \mathbf{z} \in D(S(U))$, $\mathbf{Z}^* P \mathbf{Y} = 0$.

It is obvious that (1) and (2) are equivalent. The equivalence of (2) and (3) follows from Lemma 5.3.1. We now prove the equivalence of (3) and (4).
(3)$\Rightarrow$ (4): For any $Y \in \mathcal{N}(U)$, it follows from $P(\mathcal{N}(U)) \subset \mathcal{R}(U^*)$ that $PY \in \mathcal{R}(U^*)$ and then $Y \in \mathcal{R}(P^{-1}U^*)$.
(4) $\Rightarrow$ (3): For any $Y \in P(\mathcal{N}(U))$, it follows from $\mathcal{N}(U) \subset \mathcal{R}(P^{-1}U^*)$ that $Y \in \mathcal{R}(U^*)$. Therefore $P(\mathcal{N}(U)) \subset \mathcal{R}(U^*)$. $\square$

THEOREM 12.3.3. *Let U be an $l\times 2d$ matrix with* rank $U=l$, *where $d\le l\le 2d$. Then the operator $S(U)$ is symmetric if and only if*

$$\mathcal{N}(U)\subset\mathcal{R}(P^{-1}U^*),$$

where P is defined by (12.3.3).

PROOF. This follows from Theorem 12.3.2 and Lemma 12.3.2. □

THEOREM 12.3.4. *Suppose $U\in M_{l,2d}$. Let $U=(A_1:B_1:A_2:B_2)$ where $A_1\in M_{l,m_{a_1}}$ consists of the first m_{a_1} columns of U in the same order as they are in U; $B_1\in M_{l,m_{b_1}}\ldots$; $A_2\in M_{l,m_{a_2}}\ldots$ and $B_2\in M_{l,m_{b_2}}\ldots$; (recall that $m_{a_1}+m_{b_1}+m_{a_2}+m_{b_2}=2d$) and assume that* rank $U=l$. *Then the operator $S(U)$ is self-adjoint if and only if*

$$l=d\quad\text{and}\quad kA_1E_{m_{a_1}}A_1^*-kB_1E_{m_{b_1}}B_1^*+hA_2E_{m_{a_2}}A_2^*-hB_2E_{m_{b_2}}B_2^*=0.$$

PROOF. It follows from Theorem 4.4.2 and Theorem 12.3.3 that $S(U)$ is self-adjoint if and only if $S(U)$ is a d dimensional symmetric extension of the minimal operator $S_{\min}$, i.e. if and only if $l=d$ and $\mathcal{N}(U)\subset\mathcal{R}(P^{-1}U^*)$. When $l=d$, we have $\dim(\mathcal{N}(U))=d$ and $\dim(\mathcal{R}(P^{-1}U^*))=d$. Hence $\mathcal{N}(U)\subset\mathcal{R}(P^{-1}U^*)$ is equivalent to $\mathcal{R}(P^{-1}U^*)\subset\mathcal{N}(U)$, and this is equivalent to $UP^{-1}U^*=0$, i.e.

$$UP^{-1}U^*=\frac{1}{h}A_1E_{m_{a_1}}A_1^*-\frac{1}{h}B_1E_{m_{b_1}}B_1^*+\frac{1}{k}A_2E_{m_{a_2}}A_2^*-\frac{1}{k}B_2E_{m_{b_2}}B_2^*=0.$$

This completes the proof. □

In the following, we consider boundary condition matrices U such that $(S(U))^*$ is symmetric.

THEOREM 12.3.5. *Let $U\in M_{l,2d}$, $0\le l\le 2d$ and assume that* rank $U=l$. *Then*

$$D((S(U))^*)=\{\mathbf{z}=\{z_1,z_2\}\in D_{\max}:\ \ \mathbf{Z}\in\mathcal{R}(P^{-1}U^*)\},$$

where $\mathbf{Z}$ is defined by

$$\mathbf{Z}=\begin{pmatrix}[z_1,u_{11}](a_1)\\ \vdots\\ [z_1,u_{1m_{a_1}}](a_1)\\ [z_1,v_{11}](b_1)\\ \vdots\\ [z_1,v_{1m_{b_1}}](b_1)\\ [z_2,u_{21}](a_2)\\ \vdots\\ [z_2,u_{2m_{a_2}}](a_2)\\ [z_2,v_{21}](b_2)\\ \vdots\\ [z_2,v_{2m_{b_2}}](b_2)\end{pmatrix}. \tag{12.3.5}$$

PROOF. Let $\mathbf{z} \in D_{\max}$. Then $\mathbf{z} \in D((S(U))^*)$ if and only if

$$\langle S_{\max}\mathbf{y}, \mathbf{z}\rangle = \langle \mathbf{y}, S_{\max}\mathbf{z}\rangle, \quad \textit{for all} \quad \mathbf{y} \in D(S(U)).$$

By the two-interval Green's formula (12.2.1), this is equivalent to $\mathbf{Z}^*P\mathbf{Y} = 0$ for all $\mathbf{y} \in D(S(U))$. Therefore $\mathbf{z} \in D((S(U))^*)$ if and only if $\mathbf{Y}^*P^*\mathbf{Z} = 0$, i.e. $P^*\mathbf{Z} \in \mathcal{N}(U)^\perp = \mathcal{R}(U^*)$. Therefore $\mathbf{Z} \in \mathcal{R}(P^{-1}U^*)$. This completes the proof. □

LEMMA 12.3.3. *Let $U \in M_{l,2d}$ and assume* rank $U = l$ *and* $0 \le l \le d$. *Then the following assertions are equivalent:*

(1) $(S(U))^*$ *is symmetric;*
(2) $\mathcal{N}(U) \supset \mathcal{R}(P^{-1}U^*)$;
(3) $UP^{-1}U^* = 0$.

PROOF. From Theorem 12.3.2 and Theorem 12.3.5, it follows that $(S(U))^*$ is symmetric if and only if

$$\mathbf{Z}^*P\mathbf{Y} = 0, \quad \textit{for all } \mathbf{y}, \mathbf{z} \in D((S(U))^*), \tag{12.3.6}$$

where $\mathbf{Y}$, $\mathbf{Z} \in \mathcal{R}(P^{-1}U^*)$ are defined by (12.3.5). By Lemma 12.3.1 and Theorem 12.3.5, (12.3.6) is equivalent to

$$Z^*PY = 0, \quad \textit{for all } Y, Z \in \mathcal{R}(P^{-1}U^*),$$

This is equivalent to $\mathcal{R}(P^{-1}U^*) \perp \mathcal{R}(U^*)$, i.e. $\mathcal{R}(P^{-1}U^*) \perp (N(U))^\perp$. Therefore $(S(U))^*$ is symmetric if and only if

$$\mathcal{R}(P^{-1}U^*) \subset \mathcal{N}(U).$$

The equivalence of (2) and (3) can be obtained immediately. □

LEMMA 12.3.4. *Let $U \in M_{l,2d}$ and assume that* rank $U = l$ *and* $d \le l \le 2d$. *Then the following statements are equivalent:*

(1) $S(U)$ *is a symmetric extension of the minimal operator* $S_{\min}$;
(2) $\mathcal{N}(U) \subset \mathcal{R}(P^{-1}U^*)$;
(3) *There exists a* $d \times 2d$ *matrix* $\widetilde{U}$ *satisfying* rank $\widetilde{U} = d$, $\mathcal{N}(U) \subset \mathcal{N}(\widetilde{U})$ *and* $\widetilde{U}P^{-1}\widetilde{U}^* = 0$;
(4) *There exists a* $d \times l$ *matrix* $\widetilde{V}$ *satisfying* rank $\widetilde{V} = d$ *and* $\widetilde{V}UP^{-1}U^*\widetilde{V}^* = 0$;
(5) $\operatorname{rank}(UP^{-1}U^*) = 2(l-d)$;
(6) $\operatorname{rank}(UP^{-1}U^*) \le 2(l-d)$;
(7) $\mathcal{N}(U) = P^{-1}U^*(\mathcal{N}(UP^{-1}U^*))$.

PROOF. The equivalence of (1) and (2) is provided in Theorem 12.3.3.

(1) $\Rightarrow$ (3): Let (1) hold, i.e. $S(U)$ is a symmetric extension of $S_{\min}$. Note that every symmetric extension of $S_{\min}$ is a restriction of a self-adjoint extension of $S_{\min}$. By Theorem 12.3.4, $S(\widetilde{U})$ is self-adjoint extension of $S_{\min}$. Therefore (3) holds.

(3) $\Rightarrow$ (2): It follows from $\mathcal{N}(U) \subset \mathcal{N}(\widetilde{U})$ that

$$\mathcal{R}(\widetilde{U}^*) = \mathcal{N}(\widetilde{U})^\perp \subset \mathcal{N}(U)^\perp = \mathcal{R}(U^*).$$

Hence $\mathcal{R}(P^{-1}\widetilde{U}^*) \subset \mathcal{R}(P^{-1}U^*)$. By Theorem 12.3.4 and (3), we have $\mathcal{N}(\widetilde{U}) = \mathcal{R}(P^{-1}\widetilde{U}^*)$. Therefore

$$\mathcal{N}(U) \subset \mathcal{N}(\widetilde{U}) \subset \mathcal{R}(P^{-1}U^*).$$

So (2) holds.

(3) $\Rightarrow$ (4): Let (3) hold. Since $\mathcal{N}(U) \subset \mathcal{N}(\widetilde{U})$, we have $\mathcal{R}(U^*) \supset \mathcal{R}(\widetilde{U}^*)$. Therefore there exists a $d \times l$ matrix $\widetilde{V}$ such that $\widetilde{U}^* = U^*\widetilde{V}^*$, i.e. $\widetilde{U} = \widetilde{V}U$. Hence

$$\widetilde{V}UP^{-1}U^*\widetilde{V}^* = \widetilde{U}P^{-1}\widetilde{U}^* = 0$$

and

$$\operatorname{rank} \widetilde{V} = \operatorname{rank}(\widetilde{V}U) = \operatorname{rank} \widetilde{U} = d.$$

(4) $\Rightarrow$ (3): Let $\widetilde{U} = \widetilde{V}U$. Then $\widetilde{U}P^{-1}\widetilde{U}^* = \widetilde{V}UP^{-1}U^*\widetilde{V}^* = 0$. It follows from $\operatorname{rank} U = l$ that $\operatorname{rank} \widetilde{U} = \operatorname{rank}(\widetilde{V}U) = \operatorname{rank} \widetilde{V} = d$. For any $Y \in \mathcal{N}(U)$, $\widetilde{U}Y = \widetilde{V}UY = 0$ which shows that $\mathcal{N}(U) \subset \mathcal{N}(\widetilde{U})$.

The equivalence of (2), (5), (6) and (7) can be obtained by Lemma 5.3.2. $\square$

THEOREM 12.3.6. *For $r = 1, 2$, suppose M_r is a symmetric differential expression on the interval (a_r, b_r), $-\infty \le a_r < b_r \le \infty$, of order $n \in \mathbb{N}_2$ and w_r are weight functions. Let $a_r < c_r < b_r$. Assume that the deficiency indices of M_r on (a_r, c_r), (c_r, b_r) are d_{a_r}, d_{b_r}, respectively. Then the deficiency index of M_r on (a_r, b_r) is $d_r = d_{a_r} + d_{b_r} - n$ and the the deficiency index of two-interval minimal operator $S_{\min}$ is $d = d_1 + d_2$.*

Let $u_{r1}, u_{r2}, \cdots, u_{rm_{a_r}}$, $m_{a_r} = 2d_{a_r} - n$, and $v_{r1}, v_{r2}, \cdots, v_{rm_{b_r}}$, $m_{b_r} = 2d_{b_r} - n$ be defined in Theorem 12.2.1. Define $\mathbf{Y}$ by (12.3.1). Assume $U \in M_{l,2d}$ with rank $U = l$, *$0 \le l \le 2d$ and let $U = (A_1 : B_1 : A_2 : B_2)$ where $A_1 \in M_{l,m_{a_1}}$ consists of the first m_{a_1} columns of U in the same order as they are in U; $B_1 \in M_{l,m_{b_1}}\cdots$; $A_2 \in M_{l,m_{a_2}}\cdots$ and $B_2 \in M_{l,m_{b_2}}\cdots$; (recall that $m_{a_1} + m_{b_1} + m_{a_2} + m_{b_2} = 2d$). Define the operator $S(U)$ in H by (12.3.2) and let*

$$C = C(A_1, B_1, A_2, B_2) = kA_1E_{m_{a_1}}A_1^* - kB_1E_{m_{b_1}}B_1^* + hA_2E_{m_{a_2}}A_2^* - hB_2E_{m_{b_2}}B_2^*.$$

Then we have

1. *If $l < d$, then $S(U)$ is not symmetric.*
2. *If $l = d$, then $S(U)$ is self-adjoint (and hence also symmetric) if and only if* rank $C = 0$.
3. *Let $l = d + s$, $0 < s \le d$. Then $S(U)$ is symmetric if and only if* rank $C = 2s$.

PROOF. Part (1) follows from Theorem 4.4.2.

Part (2) is given by Theorem 12.3.4.

Part (3): $d < l \le 2d$. It follows from Lemma 12.3.4 that $S(U)$ is symmetric if and only if

$$\operatorname{rank} C = \operatorname{rank}(hkUP^{-1}U^*) = \operatorname{rank}(UP^{-1}U^*) = 2(l - d) = 2s.$$

$\square$

4. Discontinuous Symmetric and Self-Adjoint Boundary Conditions

In this section we show that the special case of the 2-interval theory when the intervals have a common endpoint:

$$(12.4.1)\qquad \begin{gathered} -\infty \le a_1 < b_1 = a_2 < b_2 \le +\infty, \quad J_1 = (a_1, b_1) = (a, c), \\ J_2 = (a_2, b_2) = (c, b), \quad J = (a, b) \end{gathered}$$

developed in Sections 12.1, 12.2 and 12.3 can be applied to study *symmetric domain problems* with discontinuous boundary conditions specified at an interior point of the underlying interval J. Such *self-adjoint* conditions have been studied in the second order case and are known by various names including transmission conditions, interface conditions, multi-point conditions, point interactions (in the Physics literature) etc. Using the framework consisting of the Hilbert space H (12.1.2) with inner product (12.1.1) we generate symmetric and self-adjoint operators with regular and singular discontinuous boundary conditions. This framework provides a unified theory for the study of such problems and enlarges this class of problems.

We start with the following simple but important observations which help to illustrate how the special case (12.4.1) of the 2-interval theory produces regular and singular transmission and interface conditions.

REMARK 12.4.1. To connect the general 2-interval theory in Sections 12.2 and 12.3 to the special case (12.4.1) with the transmission and interface conditions, a key observation is that the direct sum Hilbert space $L^2(J_1, w_1) \dotplus L^2(J_2, w_2)$ can be identified with the space $L^2(J, w)$ where $w = w_1$ on J_1 and $w = w_2$ on J_2. Note that even though $b_1 = a_2 = c$ there are still four endpoint classifications since the endpoint c may have different classifications on (a, c) than on (c, b). To emphasize this point as well as to relate to the notation commonly used for regular transmission and interface conditions we use the notation c^- when c is a right endpoint i.e. for the interval (a, c) and c^+ for c as an endpoint of the interval (c, b).

REMARK 12.4.2. For any point c in the interval $J = (a, b)$, all functions in the maximal domain $D_{\max}(a, b)$ as well as their quasi-derivatives are continuous at c. Note that c is the right endpoint of the interval J_1 and the left endpoint of the interval J_2 and c is a regular endpoint for both intervals (a, c) and (c, b). Thus the 1-interval theory of Section 6.3 can be applied to the interval $J = (a, b)$. Note that this theory does not generate any symmetric operator S in $L^2(J, w)$ whose domain $D(S)$ contains a function y from the maximal domain $D_{\max}(a, b)$ such that y or some of its quasi-derivatives have a jump discontinuity at c since, in the 1-interval theory, all functions in the maximal domain $D_{\max}(a, b)$ together with their quasi-derivatives are continuous on the whole interval (a, b).

But, as we will see below, the 2-interval theory generates symmetric operators S in $L^2(J, w)$ determined by boundary conditions which specify jump discontinuities at at $c \in (a, b)$ for functions $y \in D(S)$ or some of their quasi-derivatives. These boundary conditions may be separated or coupled, in the separated case they are generally called 'transmission' conditions, in the coupled case they are often referred to as 'interface' conditions. But both the 'transmission' and the 'interface' conditions 'transmit' information from one interval to the other.

COROLLARY 12.4.1. *Let the hypothesis and notations of Theorem* 12.3.6 *hold. Let* $(a_1, b_1) = (a, c)$, $(a_2, b_2) = (c, b)$, *and let* $\operatorname{rank} U = l = d + s$, $0 \leq s \leq d$. *Then the boundary conditions* $U\mathbf{Y} = 0$, *i.e.*

$$A_1 Y(a) + B_1 Y(c^-) + A_2 Y(c^+) + B_2 Y(b) = 0$$

define symmetric operators in H *if and only if*

(12.4.2)

$$\operatorname{rank} C = \operatorname{rank}(kA_1 E_{m_{a_1}} A_1^* - kB_1 E_{m_{b_1}} B_1^* + hA_2 E_{m_{a_2}} A_2^* - hB_2 E_{m_{b_2}} B_2^*) = 2s.$$

Here

$$Y(c^-) = \lim_{t\to c^-} Y(t) = \lim_{t\to b_1^-} Y(t) = Y(b_1),$$

$$Y(c^+) = \lim_{t\to c^+} Y(t) = \lim_{t\to a_2^+} Y(t) = Y(a_2).$$

Recall that, since c is regular, $Y(c^+)$ and $Y(c^-)$ exist as finite limits.

COROLLARY 12.4.2. *Let the hypothesis and notations of Theorem* 12.3.6 *hold. Let* $(a_1, b_1) = (a, c)$, $(a_2, b_2) = (c, b)$. *Assume that* $m_{a_1} = 0$ *and* $m_{b_2} = 0$, *i.e.* A_1 *and* B_2 *disappear. Let* $B_1 \in M_{l,m_{b_1}}(\mathbb{C})$, $A_2 \in M_{l,m_{a_2}}(\mathbb{C})$, $l = d + s$, $0 \le s \le d$. *If* $\operatorname{rank}(B_1 : A_2) = l$ *and*

$$\operatorname{rank}(-kB_1E_{m_{b_1}}B_1^* + hA_2E_{m_{a_2}}A_2^*) = 2s, \tag{12.4.3}$$

then the boundary conditions

$$B_1Y(c^-) + A_2Y(c^+) = 0$$

define symmetric operators in H.

COROLLARY 12.4.3. *Let the hypothesis and notations of Theorem* 12.3.6 *hold. Let* $A_0 \in M_{l_1,m_{a_1}}(\mathbb{C})$, $B_0 \in M_{l_2,m_{b_1}}(\mathbb{C})$,$C_0 \in M_{l_3,m_{a_2}}(\mathbb{C})$,$D_0 \in M_{l_4,m_{b_2}}(\mathbb{C})$, *where* $0 \le l_1 \le m_{a_1}$, $0 \le l_2 \le m_{b_1}$, $0 \le l_3 \le m_{a_2}$, $0 \le l_4 \le m_{b_2}$. *Let* $(a_1, b_1) = (a, c)$, $(a_2, b_2) = (c, b)$. *Assume that*

$$\begin{aligned} &\operatorname{rank} A_0 = l_1, \quad \operatorname{rank} B_0 = l_2, \\ &\operatorname{rank} C_0 = l_3, \quad \operatorname{rank} D_0 = l_4, \\ &\sum_{j=1}^{4} l_j = l = d + s, \quad 0 \le s \le d. \end{aligned} \tag{12.4.4}$$

Then boundary conditions

$$\begin{aligned} A_0Y(a) &= 0, \quad B_0Y(c^-) = 0, \\ C_0Y(c^+) &= 0, \quad D_0Y(b) = 0. \end{aligned} \tag{12.4.5}$$

determine symmetric operators if and only if

$$\begin{aligned} \operatorname{rank}(A_0E_{m_{a_1}}A_0^*) &= 2l_1 - m_{a_1}, \quad \operatorname{rank}(B_0E_{m_{b_1}}B_0^*) = 2l_2 - m_{b_1}, \\ \operatorname{rank}(C_0E_{m_{a_2}}C_0^*) &= 2l_3 - m_{a_2}, \quad \operatorname{rank}(D_0E_{m_{b_2}}D_0^*) = 2l_4 - m_{b_2}. \end{aligned} \tag{12.4.6}$$

PROOF. Let

$$A_1 = \begin{pmatrix} A_0 \\ 0_{l_2,m_{a_1}} \\ 0_{l_3,m_{a_1}} \\ 0_{l_4,m_{a_1}} \end{pmatrix}, \quad B_1 = \begin{pmatrix} 0_{l_1,m_{b_1}} \\ B_0 \\ 0_{l_3,m_{b_1}} \\ 0_{l_4,m_{b_1}} \end{pmatrix},$$

$$A_2 = \begin{pmatrix} 0_{l_1,m_{a_2}} \\ 0_{l_2,m_{a_2}} \\ C_0 \\ 0_{l_4,m_{a_2}} \end{pmatrix}, \quad B_2 = \begin{pmatrix} 0_{l_1,m_{b_2}} \\ 0_{l_2,m_{b_2}} \\ 0_{l_3,m_{b_2}} \\ D_0 \end{pmatrix}.$$

By (12.4.4), we have

$$\operatorname{rank} U = \operatorname{rank}(A_1 : B_1 : A_2 : B_2) = l = d + s.$$

By calculation, we have

$$\begin{aligned} C =& C(A_1, B_1, A_2, B_2) \\ =& \begin{pmatrix} kA_0E_{m_{a_1}}A_0^* & 0 & 0 & 0 \\ 0 & -kB_0E_{m_{b_1}}B_0^* & 0 & 0 \\ 0 & 0 & hC_0E_{m_{a_2}}C_0^* & 0 \\ 0 & 0 & 0 & -hD_0E_{m_{b_2}}D_0^* \end{pmatrix}. \end{aligned}$$

Therefore

$$\operatorname{rank} C \geq 2(l-d) = 2s.$$

By Theorem 12.3.6, $U\mathbf{Y} = 0$, i.e. (12.4.5) determines symmetric operators if and only if rank $C = 2s$, i.e. (12.4.6) holds. □

REMARK 12.4.3. Note that for these separated boundary conditions the symmetric domain is independent of h and k.

COROLLARY 12.4.4. *Let* $\gamma_1 = \frac{m_{a_1}+m_{b_2}}{2} = d_{a_1} + d_{b_2} - n$ *and* $\gamma_2 = \frac{m_{b_1}+m_{a_2}}{2} = d_{b_1} + d_{a_2} - n$. *Then* $\gamma_1 + \gamma_2 = d$. *Let* $(a_1, b_1) = (a, c)$, $(a_2, b_2) = (c, b)$. *Assume that matrices* $A_0 \in M_{\gamma_1, m_{a_1}}(\mathbb{C})$, $D_0 \in M_{\gamma_1, m_{b_2}}(\mathbb{C})$, $B_0 \in M_{\gamma_2, m_{b_1}}(\mathbb{C})$, $C_0 \in M_{\gamma_2, m_{a_2}}(\mathbb{C})$, *and assume*

$$\operatorname{rank}(A_0 : D_0) = \gamma_1, \quad \operatorname{rank}(B_0 : C_0) = \gamma_2. \tag{12.4.7}$$

Let

$$\begin{aligned} A_0Y(a) + D_0Y(b) = 0, \\ B_0Y(c^-) + C_0Y(c^+) = 0. \end{aligned} \tag{12.4.8}$$

Then the operator $S(U)$ *determined by boundary conditions* (12.4.8) *is self-adjoint if and only if the following conditions hold:*

$$\begin{aligned} kA_0E_{m_{a_1}}A_0^* - hD_0E_{m_{b_2}}D_0^* = 0, \\ kB_0E_{m_{b_1}}B_0^* - hC_0E_{m_{a_2}}C_0^* = 0. \end{aligned} \tag{12.4.9}$$

PROOF. Let

$$A_1 = \begin{pmatrix} A_0 \\ 0_{\gamma_2 \times m_{a_1}} \end{pmatrix}, \quad B_1 = \begin{pmatrix} 0_{\gamma_1 \times m_{b_1}} \\ B_0 \end{pmatrix},$$

$$A_2 = \begin{pmatrix} 0_{\gamma_1 \times m_{a_2}} \\ C_0 \end{pmatrix}, \quad B_2 = \begin{pmatrix} D_0 \\ 0_{\gamma_2 \times m_{b_2}} \end{pmatrix}.$$

Set $U = (A_1 : B_1 : A_2 : B_2)$. It follows from (12.4.7) that rank $U = \gamma_1 + \gamma_2 = d$. By Theorem 12.3.6, $S(U)$ determined by $U\mathbf{Y} = 0$ is self-adjoint if and only if

$$kA_1E_{m_{a_1}}A_1^* - kB_1E_{m_{b_1}}B_1^* + hA_2E_{m_{a_2}}A_2^* - hB_2E_{m_{b_2}}B_2^* = 0.$$

This completes the proof. □

REMARK 12.4.4. Note that, in Corollaries 12.4.1 and 12.4.2, the boundary conditions (12.4.2) and (12.4.3) depend on the constants h and k used in the definition of the inner product of the Hilbert space H. These Corollaries illustrate how the use of the constants h and k enlarges the set of boundary conditions which generate symmetric operators in H. In Corollary 12.4.4 the self-adjointness condition (12.4.9) demonstrates the dependence of self-adjoint boundary conditions on h and k; this enlarges the set of self-adjoint operators with discontinuous boundary conditions specified at an interior point of the underlying interval.

5. Examples

Theorem 12.3.6 characterizes the two-interval symmetric boundary conditions in terms of complex solutions $u_{r1}, u_{r2}, \cdots, u_{rm_{a_r}}$ and $v_{r1}, v_{r2}, \cdots, v_{rm_{b_r}}$ and matrices $E_{m_{a_r}}$, $E_{m_{b_r}}$, $r = 1, 2$, which are given in Theorem 12.2.1; this is based on the decomposition of the maximal domain $D_{\max}$ given by Theorem 4.4.4.

In this section we use the real λ decomposition of $D_{\max}$ given by Theorem 7.2.4 and the much simpler symplectic matrices $E_{m_{a_r}}$, $E_{m_{b_r}}$. This decomposition in Chapter 7 uses the hypothesis **EH** and LC solutions, see Theorems 7.2.3 and 7.2.4. These matrices are used to construct the examples given in this Section.

In this whole section we use the notation $u_{r1}, u_{r2}, \cdots, u_{rm_{a_r}}$ and $v_{r1}, v_{r2}, \cdots, v_{rm_{b_r}}$ for the real λ solutions used in Chapter 7 to avoid introducing more complicated notations. (This notation is also used in Theorem 12.3.6 for complex λ solutions.)

THEOREM 12.5.1. *For $r = 1, 2$, suppose M_r is a symmetric differential expression on the interval (a_r, b_r), $-\infty \leq a_r < b_r \leq \infty$, of order $n = 2k$, $k \in \mathbb{N}$, with real coefficients. Let $a_r < c_r < b_r$. Assume that the deficiency indices of M_r on (a_r, c_r), (c_r, b_r) are d_{a_r}, d_{b_r}, respectively. Assume that the hypothesis* ***EH*** *holds: the equation $M_r y = \lambda w_r y$ has d_{a_r} linearly independent solutions in $L^2((a_r, c_r), w_r)$ for some real $\lambda = \lambda_{a_r}$ and on (c_r, b_r) has d_{b_r} linearly independent solutions in $L^2((c_r, b_r), w_r)$ for some real $\lambda = \lambda_{b_r}$.*

Then the deficiency index of M_r on (a_r, b_r) is $d_r = d_{a_r} + d_{b_r} - n$ and the the deficiency index of two-interval minimal operator $S_{\min}$ is $d = d_1 + d_2$. Let $u_{r1}, u_{r2}, \cdots, u_{rm_{a_r}}$, $m_{a_r} = 2d_{a_r} - n$, and $v_{r1}, v_{r2}, \cdots, v_{rm_{b_r}}$, $m_{b_r} = 2d_{b_r} - n$ be LC solutions constructed on (a_r, c_r), (c_r, b_r), respectively, and extended to maximal domain functions in $D_{r\max} = D_{r\max}(a_r, b_r)$ by Theorem 7.2.3.

Using these LC solutions u_{rj} and v_{rj}, we define

$$Y_{a_r,b_r} = \begin{pmatrix} Y(a_r) \\ Y(b_r) \end{pmatrix}, \quad Y(a_r) = \begin{pmatrix} [y_r, u_{r1}]_r(a_r) \\ \vdots \\ [y_r, u_{rm_{a_r}}]_r(a_r) \end{pmatrix},$$

$$Y(b_r) = \begin{pmatrix} [y_r, v_{r1}]_r(b_r)) \\ \vdots \\ [y_r, v_{rm_{b_r}}]_r(b_r) \end{pmatrix}, \quad \mathbf{Y} = \begin{pmatrix} Y_{a_1,b_1} \\ Y_{a_2,b_2} \end{pmatrix}.$$

Assume $U \in M_{l,2d}$ with rank $U = l$, *$0 \leq l \leq 2d$ and let $U = (A_1 : B_1 : A_2 : B_2)$ where $A_1 \in M_{l,m_{a_1}}$ consists of the first m_{a_1} columns of U in the same order as they are in U; $B_1 \in M_{l,m_{b_1}} \cdots$; $A_2 \in M_{l,m_{a_2}} \cdots$ and $B_2 \in M_{l,m_{b_2}} \cdots$.*

Define the operator $S(U)$ in H by

$$D(S(U)) = \{\mathbf{y} = \{y_1, y_2\} \in D_{\max} \; : \; U\mathbf{Y} = 0\},$$
$$S(U)\mathbf{y} = S_{\max}\mathbf{y} \quad \textit{for} \quad \mathbf{y} \in D(S(U)).$$

Let

$$C = C(A_1, B_1, A_2, B_2) = kA_1 E_{m_{a_1}} A_1^* - kB_1 E_{m_{b_1}} B_1^* + hA_2 E_{m_{a_2}} A_2^* - hB_2 E_{m_{b_2}} B_2^*,$$

where $E_{m_{a_r}}$ *and* $E_{m_{b_r}}$ *are the simple symplectic matrices*

$$E_{m_{a_r}} = ((-1)^j \delta_{j,m_{a_r}+1-s})_{j,s=1}^{m_{a_r}},$$
$$E_{m_{b_r}} = ((-1)^j \delta_{j,m_{b_r}+1-s})_{j,s=1}^{m_{b_r}}.$$

Then we have

(1) *If* $l < d$*, then* $S(U)$ *is not symmetric.*
(2) *If* $l = d$*, then* $S(U)$ *is self-adjoint (and hence also symmetric) if and only if* rank $C = 0$.
(3) *Let* $l = d + s$*,* $0 < s \leq d$*. Then* $S(U)$ *is symmetric if and only if* rank $C = 2s$.

PROOF. The proof is similar to the proof of Theorem 12.3.6 and hence omitted. □

First we give some examples for the self-adjoint case.

EXAMPLE 12.5.1. Let the hypothesis and notations of Theorem 12.5.1 hold. Consider $M_r y = \lambda w_r y$ on (a_r, b_r), for $r = 1, 2$. Let M_r be a symmetric expression of order $n = 4$ with real coefficients.

Let $d_{a_r} = d_{b_r} = 3$. Then $m_{a_r} = 2d_{a_r} - n = 2$, $m_{b_r} = 2d_{b_r} - n = 2$, $d_r = d_{a_r} + d_{b_r} - n = 2$, $d = d_1 + d_2 = 4$ and $E_{m_{a_r}} = E_{m_{b_r}} = \begin{pmatrix} 0 & -1 \\ 1 & 0 \end{pmatrix}$.

In the following we assume that

$$\begin{aligned} C_1, C_2 \in \mathbb{R}, &\quad (C_1, C_2) \neq (0,0); \\ D_1, D_2 \in \mathbb{R}, &\quad (D_1, D_2) \neq (0,0); \\ G_1, G_2 \in \mathbb{R}, &\quad (G_1, G_2) \neq (0,0); \\ F_1, F_2 \in \mathbb{R}, &\quad (F_1, F_2) \neq (0,0); \end{aligned}$$
$$K = (k_{ij}),\ k_{ij} \in \mathbb{R},\ i, j = 1, 2; \quad \det K \neq 0.$$

Case I.
Let

$$A_1 = \begin{pmatrix} 0 & 0 \\ G_1 & G_2 \\ 0 & 0 \\ 0 & 0 \end{pmatrix}, \quad B_1 = \begin{pmatrix} C_1 & C_2 \\ 0 & 0 \\ 0 & 0 \\ 0 & 0 \end{pmatrix},$$
$$A_2 = \begin{pmatrix} 0 & 0 \\ 0 & 0 \\ F_1 & F_2 \\ 0 & 0 \end{pmatrix}, \quad B_2 = \begin{pmatrix} 0 & 0 \\ 0 & 0 \\ 0 & 0 \\ D_1 & D_2 \end{pmatrix}.$$

Clearly rank$(A_1 : B_1 : A_2 : B_2) = d = 4$.

By computation, we have

$$A_1 E_{m_{a_1}} A_1^* = B_1 E_{m_{b_1}} B_1^* = A_2 E_{m_{a_2}} A_2^* = B_2 E_{m_{b_2}} B_2^* = 0.$$

Consider

$$\begin{aligned} &G_1[y_1, u_{11}]_1(a_1) + G_2[y_1, u_{12}]_1(a_1) = 0; \\ &C_1[y_1, v_{11}]_1(b_1) + C_2[y_1, v_{12}]_1(b_1) = 0; \\ &F_1[y_2, u_{21}]_2(a_2) + F_2[y_2, u_{22}]_2(a_2) = 0; \\ &D_1[y_2, v_{21}]_2(b_2) + D_2[y_2, v_{22}]_2(b_2) = 0. \end{aligned} \tag{12.5.1}$$

By Theorem 12.5.1, the separated boundary conditions (12.5.1) are self-adjoint. Note that this case is independent of the inner product constants h, k.

Case II.

Let

$$A_1 = \begin{pmatrix} 0 & 0 \\ k_{11} & k_{12} \\ k_{21} & k_{22} \\ 0 & 0 \end{pmatrix}, \quad B_1 = \begin{pmatrix} C_1 & C_2 \\ 0 & 0 \\ 0 & 0 \\ 0 & 0 \end{pmatrix},$$

$$A_2 = \begin{pmatrix} 0 & 0 \\ -1 & 0 \\ 0 & -1 \\ 0 & 0 \end{pmatrix}, \quad B_2 = \begin{pmatrix} 0 & 0 \\ 0 & 0 \\ 0 & 0 \\ D_1 & D_2 \end{pmatrix},$$

where clearly $\operatorname{rank}(A_1 : B_1 : A_2 : B_2) = d = 4$.

By computation, we have

$$B_1 E_{m_{b_1}} B_1^* = 0, \quad B_2 E_{m_{b_2}} B_2^* = 0,$$

$$A_1 E_{m_{a_1}} A_1^* = \begin{pmatrix} 0 & 0 & 0 & 0 \\ 0 & 0 & -\det K & 0 \\ 0 & \det K & 0 & 0 \\ 0 & 0 & 0 & 0 \end{pmatrix}, \quad A_2 E_{m_{a_2}} A_2^* = \begin{pmatrix} 0 & 0 & 0 & 0 \\ 0 & 0 & -1 & 0 \\ 0 & 1 & 0 & 0 \\ 0 & 0 & 0 & 0 \end{pmatrix}.$$

Consider

$$\begin{aligned} &C_1[y_1, v_{11}]_1(b_1) + C_2[y_1, v_{12}]_1(b_1) = 0; \\ &D_1[y_2, v_{21}]_2(b_2) + D_2[y_2, v_{22}]_2(b_2) = 0; \\ &\begin{pmatrix} [y_2, u_{21}]_2(a_2) \\ [y_2, u_{22}]_2(a_2) \end{pmatrix} = \begin{pmatrix} k_{11} & k_{12} \\ k_{21} & k_{22} \end{pmatrix} \begin{pmatrix} [y_1, u_{11}]_1(a_1) \\ [y_1, u_{12}]_1(a_1) \end{pmatrix}. \end{aligned} \tag{12.5.2}$$

By Theorem 12.5.1 the boundary conditions (12.5.2) are self-adjoint if and only if

$$k\, A_1 E_{m_{a_1}} A_1^* + h\, A_2 E_{m_{a_2}} A_2^* = 0,$$

i.e. $\det K = -\frac{h}{k} < 0$. If $k = 1$, $h > 0$, then the conditions (12.5.2) are self-adjoint if and only if $\det K = -h < 0$.

Case III.

Let

$$A_1 = \begin{pmatrix} C_1 & C_2 \\ 0 & 0 \\ 0 & 0 \\ 0 & 0 \end{pmatrix}, \quad B_1 = \begin{pmatrix} 0 & 0 \\ k_{11} & k_{12} \\ k_{21} & k_{22} \\ 0 & 0 \end{pmatrix},$$

$$A_2 = \begin{pmatrix} 0 & 0 \\ -1 & 0 \\ 0 & -1 \\ 0 & 0 \end{pmatrix}, \quad B_2 = \begin{pmatrix} 0 & 0 \\ 0 & 0 \\ 0 & 0 \\ D_1 & D_2 \end{pmatrix}.$$

It is obvious that $\operatorname{rank}(A_1 : B_1 : A_2 : B_2) = 4$, and

$$A_1 E_{m_{a_1}} A_1^* = 0, \quad B_2 E_{m_{b_2}} B_2^* = 0,$$

$$B_1 E_{m_{b_1}} B_1^* = \begin{pmatrix} 0 & 0 & 0 & 0 \\ 0 & 0 & -\det K & 0 \\ 0 & \det K & 0 & 0 \\ 0 & 0 & 0 & 0 \end{pmatrix}, \quad A_2 E_{m_{a_2}} A_2^* = \begin{pmatrix} 0 & 0 & 0 & 0 \\ 0 & 0 & -1 & 0 \\ 0 & 1 & 0 & 0 \\ 0 & 0 & 0 & 0 \end{pmatrix}.$$

Consider

$$
(12.5.3)\quad
\begin{aligned}
&C_1[y_1, u_{11}]_1(a_1) + C_2[y_1, u_{12}]_1(a_1) = 0;\\
&D_1[y_2, v_{21}]_2(b_2) + D_2[y_2, v_{22}]_2(b_2) = 0;\\
&\begin{pmatrix} [y_2, u_{21}]_2(a_2) \\ [y_2, u_{22}]_2(a_2) \end{pmatrix} = \begin{pmatrix} k_{11} & k_{12} \\ k_{21} & k_{22} \end{pmatrix} \begin{pmatrix} [y_1, v_{11}]_1(b_1) \\ [y_1, v_{12}]_1(b_1) \end{pmatrix}.
\end{aligned}
$$

By Theorem 12.5.1, the boundary conditions (12.5.3) are self-adjoint if and only if

$$-kB_1 E_{m_{b_1}} B_1^* + hA_2 E_{m_{a_2}} A_2^* = 0,$$

i.e. $\det K = \frac{h}{k} > 0$. If $k = 1$, $h > 0$, then the conditions (12.5.3) are self-adjoint if and only if $\det K = h > 0$.

REMARK 12.5.1. Let $(a_1, b_1) = (a, c)$, $(a_2, b_2) = (c, b)$. For Example 12.5.1, we may obtain the following self-adjoint operators with discontinuous boundary conditions (note that the self-adjointness conditions are not changed).

Boundary conditions (12.5.1) can be written as:

$$
\begin{aligned}
&G_1[y_1, u_{11}](a) + G_2[y_1, u_{12}](a) = 0;\\
&C_1[y_1, v_{11}](c^-) + C_2[y_1, v_{12}](c^-) = 0;\\
&F_1[y_2, u_{21}](c^+) + F_2[y_2, u_{22}](c^+) = 0;\\
&D_1[y_2, v_{21}](b) + D_2[y_2, v_{22}](b) = 0.
\end{aligned}
$$

Boundary conditions (12.5.3) can be written as:

$$
\begin{aligned}
&C_1[y_1, u_{11}](a) + C_2[y_1, u_{12}](a) = 0;\\
&D_1[y_2, v_{21}](b) + D_2[y_2, v_{22}](b) = 0;\\
&\begin{pmatrix} [y_2, u_{21}](c^+) \\ [y_2, u_{22}](c^+) \end{pmatrix} = \begin{pmatrix} k_{11} & k_{12} \\ k_{21} & k_{22} \end{pmatrix} \begin{pmatrix} [y_1, v_{11}](c^-) \\ [y_1, v_{12}](c^-) \end{pmatrix}.
\end{aligned}
$$

EXAMPLE 12.5.2. Let the hypothesis and notations of Theorem 12.5.1 hold. Consider $M_r y = \lambda w_r y$ on (a_r, b_r), for $r = 1, 2$. Let M_r be a symmetric expression of order $n = 4$ with real coefficients.

Let $d_{a_1} = 3$, $d_{b_1} = 4$, $d_{a_2} = 4$, $d_{b_2} = 3$. Then $m_{a_1} = 2d_{a_1} - n = 2$, $m_{b_1} = 2d_{b_1} - n = 4$, $m_{a_2} = 2d_{a_2} - n = 4$, $m_{b_2} = 2d_{b_2} - n = 2$, $d_1 = 3$, $d_2 = 3$, $d = d_1 + d_2 = 6$, and

$$E_{m_{a_1}} = E_{m_{b_2}} = \begin{pmatrix} 0 & -1 \\ 1 & 0 \end{pmatrix}, \quad E_{m_{a_2}} = E_{m_{b_1}} = \begin{pmatrix} 0 & 0 & 0 & -1 \\ 0 & 0 & 1 & 0 \\ 0 & -1 & 0 & 0 \\ 1 & 0 & 0 & 0 \end{pmatrix}.$$

Let

$$A_1 = \begin{pmatrix} 0 & 0 \\ 0 & 0 \\ 0 & 0 \\ 0 & 0 \\ C_1 & C_2 \\ 0 & 0 \end{pmatrix}, \quad B_1 = \begin{pmatrix} k_{11} & k_{12} & k_{13} & k_{14} \\ k_{21} & k_{22} & k_{23} & k_{24} \\ k_{31} & k_{32} & k_{33} & k_{34} \\ k_{41} & k_{42} & k_{43} & k_{44} \\ 0 & 0 & 0 & 0 \\ 0 & 0 & 0 & 0 \end{pmatrix},$$

$$A_2 = \begin{pmatrix} -1 & 0 & 0 & 0 \\ 0 & -1 & 0 & 0 \\ 0 & 0 & -1 & 0 \\ 0 & 0 & 0 & -1 \\ 0 & 0 & 0 & 0 \\ 0 & 0 & 0 & 0 \end{pmatrix}, \quad B_2 = \begin{pmatrix} 0 & 0 \\ 0 & 0 \\ 0 & 0 \\ 0 & 0 \\ 0 & 0 \\ D_1 & D_2 \end{pmatrix},$$

$$C_1, C_2 \in \mathbb{R}, \quad (C_1, C_2) \neq (0,0);$$
$$D_1, D_2 \in \mathbb{R}, \quad (D_1, D_2) \neq (0,0);$$
$$K = (k_{ij}),\ k_{ij} \in \mathbb{R},\ i,j = 1,2,3,4;\ \ \det K \neq 0.$$

It is obvious that $\operatorname{rank}(A_1 : B_1 : A_2 : B_2) = 6$, and

$$A_1 E_{m_{a_1}} A_1^* = 0, \quad B_2 E_{m_{b_2}} B_2^* = 0,$$

$$B_1 E_{m_{b_1}} B_1^* = \begin{pmatrix} K E_{m_{b_1}} K^* & 0 \\ 0 & 0 \end{pmatrix}, \quad A_2 E_{m_{a_2}} A_2^* = \begin{pmatrix} E_{m_{a_2}} & 0 \\ 0 & 0 \end{pmatrix}.$$

Consider the boundary conditions:

$$C_1[y, u_{11}]_1(a_1) + C_2[y, u_{12}]_1(a_1) = 0;$$
$$D_1[y, v_{21}]_2(b_2) + D_2[y, v_{22}]_2(b_2) = 0;$$

$$\begin{pmatrix} [y, u_{21}]_2(a_2) \\ [y, u_{22}]_2(a_2) \\ [y, u_{23}]_2(a_2) \\ [y, u_{24}]_2(a_2) \end{pmatrix} = K \begin{pmatrix} [y, v_{11}]_1(b_1) \\ [y, v_{12}]_1(b_1) \\ [y, v_{13}]_1(b_1) \\ [y, v_{14}]_1(b_1) \end{pmatrix}. \tag{12.5.4}$$

By Theorem 12.5.1 we have that the boundary conditions (12.5.4) are self-adjoint if and only if

$$-kKE_4K^* + hE_4 = 0.$$

REMARK 12.5.2. Let the hypothesis of Example 12.5.2 hold. Let $(a_1, b_1) = (a, c)$, $(a_2, b_2) = (c, b)$, and let b_1 and a_2 be LC endpoints, i.e. c is a LC singular endpoint for both intervals (a, c) and (c, b). Then the following discontinuous boundary conditions:

$$C_1[y, u_{11}](a) + C_2[y, u_{12}](a) = 0, \quad C_1, C_2 \in \mathbb{R}, \quad (C_1, C_2) \neq (0,0);$$
$$D_1[y, v_{21}](b) + D_2[y, v_{22}](b) = 0, \quad D_1, D_2 \in \mathbb{R}, \quad (D_1, D_2) \neq (0,0);$$

$$\begin{pmatrix} [y, u_{21}](c^+) \\ [y, u_{22}](c^+) \\ [y, u_{23}](c^+) \\ [y, u_{24}](c^+) \end{pmatrix} = K \begin{pmatrix} [y, v_{11}](c^-) \\ [y, v_{12}](c^-) \\ [y, v_{13}](c^-) \\ [y, v_{14}](c^-) \end{pmatrix}, \tag{12.5.5}$$

$$K = (k_{ij}),\ k_{ij} \in \mathbb{R},\ i,j = 1,2,3,4;\ \ \det K \neq 0,$$

determine self-adjoint operators if and only if $-kKE_4K^* + hE_4 = 0$.

If c is a regular endpoint for both intervals (a, c) and (c, b). The discontinuous condition (12.5.5) can be written as

$$\begin{pmatrix} y(c^+) \\ y^{[1]}(c^+) \\ y^{[2]}(c^+) \\ y^{[3]}(c^+) \end{pmatrix} = K \begin{pmatrix} y(c^-) \\ y^{[1]}(c^-) \\ y^{[2]}(c^-) \\ y^{[3]}(c^-) \end{pmatrix}.$$

Next we give some examples of discontinuous symmetric boundary conditions which are not self-adjoint.

EXAMPLE 12.5.3. Let the hypothesis and notations of Theorem 12.5.1 hold. Consider $M_r y = \lambda w_r y$ on (a_r, b_r), for $r = 1, 2$. Let M_r be a symmetric expression of order $n = 4$ with real coefficients.

Let $d_{a_r} = d_{b_r} = 3$. Then $m_{a_r} = 2d_{a_r} - n = 2$, $m_{b_r} = 2d_{b_r} - n = 2$, $d_r = d_{a_r} + d_{b_r} - n = 2$, $d = d_1 + d_2 = 4$, $E_{m_{a_r}} = E_{m_{b_r}} = E_2 = \begin{pmatrix} 0 & -1 \\ 1 & 0 \end{pmatrix}$.

By Theorem 12.5.1, if $l = d + s = 4 + s$, $0 < s \leq 4$, then $S(U)$ is symmetric if and only if rank $C = 2s$.

Here we only construct examples for the following two cases:

Case I. Let $s = 1$. Then $l = 5$.

(1) Let

$$A_1 = \begin{pmatrix} 1 & 0 \\ 0 & 0 \\ 0 & 0 \\ 0 & 0 \\ 0 & 0 \end{pmatrix}, \quad B_1 = \begin{pmatrix} 0 & 0 \\ 0 & 0 \\ 0 & 0 \\ 0 & 1 \\ 1 & 0 \end{pmatrix}, \quad A_2 = \begin{pmatrix} 0 & 0 \\ 1 & 0 \\ 0 & 0 \\ 0 & 0 \\ 0 & 0 \end{pmatrix}, \quad B_2 = \begin{pmatrix} 0 & 0 \\ 0 & 0 \\ 1 & 0 \\ 0 & 0 \\ 0 & 0 \end{pmatrix}.$$

Then rank $U = \operatorname{rank}(A_1 : B_1 : A_2 : B_2) = l = 5$. By computation, we have

$$A_1 E_2 A_1^* = A_2 E_2 A_2^* = B_2 E_2 B_2^* = 0,$$

$$B_1 E_2 B_1^* = \begin{pmatrix} 0 & 0 & 0 & 0 & 0 \\ 0 & 0 & 0 & 0 & 0 \\ 0 & 0 & 0 & 0 & 0 \\ 0 & 0 & 0 & 0 & 1 \\ 0 & 0 & 0 & -1 & 0 \end{pmatrix}.$$

So

$$\operatorname{rank} C = \operatorname{rank}(-k B_1 E_{m_{b_1}} B_1^*) = 2.$$

Therefore, by Theorem 12.5.1, the operator $S(U)$ determined by the following boundary conditions is symmetric:

$$\begin{aligned} &[y, u_{11}]_1(a_1) = 0, \quad [y, v_{12}]_1(b_1) = 0, \quad [y, v_{11}]_1(b_1) = 0, \\ &[y, u_{21}]_2(a_2) = 0, \quad [y, v_{21}]_2(b_2) = 0. \end{aligned} \tag{12.5.6}$$

(2) Let

$$A_1 = \begin{pmatrix} 1 & 0 \\ 0 & 0 \\ 0 & 0 \\ 0 & 0 \\ 0 & 0 \end{pmatrix}, \quad B_1 = \begin{pmatrix} 0 & 0 \\ 0 & 0 \\ 0 & 0 \\ 0 & 1 \\ 1 & 0 \end{pmatrix}, \quad A_2 = \begin{pmatrix} 0 & 0 \\ 1 & 0 \\ 0 & 0 \\ 0 & -i \\ 1 & 0 \end{pmatrix}, \quad B_2 = \begin{pmatrix} 0 & 0 \\ 0 & 0 \\ 1 & 0 \\ 0 & 0 \\ 0 & 0 \end{pmatrix}.$$

Then rank $U = \operatorname{rank}(A_1 : B_1 : A_2 : B_2) = l = 5$. By computation, we have

$$A_1 E_2 A_1^* = 0 = B_2 E_2 B_2^*,$$

$$B_1 E_2 B_1^* = \begin{pmatrix} 0 & 0 & 0 & 0 & 0 \\ 0 & 0 & 0 & 0 & 0 \\ 0 & 0 & 0 & 0 & 0 \\ 0 & 0 & 0 & 0 & 1 \\ 0 & 0 & 0 & -1 & 0 \end{pmatrix}, \quad A_2 E_2 A_2^* = \begin{pmatrix} 0 & 0 & 0 & 0 & 0 \\ 0 & 0 & 0 & -i & 0 \\ 0 & 0 & 0 & 0 & 0 \\ 0 & -i & 0 & 0 & -i \\ 0 & 0 & 0 & -i & 0 \end{pmatrix}.$$

So

$$\operatorname{rank} C = \operatorname{rank}(-k B_1 E_2 B_1^* + h A_2 E_2 A_2^*) = 2.$$

Therefore, by Theorem 12.5.1, the operator $S(U)$ determined by the following boundary conditions is symmetric:

$$\begin{aligned} &[y,u_{11}]_1(a_1)=0, \quad [y,u_{21}]_2(a_2)=0,\\ &[y,v_{11}]_1(b_1)=0, \quad [y,v_{21}]_2(b_2)=0,\\ &[y,v_{12}]_1(b_1)-i[y,u_{22}]_2(a_2)=0. \end{aligned} \tag{12.5.7}$$

Case II. Let $s=3$, then $l=7$.

Let

$$A_1=\begin{pmatrix} 0&0\\0&0\\0&0\\0&0\\0&0\\1&0\\0&1 \end{pmatrix}, \quad B_1=\begin{pmatrix} 1&0\\0&i\\0&0\\0&0\\0&0\\0&0\\0&0 \end{pmatrix}, \quad A_2=\begin{pmatrix} 0&0\\0&1\\1&0\\0&0\\0&0\\0&0\\0&0 \end{pmatrix}, \quad B_2=\begin{pmatrix} 0&0\\0&0\\0&0\\1&0\\0&1\\0&0\\0&0 \end{pmatrix}.$$

Then $\operatorname{rank} U=\operatorname{rank}(A_1:B_1:A_2:B_2)=l=7$. By computation, we have

$$\operatorname{rank} C=\operatorname{rank}(kA_1E_2A_1^*-kB_1E_2B_1^*+hA_2E_2A_2^*-hB_2E_2B_2^*)=2s=6.$$

Therefore, by Theorem 12.5.1, the operator $S(U)$ determined by the following boundary conditions is symmetric:

$$\begin{aligned} &[y,u_{11}]_1(a_1)=0, \quad [y,u_{12}]_1(a_1)=0,\\ &[y,v_{21}]_2(b_2)=0, \quad [y,v_{22}]_2(b_2)=0,\\ &[y,u_{21}]_2(a_2)=0, \quad [y,v_{11}]_1(b_1)=0,\\ &i[y,v_{12}]_1(b_1)+[y,u_{22}]_2(a_2)=0. \end{aligned} \tag{12.5.8}$$

REMARK 12.5.3. For Example 12.5.3, let $(a_1,b_1)=(a,c)$, $(a_2,b_2)=(c,b)$. Then boundary conditions (12.5.6) can be written as

$$\begin{aligned} &[y,u_{11}]_1(a)=0, \quad [y,v_{12}]_1(c^-)=0, \quad [y,v_{11}]_1(c^-)=0,\\ &[y,u_{21}]_2(c^+)=0, \quad [y,v_{21}]_2(b)=0. \end{aligned} \tag{12.5.9}$$

Conditions (12.5.7) can be written as

$$\begin{aligned} &[y,u_{11}]_1(a)=0, \quad [y,u_{21}]_2(c^+)=0,\\ &[y,v_{11}]_1(c^-)=0, \quad [y,v_{21}]_2(b)=0,\\ &[y,v_{12}]_1(c^-)-i[y,u_{22}]_2(c^+)=0. \end{aligned} \tag{12.5.10}$$

Condition (12.5.8) can be written as

$$\begin{aligned} &[y,u_{11}]_1(a)=0, \quad [y,u_{12}]_1(a)=0,\\ &[y,v_{21}]_2(b)=0, \quad [y,v_{22}]_2(b)=0,\\ &[y,u_{21}]_2(c^+)=0, \quad [y,v_{11}]_1(c^-)=0,\\ &i[y,v_{12}]_1(c^-)+[y,u_{22}]_2(c^+)=0. \end{aligned} \tag{12.5.11}$$

The operators determined by the discontinuous boundary conditions (12.5.9), (12.5.10) and (12.5.11) are symmetric.

EXAMPLE 12.5.4. For $r = 1, 2$, suppose M_r is a symmetric differential expression on the interval (a_r, b_r), $-\infty \leq a_r < b_r \leq \infty$, of order $n = 2$ with real coefficients. Let $a_r < c_r < b_r$. Assume that the deficiency indices of M_r on (a_r, c_r), (c_r, b_r) are d_{a_r}, d_{b_r}, respectively. Consider

$$M_r y = -(p_r y')' + q_r y,$$

where p_r^{-1}, $q_r \in L_{loc}(J_r, \mathbb{R})$. Let $d_{a_1} = d_{b_2} = 1$, $d_{b_1} = d_{a_2} = 2$. Then $m_{a_1} = m_{b_2} = 0$, $m_{b_1} = m_{a_2} = 2$, $d_r = d_{a_r} + d_{b_r} - 2 = 1$, $r = 1, 2$, $d = d_1 + d_2 = 2$ and $E_{m_{a_1}} = E_{m_{b_2}} = 0$, $E_{m_{b_1}} = E_{m_{a_2}} = E_2 = \begin{pmatrix} 0 & -1 \\ 1 & 0 \end{pmatrix}$.

By Theorem 12.5.1, if $l = d + s = 2 + s$, $0 < s \leq 2$, then $S(U)$ is symmetric if and only if rank $C = 2s$.

Consider $s = 1$. Then $l = 3$. Note that A_1 and B_2 are 0 since $m_{a_1} = m_{b_2} = 0$. Let

$$B_1 = \begin{pmatrix} 1 & 0 \\ 0 & 1 \\ 0 & 0 \end{pmatrix}, \quad A_2 = \begin{pmatrix} 0 & 0 \\ i & 0 \\ 0 & 1 \end{pmatrix}.$$

Then $\operatorname{rank}(B_1 : A_2) = l = 3$ and

$$\operatorname{rank} C = \operatorname{rank}(-kB_1 E_{m_{b_1}} B_1^* + hA_2 E_{m_{a_2}} A_2^*) = 2 = 2s.$$

Therefore the operator $S(U)$ determined by the following conditions

$$\begin{aligned} &[y, v_{11}](b_1) = 0, \\ &[y, v_{12}](b_1) + i[y, u_{21}](a_2) = 0, \\ &[y, u_{22}](a_2) = 0 \end{aligned}$$

is symmetric.

REMARK 12.5.4. For Example 12.5.4, let $(a_1, b_1) = (a, c)$, $(a_2, b_2) = (c, b)$. Then the following complex discontinuous boundary condition

$$\begin{aligned} &[y, v_{11}](c^-) = 0, \\ &[y, v_{12}](c^-) + i[y, u_{21}](c^+) = 0, \\ &[y, u_{22}](c^+) = 0 \end{aligned} \tag{12.5.12}$$

determines a symmetric operator $S(U)$ in the Hilbert space $L^2(J, w)$, $J = (a, b)$ whose domain consists of functions which have a discontinuity at the interior point c of the underlying interval J.

6. Comments

The two-interval characterization of symmetric boundary conditions given by Theorem 12.3.6 with the Mukhtarov-Yakubov constants h, k is published here for the first time. A one-interval version of this characterization was published by the authors in 2018 [**562**]. (Of course the one-interval characterization does not involve h, k.)

We believe the operators determined by (12.5.10) and (12.5.11) may be the first known examples of symmetric non-self-adjoint operators generated by singular nonreal discontinuous boundary conditions. They are also published here for the first time.

As an application of the two-interval theory, the authors study self-adjoint Sturm-Liouville problems with discontinuous boundary conditions in [**561**], and eigenvalues of such problems in [**558**]. See also the papers by Sun-Wang [**530**] and by Wang-Sun-Hao-Yao [**552**].

CHAPTER 13

Two-Interval Symmetric Domain Characterization with Maximal Domain Functions

1. Introduction and Main Theorem

In this chapter we characterize the two-interval symmetric boundary conditions in terms of functions from the maximal domain. The proof uses the maximal domain decomposition from Chapter 4 on each interval (a_r, b_r) :
(13.1.1)
$$D_{\max}(a_r, b_r) = D_{\min}(a_r, b_r) \dotplus \text{span}\{u_{r1}, \cdots, u_{rm_{a_r}}\} \dotplus \text{span}\{v_{r1}, \cdots, v_{rm_{b_r}}\}, \quad r = 1, 2.$$

Recall from Chapter 4 that in this maximal domain decomposition, the solutions u_{ri}, $i = 1, ..., m_{a_r}$ are solutions on the interval (a_r, c_r) for some $\lambda_{a_r} \in \mathbb{C}$ with $\text{Im}(\lambda_{a_r}) \neq 0$ which are identically 0 in a neighborhood of b_r, and v_{ri}, $i = 1, ..., m_{b_r}$ are solutions on the interval (c_r, b_r) for some $\lambda_{b_r} \in \mathbb{C}$ with $\text{Im}(\lambda_{b_r}) \neq 0$ which are identically 0 in a neighborhood of a_r.

The next theorem characterizes the 2-interval boundary conditions which generate the symmetric operators in the direct sum Hilbert space

$$H = (L^2(J_1, w_1) \dotplus L^2(J_2, w_2), \langle \cdot, \cdot \rangle)$$

with inner product

$$\langle \mathbf{f}, \mathbf{g} \rangle = h\ (f_1, g_1)_1 + k\ (f_2, g_2)_2, \quad h > 0, \ \ k > 0.$$

In other words the operators S in H which satisfy

$$S_{\min} \subset S \subset S^* \subset S_{\max}.$$

The self-adjoint operators $S = S^*$ are a special case.

THEOREM 13.1.1. *Suppose M_r, $r = 1, 2$, are symmetric expressions of order n, $n > 1$, even or odd, with complex coefficients. Let $a_r < c_r < b_r$, the deficiency indices of M_r on (a_r, c_r), (c_r, b_r) be d_{a_r}, d_{b_r}, respectively, and the deficiency index of M_r on (a_r, b_r) be $d_r = d_{a_r} + d_{b_r} - n$. Then the deficiency index d of the two-interval minimal operator $S_{\min}$ is $d = d_1 + d_2$. A linear submanifold $D(S)$ of $D_{\max}$ is the domain of a symmetric extension S of $S_{\min}$ if and only if there exist functions $\mathbf{w}_1, \mathbf{w}_2, \cdots, \mathbf{w}_l$ $(d \leq l \leq 2d)$ in $D_{\max}$ satisfying the following conditions:*

(i) $\mathbf{w}_1, \mathbf{w}_2, \cdots, \mathbf{w}_l$ *are linearly independent modulo* $D_{\min}$*;*
(ii) $\text{rank}\,\mathbf{W} = 2(l - d)$, *where*

$$\mathbf{W} = \begin{pmatrix} [\mathbf{w}_1, \mathbf{w}_1] & [\mathbf{w}_2, \mathbf{w}_1] & \cdots & [\mathbf{w}_l, \mathbf{w}_1] \\ \cdots & \cdots & \cdots & \cdots \\ [\mathbf{w}_1, \mathbf{w}_l] & [\mathbf{w}_2, \mathbf{w}_l] & \cdots & [\mathbf{w}_l, \mathbf{w}_l] \end{pmatrix};$$

(iii)

$$D(S)=\{\mathbf{y}=\{y_1,y_2\}\in D_{\max}:\ [\mathbf{y},\mathbf{w}_j]=0,\quad j=1,2,\cdots,l\}. \tag{13.1.2}$$

PROOF. *Sufficiency.* Assume that there exist $\mathbf{w}_1,\mathbf{w}_2,\cdots,\mathbf{w}_l$ $(d\le l\le 2d)$ in $D_{\max}$ satisfying conditions (i) and (ii). Now we prove that $D(S)$ given by (13.1.2) is the domain of a symmetric extension S of $S_{\min}$.

By the decomposition of the maximal domain given in (13.1.1), for $j=1,2,\ldots,l$, $\mathbf{w}_j=\{w_{1j},w_{2j}\}$ can be represented as

$$w_{rj}=w_{rj,0}+a_{rj,1}u_{r1}+a_{rj,2}u_{r2}+\cdots+a_{rj,m_{a_r}}u_{rm_{a_r}}+b_{rj,1}v_{r1}+\cdots+b_{rj,m_{b_r}}v_{rm_{b_r}},\quad r=1,2, \tag{13.1.3}$$

where $a_{rj,1},\cdots,a_{rj,m_{a_r}},b_{rj,1},\cdots,b_{rj,m_{b_r}}\in\mathbb{C}$ and $w_{rj,0}\in D_{r\min}$.

For $r=1,2$, $j=1,2,\cdots,l$, and any $\mathbf{y}=\{y_1,y_2\}\in D(S)$,

$$[y_r,w_{rj}]_r(a_r)=\overline{a}_{rj,1}[y_r,u_{r1}]_r(a_r)+\overline{a}_{rj,2}[y_r,u_{r2}]_r(a_r)+\cdots+\overline{a}_{rj,m_{a_r}}[y_r,u_{rm_{a_r}}]_r(a_r),$$

$$[y_r,w_{rj}]_r(b_r)=\overline{b}_{rj,1}[y_r,v_{r1}]_r(b_r)+\overline{b}_{rj,2}[y_r,v_{r2}]_r(b_r)+\cdots+\overline{b}_{rj,m_{b_r}}[y_r,v_{rm_{b_r}}]_r(b_r).$$

By $\mathbf{y}\in D(S)$, i.e. $[\mathbf{y},\mathbf{w}_j]=0,\quad j=1,2,\cdots,l$, we have

$$\begin{aligned}
&h\begin{pmatrix}[y_1,w_{11}]_1(b_1)\\ [y_1,w_{12}]_1(b_1)\\ \vdots\\ [y_1,w_{1l}]_1(b_1)\end{pmatrix}-h\begin{pmatrix}[y_1,w_{11}]_1(a_1)\\ [y_1,w_{12}]_1(a_1)\\ \vdots\\ [y_1,w_{1l}]_1(a_1)\end{pmatrix}\\
&\quad+k\begin{pmatrix}[y_2,w_{21}]_2(b_2)\\ [y_2,w_{22}]_2(b_2)\\ \vdots\\ [y_2,w_{2l}]_2(b_2)\end{pmatrix}-k\begin{pmatrix}[y_2,w_{21}]_2(a_2)\\ [y_2,w_{22}]_2(a_2)\\ \vdots\\ [y_2,w_{2l}]_2(a_2)\end{pmatrix}\\
&=B_1\begin{pmatrix}[y_1,v_{11}](b_1)\\ \vdots\\ [y_1,v_{1m_{b_1}}](b_1)\end{pmatrix}+A_1\begin{pmatrix}[y_1,u_{11}](a_1)\\ \vdots\\ [y_1,u_{1m_{a_1}}](a_1)\end{pmatrix}\\
&\quad+B_2\begin{pmatrix}[y_2,v_{21}](b_2)\\ \vdots\\ [y_2,v_{2m_{b_2}}](b_2)\end{pmatrix}+A_2\begin{pmatrix}[y_2,u_{21}](a_2)\\ \vdots\\ [y_2,u_{2m_{a_2}}](a_2)\end{pmatrix}=0,
\end{aligned}$$

where

$$A_1=-h\begin{pmatrix}\overline{a}_{11,1}&\cdots&\overline{a}_{11,m_{a_1}}\\ \overline{a}_{12,1}&\cdots&\overline{a}_{12,m_{a_1}}\\ \cdots&\cdots&\cdots\\ \overline{a}_{1l,1}&\cdots&\overline{a}_{1l,m_{a_1}}\end{pmatrix},\quad B_1=h\begin{pmatrix}\overline{b}_{11,1}&\cdots&\overline{b}_{11,m_{b_1}}\\ \overline{b}_{12,1}&\cdots&\overline{b}_{12,m_{b_1}}\\ \cdots&\cdots&\cdots\\ \overline{b}_{1l,1}&\cdots&\overline{b}_{1l,m_{b_1}}\end{pmatrix},$$

$$A_2=-k\begin{pmatrix}\overline{a}_{21,1}&\cdots&\overline{a}_{21,m_{a_2}}\\ \overline{a}_{22,1}&\cdots&\overline{a}_{22,m_{a_2}}\\ \cdots&\cdots&\cdots\\ \overline{a}_{2l,1}&\cdots&\overline{a}_{2l,m_{a_2}}\end{pmatrix},\quad B_2=k\begin{pmatrix}\overline{b}_{21,1}&\cdots&\overline{b}_{21,m_{b_2}}\\ \overline{b}_{22,1}&\cdots&\overline{b}_{22,m_{b_2}}\\ \cdots&\cdots&\cdots\\ \overline{b}_{2l,1}&\cdots&\overline{b}_{2l,m_{b_2}}\end{pmatrix}.$$

Therefore the boundary condition (13.1.2) of $D(S)$ is equivalent to

$$A_1Y(a_1)+B_1Y(b_1)+A_2Y(a_2)+B_2Y(b_2)=0,$$

i.e.

$$(A_1 : B_1 : A_2 : B_2)\mathbf{Y} = U\mathbf{Y} = 0,$$

where $Y(a_r)$, $Y(b_r)$ and $\mathbf{Y}$ are defined in Definition 12.3.1. Using the symmetric Theorem 12.3.6, we need to prove that matrices A_1, A_2, B_1 and B_2 satisfy the following two conditions:

(1) $\text{rank}U = \text{rank}(A_1 : B_1 : A_2 : B_2) = l$;

(2)

$$\begin{aligned}\text{rank}C &= \text{rank}(hkUP^{-1}U^*)\\ &= \text{rank}\,(kA_1E_{m_{a_1}}A_1^* - kB_1E_{m_{b_1}}B_1^* + hA_2E_{m_{a_2}}A_2^* - hB_2E_{m_{b_2}}B_2^*)\\ &= 2(l-d).\end{aligned}$$

It is obvious that $\text{rank}U \leq l$. If $\text{rank}U < l$, then there exist constants $h_1, h_2, \cdots, h_l$, not all zero, such that

$$(h_1, h_2, \cdots, h_l)U = 0.$$

Hence

$$(h_1, h_2, \cdots, h_l)A_r = 0, \quad r = 1, 2, \tag{13.1.4}$$

$$(h_1, h_2, \cdots, h_l)B_r = 0, \quad r = 1, 2. \tag{13.1.5}$$

Let $\mathbf{z} = \{z_1, z_2\} = \overline{h}_1\mathbf{w}_1 + \overline{h}_2\mathbf{w}_2 + \cdots + \overline{h}_l\mathbf{w}_l$. It follows from (13.1.3) that

$$z_r = \sum_{j=1}^{l} \overline{h}_j w_{rj,0} + \sum_{j=1}^{m_{a_r}}\sum_{i=1}^{l} \overline{h}_i a_{ri,j} u_{rj} + \sum_{j=1}^{m_{b_r}}\sum_{i=1}^{l} \overline{h}_i b_{ri,j} v_{rj}, \quad r = 1, 2.$$

By (13.1.4) and (13.1.5), we have

$$\sum_{j=1}^{m_{a_r}}\sum_{i=1}^{l} \overline{h}_i a_{ri,j} u_{rj} = 0, \quad \sum_{j=1}^{m_{b_r}}\sum_{i=1}^{l} \overline{h}_i b_{ri,j} v_{rj} = 0, \quad r = 1, 2.$$

Therefore $z_r = \sum_{j=1}^{l} \overline{h}_j w_{rj,0}$ and then $\mathbf{z} \in D_{\min}$ which contradicts that $\mathbf{w}_1, \mathbf{w}_2, \cdots, \mathbf{w}_l$ are linearly independent modulo $D_{\min}$. So $\text{rank}\,U = l$.

It follows from (13.1.3) that for any $i, j = 1, 2, \cdots, l$,

$$\begin{aligned}&[w_{ri}, w_{rj}]_r(b_r) = (b_{ri,1}, b_{ri,2}, \cdots, b_{ri,m_{b_r}})\\ &\quad\cdot\begin{pmatrix} [v_{r1}, v_{r1}](b_r) & \cdots & [v_{r1}, v_{rm_{b_r}}](b_r) \\ \cdots & \cdots & \cdots \\ [v_{rm_{b_r}}, v_{r1}](b_r) & \cdots & [v_{rm_{b_r}}, v_{rm_{b_r}}](b_r) \end{pmatrix} \begin{pmatrix} \overline{b}_{rj,1} \\ \overline{b}_{rj,2} \\ \vdots \\ \overline{b}_{rj,m_{b_r}} \end{pmatrix}.\end{aligned}$$

Therefore

$$\begin{aligned}&\begin{pmatrix} [w_{11}, w_{11}](b_1) & \cdots & [w_{1l}, w_{11}](b_1) \\ \cdots & \cdots & \cdots \\ [w_{11}, w_{1l}](b_1) & \cdots & [w_{1l}, w_{1l}](b_1) \end{pmatrix}\\ =&\frac{B_1}{h} \begin{pmatrix} [v_{11}, v_{11}](b_1) & \cdots & [v_{1m_{b_1}}, v_{11}](b_1) \\ \cdots & \cdots & \cdots \\ [v_{11}, v_{1m_{b_1}}](b_1) & \cdots & [v_{1m_{b_1}}, v_{1m_{b_1}}](b_1) \end{pmatrix} \frac{B_1^*}{h}\\ =&\frac{1}{h^2}B_1E_{m_{b_1}}B_1^*.\end{aligned}$$

Similarly, we have

$$\begin{pmatrix} [w_{21},w_{21}](b_2) & \cdots & [w_{2l},w_{21}](b_2) \\ \cdots & \cdots & \cdots \\ [w_{21},w_{2l}](b_2) & \cdots & [w_{2l},w_{2l}](b_2) \end{pmatrix} = \frac{1}{k^2} B_2 E_{m_{b_2}} B_2^*.$$

$$\begin{pmatrix} [w_{11},w_{11}](a_1) & \cdots & [w_{1l},w_{11}](a_1) \\ \cdots & \cdots & \cdots \\ [w_{11},w_{1l}](a_1) & \cdots & [w_{1l},w_{1l}](a_1) \end{pmatrix} = \frac{1}{h^2} A_1 E_{m_{a_1}} A_1^*,$$

$$\begin{pmatrix} [w_{21},w_{21}](a_2) & \cdots & [w_{2l},w_{21}](a_2) \\ \cdots & \cdots & \cdots \\ [w_{21},w_{2l}](a_2) & \cdots & [w_{2l},w_{2l}](a_2) \end{pmatrix} = \frac{1}{k^2} A_2 E_{m_{a_2}} A_2^*.$$

Hence the matrix

$$\begin{aligned} \mathbf{W} &= \begin{pmatrix} [\mathbf{w}_1,\mathbf{w}_1] & \cdots & [\mathbf{w}_l,\mathbf{w}_1] \\ \cdots & \cdots & \cdots \\ [\mathbf{w}_1,\mathbf{w}_l] & \cdots & [\mathbf{w}_l,\mathbf{w}_l] \end{pmatrix} \\ &= \frac{1}{h} B_1 E_{m_{b_1}} B_1^* - \frac{1}{h} A_1 E_{m_{a_1}} A_1^* + \frac{1}{k} B_2 E_{m_{b_2}} B_2^* - \frac{1}{k} A_2 E_{m_{a_2}} A_2^*. \end{aligned}$$

Note that $\mathbf{w}_1, \mathbf{w}_2, \cdots, \mathbf{w}_l$ satisfy condition (ii). Thus we have

$$\operatorname{rank}[kA_1E_{m_{a_1}}A_1^* - kB_1E_{m_{b_1}}B_1^* + hA_2E_{m_{a_2}}A_2^* - hB_2E_{m_{b_2}}B_2^*] = \operatorname{rank}\mathbf{W} = 2(l-d).$$

By Theorem 12.3.6, the operator $S(U)$ determined by boundary conditions $(A_1 : B_1 : A_2 : B_2)\mathbf{Y} = 0$ is symmetric. By the equivalence of (13.1.2) and $(A_1 : B_1 : A_2 : B_2)\mathbf{Y} = 0$, we know that $D(S)$ determined by (13.1.2) is the domain of a symmetric extension S of $S_{\min}$.

Necessity. Let a linear submanifold $D(S)$ of $D_{\max}$ be the domain of a symmetric extension S of $S_{\min}$. In the following we prove that there exist $\mathbf{w}_1, \mathbf{w}_2, \cdots, \mathbf{w}_l$ $(d \le l \le 2d)$ in $D_{\max}$ satisfying conditions (i), (ii) and $D(S)$ can be characterized by (13.1.2).

Since S is a symmetric operator, by Theorem 12.3.6, there exists a matrix $U = (A_1 : B_1 : A_2 : B_2)$ of order $l \times 2d$ such that $\operatorname{rank} U = l$ and

(13.1.6) $$\operatorname{rank}\left(kA_1E_{m_{a_1}}A_1^* - kB_1E_{m_{b_1}}B_1^* + hA_2E_{m_{a_2}}A_2^* - hB_2E_{m_{b_2}}B_2^*\right) = 2(l-d),$$

where $d \le l \le 2d$ and matrices A_r, B_r are of order $l \times m_{a_r}$, $l \times m_{b_r}$, $r = 1, 2$, respectively. And then the symmetric domain $D(S)$ can be characterized by

(13.1.7) $$D(S) = \{\mathbf{y} \in D_{\max} : U\mathbf{Y} = 0\}.$$

Let

$$A_1 = -h \begin{pmatrix} \bar{a}_{11,1} & \cdots & \bar{a}_{11,m_{a_1}} \\ \cdots & \cdots & \cdots \\ \bar{a}_{1l,1} & \cdots & \bar{a}_{1l,m_{a_1}} \end{pmatrix}, \quad B_1 = h \begin{pmatrix} \bar{b}_{11,1} & \cdots & \bar{b}_{11,m_{b_1}} \\ \cdots & \cdots & \cdots \\ \bar{b}_{1l,1} & \cdots & \bar{b}_{1l,m_{b_1}} \end{pmatrix},$$

$$A_2 = -k \begin{pmatrix} \bar{a}_{21,1} & \cdots & \bar{a}_{21,m_{a_2}} \\ \cdots & \cdots & \cdots \\ \bar{a}_{2l,1} & \cdots & \bar{a}_{2l,m_{a_2}} \end{pmatrix}, \quad B_2 = k \begin{pmatrix} \bar{b}_{21,1} & \cdots & \bar{b}_{21,m_{b_2}} \\ \cdots & \cdots & \cdots \\ \bar{b}_{2l,1} & \cdots & \bar{b}_{2l,m_{b_2}} \end{pmatrix},$$

and let

(13.1.8) $$w_{rj} = \sum_{i=1}^{m_{a_r}} a_{rj,i}u_{ri} + \sum_{i=1}^{m_{b_r}} b_{rj,i}v_{ri}, \quad j = 1, 2, \cdots, l, \quad r = 1, 2.$$

It is obvious that $\mathbf{w}_j = \{w_{1j}, w_{2j}\} \in D_{\max}$, $j = 1, 2, \cdots, l$. By a direct computation, we have

$$\begin{pmatrix} [y_1, w_{11}](a_1) \\ \vdots \\ [y_1, w_{1l}](a_1) \end{pmatrix} = -\frac{A_1}{h} \begin{pmatrix} [y_1, u_{11}](a_1) \\ \vdots \\ [y_1, u_{1m_{a_1}}](a_1) \end{pmatrix},$$

$$\begin{pmatrix} [y_1, w_{11}](b_1) \\ \vdots \\ [y_1, w_{1l}](b_1) \end{pmatrix} = \frac{B_1}{h} \begin{pmatrix} [y_1, v_{11}](b_1) \\ \vdots \\ [y_1, v_{1m_{b_1}}](b_1) \end{pmatrix},$$

$$\begin{pmatrix} [y_2, w_{21}](a_2) \\ \vdots \\ [y_2, w_{2l}](a_2) \end{pmatrix} = -\frac{A_2}{k} \begin{pmatrix} [y_2, u_{21}](a_2) \\ \vdots \\ [y_2, u_{2m_{a_2}}](a_2) \end{pmatrix},$$

$$\begin{pmatrix} [y_2, w_{21}](b_2) \\ \vdots \\ [y_2, w_{2l}](b_2) \end{pmatrix} = \frac{B_2}{k} \begin{pmatrix} [y_2, v_{21}](b_2) \\ \vdots \\ [y_2, v_{2m_{b_2}}](b_2) \end{pmatrix}.$$

Then

$$h \begin{pmatrix} [y_1, w_{11}](b_1) \\ \vdots \\ [y_1, w_{1l}](b_1) \end{pmatrix} - h \begin{pmatrix} [y_1, w_{11}](a_1) \\ \vdots \\ [y_1, w_{1l}](a_1) \end{pmatrix} = A_1 Y(a_1) + B_1 Y(b_1),$$

$$k \begin{pmatrix} [y_2, w_{21}](b_2) \\ \vdots \\ [y_2, w_{2l}](b_2) \end{pmatrix} - k \begin{pmatrix} [y_2, w_{21}](a_2) \\ \vdots \\ [y_2, w_{2l}](a_2) \end{pmatrix} = A_2 Y(a_2) + B_2 Y(b_2).$$

Therefore

$$\begin{aligned} &\begin{pmatrix} [\mathbf{y}, \mathbf{w}_1] \\ \vdots \\ [\mathbf{y}, \mathbf{w}_l] \end{pmatrix} \\ =& A_1 Y(a_1) + B_1 Y(b_1) + A_2 Y(a_2) + B_2 Y(b_2) \\ =& (A_1 : B_1 : A_2 : B_2)\mathbf{Y} = U\mathbf{Y}. \end{aligned}$$

This shows that (13.1.2) is equivalent to (13.1.7).

In fact, $\mathbf{w}_1, \mathbf{w}_2, \cdots, \mathbf{w}_l$ are linearly independent modulo $D_{\min}$. If not, there exist constants $h_1, h_2, \cdots, h_l$, not all zero, such that

$$\gamma = \{\gamma_1, \gamma_2\} = h_1 \mathbf{w}_1 + h_2 \mathbf{w}_2 + \cdots + h_l \mathbf{w}_l \in D_{\min}.$$

Then

$$[u_{rj}, \gamma_r](a_r) = 0, \quad j = 1, 2, \cdots, m_{a_r}, \quad r = 1, 2$$

and

$$[v_{rj}, \gamma_r](b_r) = 0, \quad j = 1, 2, \cdots, m_{b_r}, \quad r = 1, 2.$$

It follows from (13.1.8) that

$$\gamma_r = \sum_{j=1}^{l} h_j \Big(\sum_{i=1}^{m_{a_r}} a_{rj,i} u_{ri} + \sum_{i=1}^{m_{b_r}} b_{rj,i} v_{ri}\Big), \quad r = 1, 2.$$

It can be verified that

$$\begin{pmatrix} [u_{11},\gamma_1](a_1) \\ [u_{12},\gamma_1](a_1) \\ \vdots \\ [u_{1m_{a_1}},\gamma_1](a_1) \end{pmatrix} = 0$$

is equivalent to

$$(\overline{h}_1,\overline{h}_2,\cdots,\overline{h}_l)(-\frac{A_1}{h})E_{m_{a_1}} = 0.$$

Therefore $(\overline{h}_1,\overline{h}_2,\cdots,\overline{h}_l)A_1 = 0$. Similarly, we have $(\overline{h}_1,\overline{h}_2,\cdots,\overline{h}_l)A_2 = 0$, and $(\overline{h}_1,\overline{h}_2,\cdots,\overline{h}_l)B_r = 0$, $r = 1,2$. Therefore

$$(\overline{h}_1,\overline{h}_2,\cdots,\overline{h}_l)(A_1 : B_1 : A_2 : B_2) = (\overline{h}_1,\overline{h}_2,\cdots,\overline{h}_l)U = 0.$$

This contradicts with $\operatorname{rank} U = l$. Therefore $\mathbf{w}_1,\mathbf{w}_2,\cdots,\mathbf{w}_l$ are linearly independent modulo $D_{\min}$.

By (13.1.8), we can obtain that

$$\begin{aligned}\mathbf{W} &= \begin{pmatrix} [\mathbf{w}_1,\mathbf{w}_1] & \cdots & [\mathbf{w}_l,\mathbf{w}_1] \\ \cdots & \cdots & \cdots \\ [\mathbf{w}_1,\mathbf{w}_l] & \cdots & [\mathbf{w}_l,\mathbf{w}_l] \end{pmatrix} \\ &= \frac{1}{h}B_1E_{m_{b_1}}B_1^* - \frac{1}{h}A_1E_{m_{a_1}}A_1^* + \frac{1}{k}B_2E_{m_{b_2}}B_2^* - \frac{1}{k}A_2E_{m_{a_2}}A_2^*.\end{aligned}$$

It follows from (13.1.6) that rank $\mathbf{W} = \operatorname{rank}(-hk\mathbf{W}) = 2(l-d)$. This completes the proof. □

The self-adjoint 2-interval theorem is a special case:

COROLLARY 13.1.1. *Let the hypotheses and notation of Theorem* 13.1.1 *hold. In particular, if $l = d$, then we have the following result:*

A linear submanifold $D(S)$ of $D_{\max}$ is the domain of a self-adjoint extension S of $S_{\min}$ if and only if there exist functions $\mathbf{w}_1,\mathbf{w}_2,\cdots,\mathbf{w}_d$ in $D_{\max}$ such that the following conditions hold:

(i) *$\mathbf{w}_1,\mathbf{w}_2,\cdots,\mathbf{w}_d$ are linearly independent modulo $D_{\min}$;*
(ii) *$[\mathbf{w}_i,\mathbf{w}_j] = 0,\ \ i,j = 1,2,\cdots,d$;*
(iii) *$D(S) = \{\mathbf{y} \in D_{\max} : [\mathbf{y},\mathbf{w}_j] = 0,\ \ j = 1,2,\cdots,d\}$.*

2. Comments

As mentioned in the Preface, in the 1-interval self-adjoint case, the GKN theorem was proven first and then 'applied' to characterize the 1-interval self-adjoint boundary conditions. In the 2-interval case the characterization of the symmetric boundary conditions given by Theorem 12.3.6 was used to prove the 2-interval symmetric characterization in terms of maximal domain functiions given by Theorem 13.1.1 which includes the self-adjoint case. The authors were surprised when they discovered that the characterization of the symmetric boundary conditions in terms of Lagrange brackets evaluated at the endpoints used in Theorem 12.3.6 can be used to prove the symmetric characterization in terms of maximal domain functions given by Theorem 13.1.1. This is published here for the first time.

Part 4

Other Topics

CHAPTER 14

Green's Function and Adjoint Problems

1. Introduction

In this chapter we construct the Green's function of regular and singular 1-interval self-adjoint and non-self-adjoint boundary value problems consisting of general quasi-differential equations on an open, bounded or unbounded, interval (a, b) of the real line. In contrast to the usual construction of singular such problems as, for example, in the well known book by Coddington and Levinson [**102**], which involves a selection theorem to select a sequence of regular Green's functions on truncated intervals whose limit as the truncated endpoints approach the singular endpoints is the Green's function of the singular problem, our construction is direct, elementary and explicit in terms of solutions. Our method is based on a simple transformation of the dependent variable which leaves the underlying interval unchanged and transforms the singular problem with limit-circle (LC) endpoints into a regular problem. Necessary and sufficient conditions for two singular problems to be adjoint to each other and, in particular, self-adjoint, then follow from the regular case. As an illustration we construct the Green's function of the classical Legendre equation with arbitrary separated or coupled self-adjoint boundary conditions.

2. Adjoint Matrices and Green's Functions for Regular Systems

In this section we review the concept of 'Lagrange Adjoint' for first order systems and establish the corresponding 'Lagrange Hermitian' properties of their Green's matrices. Fundamental to this analysis is the 'Adjointness Lemma' (see below) and the above mentioned construction of the Green's matrices. The next section will contain applications of these results to scalar problems and the following sections will discuss singular systems and singular scalar problems.

For $P \in M_n(L_{loc}(J))$ recall from Theorem 1.2.1 that for each point u of J there is exactly one matrix solution X of (1.2.5) satisfying $X(u) = I_n$ where I_n denotes the $n \times n$ identity matrix. Also recall the definition of the primary fundamental matrix from Section 1.3 which we repeat here for the benefit of the reader.

DEFINITION 14.2.1 (Primary Fundamental Matrix). For each fixed $u \in J$ let $\Phi(\cdot, u)$ be the fundamental matrix of (1.2.5) satisfying

$$\Phi(u, u) = I_n\,.$$

Note that for each fixed u in J, $\Phi(\cdot, u)$ belongs to $M_n(AC_{loc}(J))$. Furthermore, if J is compact and $P \in M_n(L(J, \mathbb{C}))$, then u can be an endpoint of J and $\Phi(\cdot, u)$ belongs to $M_n(AC(J))$. By Theorem 1.2.2, $\Phi(t, u)$ is invertible for each $t, u \in J$ and we note that

$$\Phi(t, u) = Y(t)Y^{-1}(u)$$

for any fundamental matrix Y of (1.2.5).

We call Φ the primary fundamental matrix of the system $Y' = PY$, or just the primary fundamental matrix of P and also write

$$\Phi = \Phi(P) = (\Phi_{rs})_{r,s=1}^{n}, \quad \Phi(P)(t,u) = \Phi(t,u,P).$$

Observe that for any constant $n \times m$ matrix C, ΦC is also a solution of (1.2.5). If C is a constant nonsingular $n \times n$ matrix then ΦC is a fundamental matrix solution and every fundamental matrix solution has this form.

Next we discuss some properties of the primary fundamental matrix.

LEMMA 14.2.1. *Let $P \in M_n(L_{loc}(J))$ and let $\Phi(t,s)$ be the primary fundamental matrix of P. Then for any $t, s, u \in J$ we have*

$$\Phi(t,u)\,\Phi(u,s) = \Phi(t,s). \tag{14.2.1}$$

Furthermore, if $P \in L(J)$ then (14.2.1) *holds also when t, s, u are equal to a or b. Here $a = -\infty$ and $b = \infty$ are allowed.*

PROOF. This follows from the well known representation $\Phi(t,s) = Y(t)Y^{-1}(s)$ where Y is any fundamental matrix of $Y' = PY$. For the furthermore statement see Section 1.5. □

LEMMA 14.2.2. *Let $P \in M_n(L_{loc}(J))$ and let $\Phi(t,s)$ be the primary fundamental matrix of P. Suppose trace$P(t) = 0$, for $t \in J$. Then*

$$\det(\Phi(t,s)) = 1, \quad t, s \in J.$$

PROOF. Fix $s \in J$. It is well known that $(\det \Phi(t,s))' = trace(P(t)) \det(\Phi(t,s))$ and the result follows. □

Next we establish a basic relationship between the primary fundamental matrices of P and its Lagrange adjoint matrix $P^+ = -E^{-1}P^*E$. Recall the definition of the symplectic matrix E in Section 3.2 and its basic properties:

DEFINITION 14.2.2. For $k \in \mathbb{N}_2$ let E_k be defined by

$$E_k = ((-1)^r \delta_{r,k+1-s})_{r,s=1}^{k}, \tag{14.2.2}$$

where $\delta_{i,j}$ is the Kronecker δ. Note that

$$E_k^* = E_k^{-1} = (-1)^{k+1} E_k. \tag{14.2.3}$$

LEMMA 14.2.3 (Adjointness Lemma). *Suppose $P \in M_n(L_{loc}(J))$. Then $P^+ \in M_n(L_{loc}(J))$. If $\Phi(t,s) = \Phi(t,s,P)$ and $\Psi(t,s) = \Phi(t,s,P^+)$ are the primary fundamental matrices of P and P^+, respectively, then*

$$\Psi(t,s) = E^{-1}\Phi^*(s,t)E, \quad \Phi(t,s) = E^{-1}\Psi^*(s,t)E, \quad t, s \in J. \tag{14.2.4}$$

Furthermore, if $P \in L(J)$ then $P^+ \in L(J)$ and (14.2.1), (14.2.4) *hold for $a \leq s,t \leq b$, even when $a = -\infty$ or $b = +\infty$.*

PROOF. Fix $s \in J$ and let $X(t) = E^{-1*}\Phi^*(t,s)E^*\Psi(t,s)$, for $t \in J$. Then $X(s) = I$ and for all $t \in J$

$$\begin{aligned}
X'(t) &= E(\Phi^*)'(t,s)E^*\Psi(t,s) + E\Phi^*(t,s)E^*\Psi'(t,s) \\
&= E(P(t)\Phi(t,s))^*E^*\Psi(t,s) + E\Phi^*(t,s)E^*P^+(t)\Psi(t,s) \\
&= E\Phi^*(t,s)P^*(t)E^*\Psi(t,s) + E\Phi^*(t,s)E^*(-E^{-1}P^*(t)E)\Psi(t,s) \\
&= E\Phi^*(t,s)P^*(t)E^*\Psi(t,s) - E\Phi^*(t,s)(-1)^{n+1}EE^{-1}P^*(t)(-1)^{n+1}E^*\Psi(t,s) \\
&= 0.
\end{aligned}$$

Hence, for all $t \in J$, $X(t) = I$ and

$$\Psi(t,s) = E^{-1}(\Phi^*)^{-1}(t,s)E = E^{-1}(\Phi^{-1})^*(t,s)E = E^{-1}\Phi^*(s,t)E.$$

In the last step we used $\Phi^{-1}(t,s) = \Phi(s,t)$ which follows from (14.2.1). This completes the proof of the first part of the Lemma and the second part of (14.2.4) follows from the first part and (14.2.3). The furthermore statement follows by taking limits as t, s approach a or b. These limits exist and are finite by Section 1.5. □

For $\lambda \in \mathbb{C}$ consider the following vector matrix boundary value problem

$$Y' = (P - \lambda W)Y, \quad AY(a) + BY(b) = 0, \quad A, B \in M_n(\mathbb{C}). \tag{14.2.5}$$

We now construct the Green's matrix for regular systems (14.2.5).

THEOREM 14.2.1. *Assume that $P, W \in M_n(L(J))$. Let $\lambda \in \mathbb{C}$ and let $\Phi(t,s,\lambda)$ be the primary fundamental matrix of $P - \lambda W$.*

(1) *The homogeneous boundary value problem* (14.2.5) *has a nontrivial solution if and only if*

$$\det[A + B\,\Phi(b,a,\lambda)] = 0.$$

(2) *If* $\det[A + B\,\Phi(b,a,\lambda)] \neq 0$, *then for every $F \in L(J)$ the inhomogeneous problem*

$$Y' = (P - \lambda W)Y + F, \quad AY(a) + BY(b) = 0$$

has a unique solution Y given by

$$Y(t) = \int_a^b K(t,s,\lambda)F(s)\mathrm{ds}, \quad a \le t \le b,$$

where

$$K(t,s,\lambda) = \begin{cases} \Phi(t,a,\lambda)U\,\Phi(a,s,\lambda), & a \le t < s \le b, \\ \Phi(t,a,\lambda)U\,\Phi(a,s,\lambda) + \Phi(t,s,\lambda), & a \le s < t \le b, \\ \Phi(t,a,\lambda)U\,\Phi(a,s,\lambda) + \frac{1}{2}\Phi(t,s,\lambda), & a \le s = t \le b \end{cases} \tag{14.2.6}$$

with

$$U = -[A + B\,\Phi(b,a,\lambda)]^{-1}B\,\Phi(b,a,\lambda). \tag{14.2.7}$$

We call $K(t,s,\lambda) = K(t,s,\lambda,P,W,A,B)$ the Green's matrix of the boundary value problem (14.2.5).

PROOF. The construction given in Section 4.11 of the book [**620**] for $n = 2$ can be directly applied here. See also Coddington and Zettl [**103**], and [**609**], [**594**], [**596**]. □

Consider the boundary value problem

$$Z' = (P - \lambda W)^+ Z, \quad CZ(a) + DZ(b) = 0, \quad C, D \in M_n(\mathbb{C}). \tag{14.2.8}$$

and its Green's matrix $L(t,s,\overline{\lambda}) = L(t,s,\overline{\lambda},P^+,W^+,C,D)$ given by

$$L(t,s,\overline{\lambda}) = \begin{cases} \Psi(t,a,\overline{\lambda})V\Psi(a,s,\overline{\lambda}), & a \le t < s \le b, \\ \Psi(t,a,\overline{\lambda})V\Psi\,(a,s,\overline{\lambda}) + \Psi(t,s,\overline{\lambda}), & a \le s < t \le b, \\ \Psi(t,a,\overline{\lambda})V\Psi(a,s,\overline{\lambda}) + \frac{1}{2}\Psi(t,s,\overline{\lambda}), & a \le s = t \le b, \end{cases} \tag{14.2.9}$$

with

$$V = -[C + D\,\Psi(b,a,\overline{\lambda})]^{-1}D\,\Psi(b,a,\overline{\lambda}). \tag{14.2.10}$$

LEMMA 14.2.4. *Let $\lambda \in \mathbb{C}$. Suppose that $P, W \in M_n(L(J))$. Let $\Phi(t,s,\lambda)$, $\Psi(t,s,\overline{\lambda})$ be the primary fundamental matrices of $(P-\lambda W)$ and $(P-\lambda W)^+$, respectively. Let $K(t,s,\lambda) = K(t,s,\lambda,P,W,A,B)$ be the Green's matrix of the boundary value problem* (14.2.5) *and let $L(t,s,\overline{\lambda}) = L(t,s,\overline{\lambda},P^+,W^+,C,D)$ be the Green's matrix of the boundary value problem* (14.2.8). *Assume that*

$$\det[A + B\,\Phi(b,a,\lambda)] \neq 0 \neq \det[C + D\,\Psi(b,a,\overline{\lambda})]. \tag{14.2.11}$$

Then the Green's matrices $K(t,s,\lambda)$ and $L(t,s,\overline{\lambda})$ exist by Theorem 14.2.1 *and we have*

$$K(t,s,\lambda) + E^{-1}L^*(s,t,\overline{\lambda})\,E = \Phi(t,a,\lambda)\,\Gamma\,\Phi(a,s,\lambda), \quad a \leq s,t \leq b \tag{14.2.12}$$

with

$$\Gamma = U + E^{-1}V^*E + I.$$

PROOF. This follows from the above construction and notation as follows.

$$\begin{aligned} &K(t,s,\lambda) + E^{-1}L^*(s,t,\overline{\lambda})\,E - \Phi(t,a,\lambda)U\,\Phi(a,s,\lambda) \\ &= E^{-1}[\Psi(s,a,\overline{\lambda})V\,\Psi(a,t,\overline{\lambda})]^*E + E^{-1}\Psi^*(s,t,\overline{\lambda})\,E \\ &= E^{-1}[\Psi^*(a,t,\overline{\lambda})V^*\Psi^*(s,a,\overline{\lambda})]E + \Phi(t,s,\lambda) \\ &= \Phi(t,a,\lambda)E^{-1}V^*E\,\Phi(a,s,\lambda) + \Phi(t,a,\lambda)\,\Phi(a,s,\lambda). \end{aligned} \tag{14.2.13}$$

Similarly (14.2.13) also holds for the cases $a \leq s < t \leq b$ and $a \leq t = s \leq b$. (The fraction $\frac{1}{2}$ in the constructions (14.2.6) and (14.2.9) is used when $s = t$.) The theorem follows using (14.2.13) for $t > s$, $t < s$ and $t = s$. □

LEMMA 14.2.5. *Let the hypotheses and notation of Lemma* 14.2.4 *hold. Then*

$$K(t,s,\lambda) = -E^{-1}L^*(s,t,\overline{\lambda})\,E, \quad \textit{for all } a \leq s,t \leq b, \tag{14.2.14}$$

if and only if

$$U = -(E^{-1}V^*E + I). \tag{14.2.15}$$

PROOF. This follows from Lemma 14.2.4 by noting that (14.2.14) holds if and only if $\Gamma = 0$ and $\Gamma = 0$ is equivalent to (14.2.15). □

LEMMA 14.2.6. *Let the hypotheses and notation of Lemma* 14.2.4 *hold. Then*

$$K(t,s,\lambda) = +E^{-1}L^*(s,t,\overline{\lambda})\,E, \quad \textit{for all } a \leq s,t \leq b,$$

if and only if

$$U = +(E^{-1}V^*E + I). \tag{14.2.16}$$

PROOF. The proofs of Lemmas 14.2.4 and 14.2.5 can be easily adapted to prove this lemma. □

THEOREM 14.2.2. *Let the hypotheses and notation of Lemma* 14.2.4 *hold. Then the Green's matrices $K(t,s,\lambda)$ and $L(t,s,\overline{\lambda})$ exist by Theorem* 14.2.1 *and* (14.2.14) *holds if and only if*

$$AEC^* = BED^*. \tag{14.2.17}$$

PROOF. By Lemma 14.2.5 we only need to show that (14.2.15) holds. From (14.2.7) and (14.2.10) we have

$$\begin{aligned} -I =& U + E^{-1}V^*E \\ =& - [A + B\,\Phi(b,a,\lambda)]^{-1}B\,\Phi(b,a,\lambda) - \\ & E^{-1}\Psi^*(b,a,\overline{\lambda})D^*[(C + D\Psi(b,a,\overline{\lambda}))^{-1}]^*E. \end{aligned} \tag{14.2.18}$$

Multiplying (14.2.18) on the left by $-[A + B\,\Phi(b,a,\lambda)]$ and on the right by $E^{-1}[C + D\Psi(b,a,\overline{\lambda})]^*$ we obtain the equivalent identity

$$\begin{aligned} &[A + B\,\Phi(b,a,\lambda)]E^{-1}[C + D\Psi(b,a,\overline{\lambda})]^* \\ &= B\,\Phi(b,a,\lambda)\,E^{-1}[C + D\,\Psi(b,a,\overline{\lambda})]^* + [A + B\,\Phi(b,a,\lambda)]E^{-1}\Psi^*(b,a,\overline{\lambda})D^*. \end{aligned}$$

This simplifies to

$$\begin{aligned} AE^{-1}C^* &= B\,\Phi(b,a,\lambda)E^{-1}\Psi^*(b,a,\overline{\lambda})EE^{-1}D^* \\ &= B\,\Phi(b,a,\lambda)\Phi(a,b,\lambda)E^{-1}D^* \\ &= B\,E^{-1}D^* \end{aligned}$$

which is equivalent to (14.2.17). This completes the proof. Special cases of this theorem were obtained by Coddington-Zettl in [**103**], and in [**594**], [**596**]. □

THEOREM 14.2.3. *Let the hypotheses and notation of Lemma* 14.2.4 *hold. If* $U = V$ *then, for each* λ *satisfying* (14.2.11), *there exist* s, t, $a \le s, t \le b$ *such that*

$$K(t,s,\lambda) \neq +E^{-1}L^*(s,t,\overline{\lambda})E. \tag{14.2.19}$$

In particular, (14.2.19) *holds if* $P = P^+$, $W = W^+$, $A = C$, $B = D$ *and* $\lambda = \overline{\lambda}$ *satisfies* (14.2.11).

PROOF. By Lemma 14.2.6 we only need to show that (14.2.16) does not hold. In this case $U = V$. Assume (14.2.16) holds with $U = (u_{ij}) = V$. Then we have $u_{11} = \overline{u}_{nn} + 1$ and $u_{nn} = \overline{u}_{11} + 1$. Hence $1 = u_{11} - \overline{u}_{nn} = \overline{u}_{11} - u_{nn} = -1$. This contradiction completes the proof. □

3. Green's Functions of Regular Scalar Boundary Value Problems

Although in this chapter our primary focus is on singular problems, we now discuss regular problems and their Green's functions. The traditional construction of the Green's function $K(t,s,\lambda)$ of regular ordinary boundary value problems involves a recipe which, among other things, prescribes a jump discontinuity of the derivative with respect to t for fixed s when $t = s$, see e.g. Coddington and Levinson [**102**]. Here we construct $K(t,s,\lambda)$ directly in such a way that this jump discontinuity does not have to be prescribed a priori but occurs naturally. This is accomplished by converting the scalar problem to a system, constructing a Green's matrix for the system as in Section 14.2, then extracting the scalar Green's function from the Green's matrix. The jump discontinuities along the diagonal $t = s$ are clearly apparent from this construction. This and a number of other features make this construction more direct and, we believe, more "natural" than the traditional one found in textbooks. This construction is a modification of a construction used previously by Neuberger [**442**] for the second order case and by Zettl [**594**], [**596**] and by Coddington and Zettl [**103**] for higher orders. It seems not to be widely known and most current textbooks still use the 'recipe' construction mentioned above. Our construction of singular Green's functions in Section 14.5 is based on our construction of regular Green's functions, thus we give it here.

Throughout the remainder of this section we assume that $P = (p_{ij}) \in Z_n(J)$ satisfies

$$p_{ij} \in L(J),\ \ 1 \leq i \leq j,\ \ j = 1, 2, ..., n; \quad p_{j,j+1}^{-1} \in L(J),\ \ j = 1, 2, ..., n-1. \tag{14.3.1}$$

Let $w \in L(J)$ and let W be the $n \times n$ matrix

$$W := \begin{pmatrix} 0 & 0 & \dots & 0 \\ \vdots & \vdots & \dots & \vdots \\ 0 & 0 & \dots & 0 \\ w & 0 & \dots & 0 \end{pmatrix}. \tag{14.3.2}$$

For any $\lambda \in \mathbb{C}$ and $f \in L(J)$ we now consider the equation

$$My = \lambda wy + f \tag{14.3.3}$$

and its equivalent systems formulation

$$Y' = (P - \lambda W)Y + F$$

where $F = (0, \cdots, 0, f)^T$, $Y = (y, y^{[1]}, \cdots, y^{[n-1]})^T$ and $M = M_P$, the quasi-differential expression generated by P, is defined by:

$$My = M_P y = i^n y^{[n]} \quad \text{for} \quad y \in D(P).$$

For $A, B \in M_n(\mathbb{C})$ consider the two point boundary value problem (14.3.3) consisting with the boundary conditions

$$AY(a) + BY(b) = 0.$$

Note that this is a well defined problem since the quasi-derivatives $y^{[r]}$ exist as finite limits at both endpoints from Lemma 3.3.3.

THEOREM 14.3.1. *Let $\lambda \in \mathbb{C}$, let W be given by* (14.3.2) *with $w \in L(J)$. Assume that $P = (p_{ij}) \in Z_n(J)$ satisfies* (14.3.1). *Let $\Phi(t, s, \lambda)$ be the primary fundamental matrix of $P - \lambda W$. Then*

(1) *The homogeneous boundary value problem*

$$Y' = (P - \lambda W)Y, \quad AY(a) + BY(b) = 0, \quad A, B \in M_n(\mathbb{C}), \tag{14.3.4}$$

has a nontrivial solution if and only if

$$\det[A + B\,\Phi(b, a, \lambda)] = 0.$$

(2) *The homogeneous boundary value problem*

$$My = \lambda wy, \quad AY(a) + BY(b) = 0,$$

has a nontrivial solution if and only if

$$\det[A + B\,\Phi(b, a, \lambda)] = 0.$$

(3) *If* $\det[A + B\,\Phi(b, a, \lambda)] \neq 0$, *then for every $F \in L(J)$ the inhomogeneous problem*

$$Y' = (P - \lambda W)Y + F, \quad AY(a) + BY(b) = 0$$

has a unique solution Y *given by*

$$Y(t) = \int_a^b K(t,s,\lambda)F(s)\mathrm{d}s, \quad a \le t \le b,$$

where $K(t,s,\lambda)$ *is given by* (14.2.6).

(4) *Let* $K(t,s,\lambda) = (K_{ij}(t,s,\lambda))_{1\le i,j\le n}$. *If* $\det[A + B\,\Phi(b,a,\lambda)] \neq 0$, *then for every* $f \in L(J)$ *the inhomogeneous problem*

$$My = \lambda wy + f, \quad AY(a) + BY(b) = 0$$

has a unique solution y *given by*

$$y(t) = \int_a^b K_{1n}(t,s,\lambda)f(s)\mathrm{d}s, \quad a \le t \le b.$$

Furthermore K_{1n} *is continuous on* $[a,b]\times[a,b]$ *and is unique.*

PROOF. All four parts follow from Theorem 14.2.1. The continuity of K_{1n} is clear from (14.2.6) since $n > 1$ and the jump discontinuities of K occur only on the diagonal $s = t$. The proof of the uniqueness of K_{1n} is standard by using compact support function for f. □

Next we consider boundary value problems for the system

$$Z' = (P - \lambda W)^+ Z + F = (P^+ - \overline{\lambda} W^+)\,Z + F = (P^+ - (-1)^n\,\overline{\lambda W})\,Z + F \tag{14.3.5}$$

and its equivalent scalar equation

$$M^+ z = (-1)^n \overline{\lambda}\overline{w} z + f, \quad M^+ = M_{P^+}$$

both with boundary condition

$$CZ(a) + DZ(b) = 0, \quad C, D \in M_n(\mathbb{C}).$$

THEOREM 14.3.2. *Let* $\lambda \in \mathbb{C}$. *Let* $A, B, C, D \in M_n(\mathbb{C})$. *Assume that* $P = (p_{ij}) \in Z_n(J)$ *satisfies* (14.3.1) *and* W *is given by* (14.3.2) *with* $w \in L(J)$. *Let* E *be given by* (14.2.2). *Then* $P^+ \in Z_n(J)$ *and satisfies* (14.3.1). *Let* $\Phi(t,s,\lambda)$ *and* $\Psi(t,s,\overline{\lambda})$ *be the primary fundamental matrices of* $P - \lambda W$ *and* $(P - \lambda W)^+$, *respectively. If*

$$\det[A + B\,\Phi(b,a,\lambda)] \neq 0 \neq \det[C + D\,\Psi(b,a,\overline{\lambda})].$$

Then the Green's matrices $K(t,s,\lambda) = (K_{ij}(t,s,\lambda))$ *of* (14.3.4) *and* $L(t,s,\overline{\lambda}) = (L_{ij}(t,s,\overline{\lambda}))$ *of* (14.3.5) *exist by Theorem* 14.2.1 *and the following three statements are equivalent:*

(1)

$$K(t,s,\lambda) = -E^{-1}L^*(s,t,\overline{\lambda})E, \quad a \le s,t \le b; \tag{14.3.6}$$

(2)

$$K_{1,n}(t,s,\lambda) = (-1)^n \overline{L}_{1,n}(s,t,\overline{\lambda}), \quad a \le s,t \le b;$$

(3)

$$AEC^* = BED^*.$$

PROOF. From (14.2.12) and the nonsingularity of the primary fundamental matrix Φ it follows that (14.3.6) holds if and only if $\Gamma = 0$. By Theorem 14.2.2, (1) and (3) are equivalent. Clearly (1) implies (2). To show that (2) implies (1) we show that $\Gamma = 0$. This follows from (14.2.12) and the linear independence of $\phi_{1j}(t,a,\lambda)$, $j = 1,\cdots,n$ as functions of t and the linear independence of $\phi_{jn}(a,s,\lambda)$, $j = 1,\cdots,n$ as functions of s. Fix s and let $C(s) = \Gamma\Phi(a,s,\lambda)$. By (14.2.12) we have

$$\sum_{j=1}^{n} \phi_{1j}(t,a,\lambda)C_{jn}(s) = 0, \quad a \le t \le b.$$

Hence $C_{jn}(s) = 0$, $j = 1,\cdots,n$ by the linear independence of $\phi_{1j}(t,a,\lambda)$, $j = 1,\cdots,n$. Thus we have

$$C_{jn}(s) = \sum_{k=1}^{n} \Gamma_{jk}\phi_{kn}(a,s,\lambda) = 0, \quad a \le s \le b,$$

and from the linear independence of $\phi_{kn}(a,s,\lambda)$, $k = 1,\cdots,n$ as functions of s (see the Adjointness Lemma) we conclude that $\Gamma_{jk} = 0$, for $k = 1,\cdots,n$. Since this holds for each j we may conclude that $\Gamma_{jk} = 0$, for $j,k = 1,\cdots,n$ and this completes the proof of Theorem 14.3.2. Special cases of this theorem were proven in [**103**], [**594**], and [**596**]. □

THEOREM 14.3.3. *Let $A,B,C,D \in M_n(\mathbb{C})$. Assume that $P \in Z_n(J)$ satisfies* (14.3.1) *and W is given by* (14.3.2) *with $w \in L(J)$ and real valued. Suppose $P = P^+$, $W = W^+$, $A = C$, $B = D$ and $\lambda \in \mathbb{R}$. If $\det[A + B\,\Phi(b,a,\lambda)] \ne 0$, then there exist t, s, $a \le t, s \le b$ such that*

$$K_{1,n}(t,s,\lambda) \ne (-1)^{n+1}\overline{K}_{1,n}(s,t,\lambda). \tag{14.3.7}$$

PROOF. Note that $(P - \lambda W)^+ = P - \lambda W$ and thus (14.3.7) follows from Theorem 14.2.3. □

REMARK 14.3.1. Together, Theorems 14.3.2 and 14.3.3 say that for scalar problems when $(P-\lambda W)^+ = P-\lambda W$, $\lambda \in \mathbb{R}$ the Green's function cannot be symmetric when n is odd and it cannot be anti-symmetric when n is even.

4. Regularization of Singular Problems

In this section we show that singular scalar equations with limit-circle endpoints can be "regularized" in the sense that they can be transformed to regular problems. The endpoints may be finite or infinite and no oscillatory restrictions on the coefficients or solutions are assumed. Here the components of P and w are in $L_{loc}(J)$ but not necessarily in $L(J)$. This transformation transforms the dependent variable and leaves the independent variable and the domain interval unchanged.

Let $P \in Z_n(J)$, let W be given by (14.3.2) with $w \in L_{loc}(J)$, let $\lambda \in \mathbb{C}$. Then $P^+ \in Z_n(J)$. Let $M = M_P$, $M^+ = M_{P^+}$ and consider the scalar equations

$$My = \lambda w y \quad \text{on } J, \tag{14.4.1}$$

$$M^+ z = (-1)^n\,\overline{\lambda}\,\overline{w}\, z \quad \text{on } J, \tag{14.4.2}$$

and their system formulations

$$\begin{aligned} Y' &= (P - \lambda W)Y \quad \text{on } J, \\ Z' &= (P - \lambda W)^+ Z \quad \text{on } J. \end{aligned} \tag{14.4.3}$$

Let Φ, Ψ be the primary fundamental matrices of $(P - \lambda W)$ and $(P - \lambda W)^+$, respectively. Fix $c \in J$ and choose $r \in \mathbb{R}$ (this r can be chosen arbitrarily but, once chosen, it remains fixed).

(14.4.4)
$$U = \Phi(\cdot, c, r), \quad V = \Psi(\cdot, c, r) = (v_{ij}),$$
$$U = (u_{ij}) = (u_i^{[j-1]})^T, \quad U^{-1} = (U_{ij}).$$

THEOREM 14.4.1. *Assume* trace$(P) = 0$. *Suppose that for $\lambda = r$ all solutions of* (14.4.1) *and* (14.4.2) *are in $L^2(J, |w|)$. For any $\lambda \in \mathbb{C}$ and any vector solution $Y = Y(\cdot, \lambda)$ of the system* (14.4.3), *let $X = X(\cdot, \lambda)$ be defined by:*

(14.4.5)
$$X(t) = U^{-1}(t)\, Y(t), \quad t \in J.$$

Then

(1)
$$X' = (r - \lambda)\, w\, Q X \quad on\ J$$

where

(14.4.6)
$$Q = \begin{pmatrix} u_1 U_{1n} & u_2 U_{1n} & \dots & u_n U_{1n} \\ u_1 U_{2n} & u_2 U_{2n} & \dots & u_n U_{2n} \\ \vdots & \vdots & \vdots & \vdots \\ u_1 U_{nn} & u_2 U_{nn} & \dots & u_n U_{nn} \end{pmatrix}.$$

(2) $w\, Q \in L^1(J)$.
(3) *Both limits*
$$X(a) = \lim_{t \to a^+} X(t), \quad X(b) = \lim_{t \to b^-} X(t)$$
exist and are finite.

PROOF.
$$\begin{aligned} X' &= -U^{-1} U' U^{-1} Y + U^{-1} Y' \\ &= U^{-1}[-(P - rW) U U^{-1} Y + (P - \lambda W) Y] \\ &= (r - \lambda)(U^{-1} W U)\, X \quad on\ J. \end{aligned}$$

Now observe that
$$U^{-1} W\, U = w\, Q$$
and the proof of part (1) is complete. For the proof of part (2) the critical observation is to note that $\overline{U}_{jn}$ is a solution of the adjoint equation (14.4.2) for each $j = 1, 2, ..., n$. Using the Adjointness Lemma, the fact that $\Phi(c, t) = \Phi^{-1}(t, c)$ which follows from the representation $\Phi(c, t) = Y(c) Y^{-1}(t)$ for any fundamental matrix Y of (14.4.3) and the hypothesis that trace$(P) = 0$ which implies that $\det \Phi(c, t) = 1$ for all $t \in J$, we have
$$\Psi(t, c) = E^{-1} \Phi^*(c, t)\, E = E^{-1} \Phi^{-1*}(t, c)\, E = E^{-1} (U^{-1})^*\, E.$$

Therefore
$$\Psi_{1,n+1-j}(t, c) = \pm \overline{U}_{j,n}, \quad j = 1, 2, \cdots, n.$$

From the hypothesis that all solutions of (14.4.1) and (14.4.2) are in $L^2(J, |w|)$ and the Schwarz inequality we get that
$$\left(\int_J |w\, u_j\, U_{kn}| \right)^2 \le \int_J |w|\, |u_j|^2 \int_J |w|\, |U_{kn}|^2 < \infty, \quad j, k = 1, 2, ..., n.$$

This completes the proof of part (2). Part (3) follows from part (2), see [**620**] and the proof of the Theorem is complete. □

COROLLARY 14.4.1. *Let the hypotheses and notation of Theorem* 14.4.1 *hold. Then all solutions of* (14.4.1) *and* (14.4.2) *are in* $L^2(J, |w|)$ *for every* $\lambda \in \mathbb{C}$.

PROOF. By (14.4.5) $Y(t,\lambda) = U(t)\,X(t,\lambda)$. By hypothesis $u_{1,j} \in L^2(J, |w|)$, for $j = 1, \cdots, n$. By part (3) of Theorem 14.4.1, each component of $X(t,\lambda)$ has a finite limit at each endpoint and is therefore bounded in a neighborhood of each endpoint and therefore the conclusion follows for equation (14.4.1). Since $(P^+)^+ = P$ this argument is symmetric with respect to P and P^+ and the conclusion follows also for equation (14.4.2). □

5. Green's Functions of Singular Boundary Value Problems

In this section we construct Green's functions for singular boundary value problems on $J = (a,b)$ for the case when each endpoint is either regular or singular limit-circle (LC). The endpoints may be finite or infinite and no oscillatory restrictions on the coefficients or solutions are assumed. Here the components of P and w are in $L_{loc}(J)$ but not necessarily in $L(J)$.

THEOREM 14.5.1. *Let* $w \in L_{loc}(J)$. *Suppose* $P \in Z_n(J)$, *trace*$(P) = 0$. *Let* $M = M_P$ *and* $M^+ = M_{P^+}$ *be the scalar n-th order differential expressions generated by* P *and* P^+ *respectively. let* U, V *be determined by* (14.4.4) , X *by* (14.4.5), *and let* Q *be given by* (14.4.6). *Let* $A, B, C, D \in M_n(\mathbb{C})$. *We consider the following boundary value problems:*

$$M_P y = \lambda w y \ \ on\ J, \ \ \lambda \in \mathbb{C}, \ \ AX(a) + BX(b) = 0, \quad where\ X = U^{-1}Y,$$

$$M_{P^+} z = (-1)^n \overline{\lambda}\overline{w} z \ \ on\ J, \ \ \overline{\lambda} \in \mathbb{C}, \ \ C\,\Xi(a) + D\,\Xi(b) = 0, \quad where\ \Xi = V^{-1}Z,$$

$$Y' = (P - \lambda W)\,Y + F \ \ on\ J, \quad AX(a) + BX(b) = 0, \tag{14.5.1}$$

$$Z' = (P^+ - (-1)^n \overline{\lambda}\,\overline{W})Z + G \ \ on\ J, \quad C\,\Xi(a) + D\,\Xi(b) = 0, \tag{14.5.2}$$

$$X' = (r - \lambda)wQX + U^{-1}F \ \ on\ J, \quad AX(a) + BX(b) = 0, \tag{14.5.3}$$

$$\Xi' = (r - \overline{\lambda})\overline{w}Q^+\Xi + V^{-1}G \ \ on\ J, \quad A\,\Xi(a) + B\,\Xi(b) = 0. \tag{14.5.4}$$

Note that if $Y' = (P - \lambda W)\,Y + F$ *and* $X = U^{-1}Y$, *then from* (14.5.1) *we have*

$$\begin{aligned} U'X + UX' &= (P - \lambda W)\,UX + F, \\ UX' &= [(P - \lambda W)\,U - U']\,X + F, \\ X' &= U^{-1}[(P - \lambda W)\,U - U']\,X + U^{-1}F \\ &= (r - \lambda)wQX + U^{-1}F. \end{aligned}$$

Similarly, (14.5.4) *follows from* (14.5.2) *and the transformation* $\Xi = V^{-1}Z$.

Assume that $r - \lambda$ *is not an eigenvalue of the boundary value problem (BVP)* (14.5.3) *and* $r - \overline{\lambda}$ *is not an eigenvalue of the BVP* (14.5.4). *Then the Green's matrices of the regular systems* (14.5.3), (14.5.4)

$$K(t,s,r-\lambda,Q,W), \quad K(t,s,r-\overline{\lambda},Q^+,\overline{W})$$

exist by Theorem 14.2.1. *Define*

$$G(t,s,\lambda,P,W,A,B) = U(t)K(t,s,r-\lambda,Q,W,A,B)\,U^{-1}(s), \quad s,t \in J,$$

$$G(t,s,\overline{\lambda},P^+,\overline{W},C,D) = V(t)K(t,s,r-\overline{\lambda},Q^+,\overline{W},C,D)\,V^{-1}(s), \quad s,t \in J,$$

where Q^+ *is defined as* Q *but with* P *replaced by* P^+. *Then for each* $U^{-1}F \in L(J)$ *the regular boundary value problem* (14.5.3) *has a unique solution* X *given by*

$$X(t) = \int_a^b K(t,s,r-\lambda,Q,W)U^{-1}(s)F(s)ds, \quad a < t < b,$$

and hence

$$U^{-1}(t)Y(t) = \int_a^b U^{-1}(t)G(t,s,\lambda,P,W,A,B)U(s)U^{-1}(s)F(s)ds, \quad a < t < b.$$

Therefore

$$Y(t) = \int_a^b G(t,s,\lambda,P,W,A,B)F(s)ds, \quad a < t < b.$$

Similarly for (14.5.2) *and* (14.5.4).

Given the above, we prove that the following statements are equivalent:

(1) $AEC^* = BED^*$,
(2) $K(t,s,Q,r-\lambda,A,B) = -E^{-1}K^*(s,t,Q^+,r-\overline{\lambda},C,D)E, \quad s,t \in J,$
(3) $G(t,s,\lambda,P,W,A,B) = -E^{-1}G^*(s,t,\overline{\lambda},P^+,\overline{W},C,D)E, \quad s,t \in J,$
(4) $K_{1n}(t,s,\lambda,P,W,A,B) = (-1)^n\overline{K}_{1n}(s,t,\overline{\lambda},P^+,\overline{W},C,D), \quad s,t \in J.$

In particular, each of (1), (2), (3), (4) implies that

$$G_{1n}(t,s,\lambda,P,W,A,B) = (-1)^n\overline{G}_{1n}(s,t,\overline{\lambda},P^+,\overline{W},C,D), \quad s,t \in J.$$

PROOF. Parts (1) and (2) are equivalent by Theorem 14.2.2. To show that (3) is equivalent to (2) we proceed as follows

$$\begin{aligned}
&-E^{-1}G^*(s,t,\overline{\lambda},P^+,\overline{W},C,D)E \\
=&-E^{-1}[V(s)\,K(s,t,r-\overline{\lambda},Q^+,\overline{W},C,D)\,V^{-1}(t)\,]^*\,E \\
=&-E^{-1}\,V^{-1*}(t)\,E\,E^{-1}\,K^*(s,t,r-\overline{\lambda},Q^+,\overline{W},C,D)EE^{-1}V^*(s)\,E \\
=&U(t)\,K(t,s,r-\lambda,Q,W,A,B)\,U^{-1}(s) \\
=&G(t,s,\lambda,P,W,A,B).
\end{aligned}$$

In the last step we used the identities:

$$U(t) = E^{-1}V^{-1*}(t)\,E, \quad V(s) = E^{-1}\,U^{-1*}(s)\,E.$$

These can be established by showing that both sides satisfy the same initial value problem. The equivalence of (3) and (4) is established similarly to the corresponding result of Theorem 14.3.2. □

6. Construction of Adjoint and Self-Adjoint Boundary Conditions

In this section we comment on the adjointness and self-adjointness conditions of Theorem 14.3.2:

$$AEC^* = BED^*,$$

$$AEA^* = BEB^*,$$

and discuss a construction for these conditions for the cases $n = 2$ and $n = 4$. This construction is based on the method used in the well known book by Coddington-Levinson [**102**] but is more explicit because our Lagrange bracket is much simpler than the classical one used in [**102**]. In particular it does not depend on the coefficients of the equation.

Since the Lagrange identity is fundamental to the study of boundary value problems we review it here for the benefit of the reader. Let $P \in Z_n(J)$, $M = M_P$, let $Q = P^+$, $M^+ = M_Q$ (this Q is not related to the Q used in Section 14.4 above). For $y \in D(P)$, $z \in D(Q)$ define the Lagrange bracket $[\cdot,\cdot]$ by

$$[y, z] = i^n \sum_{r=0}^{n-1} (-1)^{n+1-r} \overline{z}^{[n-r-1]} y^{[r]}.$$

Here we have omitted the subscript P on the quasi-derivatived of y and the subscript Q on the quasi-derivatives of z.

LEMMA 14.6.1 (Lagrange Identity). *For any $y \in D(P)$, and $z \in D(Q)$ we have*

$$\overline{z}\, My - y\,\overline{(M^+ z)} = [y, z]'.$$

PROOF. See Section 3.3 above or Lemma 3.3 in [**428**]. □

Assume that $P = (p_{ij}) \in Z_n(J)$ and p_{ij}, w satisfy

$$w,\, p_{ij} \in L(J),\ 1 \le i \le j,\ j = 1, \cdots, n; \qquad p_{j,j+1}^{-1} \in L(J),\ j = 1, \cdots, n-1, \tag{14.6.1}$$

then equations (14.4.1) and (14.4.2) are regular and therefore $y^{[r]}$ and $z^{[r]}$ are well defined at both endpoints a and b as finite limits as shown in Section 3.3.

LEMMA 14.6.2. *Assume* (14.6.1) *holds. For any $y \in D(P)$ and $z \in D(Q)$ we have*

$$\int_a^b \{\overline{z}\, My - y\,\overline{(M^+ z)}\} = [y, z](b) - [y, z](a).$$

PROOF. This follows from the Lagrange Identity by integration. □

DEFINITION 14.6.1. Given matrices $A, B \in M_n(\mathbb{C})$ satisfying $\operatorname{rank}(A : B) = n$ and the boundary condition

$$AY(a) + BY(b) = 0 \tag{14.6.2}$$

and matrices $C, D \in M_n(\mathbb{C})$ satisfying $\operatorname{rank}(C : D) = n$ and boundary condition

$$CZ(a) + DZ(b) = 0 \tag{14.6.3}$$

we say that the boundary condition (14.6.3) is adjoint to (14.6.2) if $[y, z](b) - [y, z](a) = 0$ for all $y \in D(P)$ and $z \in D(Q)$. Note that (14.6.3) is adjoint to (14.6.2) if and only if (14.6.2) is adjoint to (14.6.3).

EXAMPLE 14.6.1. Let $n = 2$. Consider the Sturm-Liouville equation

$$My = -(py')' + qy = \lambda wy \quad on \ J = (a, b), \quad -\infty \leq a < b \leq \infty$$

and its adjoint equation

$$M^+ z = -(\overline{p}z')' + \overline{q}z = \overline{\lambda}\,\overline{w}\,z \quad on \ J$$

with

$$\frac{1}{p}, \ q, \ w \in L^1(J, \mathbb{C}), \quad \lambda \in \mathbb{C}.$$

Here

$$P = \begin{bmatrix} 0 & 1/p \\ q & 0 \end{bmatrix}, \quad P^+ = \begin{bmatrix} 0 & 1/\overline{p} \\ \overline{q} & 0 \end{bmatrix}, \quad E = \begin{bmatrix} 0 & -1 \\ 1 & 0 \end{bmatrix}$$

and the Lagrange Identity is

$$\overline{z}My - \overline{yM^+z} = [y, z]', \quad where \ [y, z] = y(p\overline{z}') - \overline{z}(py')$$

for all $y \in D(M)$, $z \in D(M^+)$.

The next lemma yields a construction for adjoint and, as we will see below, also self-adjoint boundary conditions.

Lemma. Let $A, B \in M_2(\mathbb{C})$, the set of 2×2 matrices over the complex numbers, with

$$\operatorname{rank}(A : B) = 2$$

and let $Y = \begin{bmatrix} y \\ py' \end{bmatrix}$. Choose any matrices α, β such that the block matrix

$$\begin{bmatrix} A & B \\ \alpha & \beta \end{bmatrix}$$

is nonsingular; then choose 2×2 matrices F, G, H, K such that

$$\begin{bmatrix} F & G \\ H & K \end{bmatrix} \begin{bmatrix} A & B \\ \alpha & \beta \end{bmatrix} = \begin{bmatrix} -E & 0 \\ 0 & E \end{bmatrix}, \quad E = \begin{bmatrix} 0 & -1 \\ 1 & 0 \end{bmatrix}. \tag{14.6.4}$$

Then the boundary conditions

$$G^* Z(a) + K^* Z(b) = 0$$

with $Z = \begin{bmatrix} z \\ \overline{p}z' \end{bmatrix}$ are adjoint to the conditions

$$AY(a) + BY(b) = 0.$$

Proof. Let $y \in D(M)$, $z \in D(M^+)$. Then

$$\overline{z}My - \overline{yM^+z} = \overline{z}[-(py')' + qy] - y\overline{[-(\overline{p}z')' + \overline{q}z]} = [yp\overline{z}' - \overline{z}py']'.$$

Note that

$$yp\overline{z}' - \overline{z}py' = [\overline{z}, p\overline{z}'] \begin{bmatrix} 0 & -1 \\ 1 & 0 \end{bmatrix} \begin{bmatrix} y \\ py' \end{bmatrix}.$$

Hence

$$\begin{aligned}\int_a^b \{\overline{z}My - y\overline{M^+z}\} =& [yp\overline{z}' - \overline{z}py'](b) - [yp\overline{z}' - \overline{z}py'](a)\\ =& Z^*(b)EY(b) - Z^*(a)EY(a)\\ =& [Z^*(a), Z^*(b)]\begin{bmatrix} -E & 0 \\ 0 & E \end{bmatrix}\begin{bmatrix} Y(a) \\ Y(b) \end{bmatrix}\\ =& [Z^*(a), Z^*(b)]\begin{bmatrix} F & G \\ H & K \end{bmatrix}\begin{bmatrix} A & B \\ \alpha & \beta \end{bmatrix}\begin{bmatrix} Y(a) \\ Y(b) \end{bmatrix}\\ =& [Z^*(a)F + Z^*(b)H][AY(a) + BY(b)]\\ &+ [Z^*(a)G + Z^*(b)K][\alpha Y(a) + \beta Y(b)].\end{aligned}$$

And $Z^*(a)G + Z^*(b)K = 0$ if and only if $G^*Z(a) + K^*Z(b) = 0$. This completes the proof.

Illustration. $B = -I$, $\alpha = -I$, $\beta = 0$. Then the following equations must hold:

(1) $FA + G\alpha = -E$,
(2) $FB + G\beta = 0$,
(3) $HA + K\alpha = 0$,
(4) $HB + K\beta = E$.

Hence

$$C = G^* = E^* = -E, \qquad D = K^* = (-EA)^* = -A^*E^* = A^*E.$$

Checking the "Green's function identities" condition we have

$$AEC^* = AE(-E)^* = AEE = -A,$$

$$BED^* = BE(A^*E)^* = BEE^*A = -IE(-E)A = -A.$$

Thus we have constructed adjoint boundary conditions. Next we show that this construction produces all self-adjoint conditions.

Let

$$\frac{1}{p},\ q,\ w \in L^1(J, \mathbb{R}), \quad \lambda \in \mathbb{C}.$$

Case 1. All real coupled self-adjoint boundary conditions.

Set $A = A$, where $A = (a_{ij})$, $a_{ij} \in \mathbb{R}$, $\det A = 1$; $B = -I$, $\alpha = 0$, $\beta = -E$. From

$$\begin{bmatrix} F & G \\ H & K \end{bmatrix}\begin{bmatrix} A & B \\ \alpha & \beta \end{bmatrix} = \begin{bmatrix} -E & 0 \\ 0 & E \end{bmatrix},$$

we have

$$F = -EA^{-1}, \quad G = -EA^{-1}E, \quad H = 0, \quad K = -I.$$

Hence

$$AEC^* = AEG = AE(-EA^{-1}E) = E,$$
$$BED^* = BEK = -IEK = -E(-I) = E.$$

Note that

$$G = -EA^{-1}E = A^*, \quad K = B^*,$$

so

$$AEA^* = BEB^*.$$

Case 2. All complex coupled self-adjoint boundary conditions.

Set $A = e^{i\gamma}T$ where T satisfies $T = (t_{ij})$, $t_{ij} \in \mathbb{R}$, $\det T = 1$ and $-\pi < \gamma < 0$ or $0 < \gamma < \pi$; $B = -I$; $\alpha = 0$; $\beta = -E$. By

$$\begin{bmatrix} F & G \\ H & K \end{bmatrix} \begin{bmatrix} A & B \\ \alpha & \beta \end{bmatrix} = \begin{bmatrix} -E & 0 \\ 0 & E \end{bmatrix},$$

we have

$$F = -EA^{-1}, \quad G = -EA^{-1}E, \quad H = 0, \quad K = -I.$$

Hence

$$AEC^* = AEG = AE(-EA^{-1}E) = E,$$
$$BED^* = BEK = -IE(-I) = E.$$

So $AEC^* = BED^*$. Note that

$$G = -EA^{-1}E = e^{-i\gamma}T^* = A^*, \quad K = B^*,$$

Hence

$$AEA^* = BEB^*.$$

Case 3. Separated self-adjoint boundary conditions.

Set $A = \begin{pmatrix} a_1 & a_2 \\ 0 & 0 \end{pmatrix}, \quad a_1, a_2 \in \mathbb{R}, \quad a_1 \neq 0;$

$B = \begin{pmatrix} 0 & 0 \\ b_1 & b_2 \end{pmatrix}, \quad b_1, b_2 \in \mathbb{R}, \; b_1 \neq 0;$

$\alpha = \begin{pmatrix} -a_1 & \frac{1}{a_1} - a_2 \\ 0 & 0 \end{pmatrix}; \quad \beta = \begin{pmatrix} 0 & 0 \\ -b_1 & -\frac{1}{b_1} - b_2 \end{pmatrix}.$

From (14.6.4), we obtain that

$$F = \begin{pmatrix} a_1 & 0 \\ a_2 - \frac{1}{a_1} & 0 \end{pmatrix}, \quad G = \begin{pmatrix} a_1 & 0 \\ a_2 & 0 \end{pmatrix},$$
$$H = \begin{pmatrix} 0 & b_1 \\ 0 & \frac{1}{b_1} + b_2 \end{pmatrix}, \quad K = \begin{pmatrix} 0 & b_1 \\ 0 & b_2 \end{pmatrix}.$$

By a computation we obtain

$$AEC^* = AEG = 0, \quad BED^* = BEK = 0.$$

Note that $G = A^*$, $K = B^*$, so

$$AEA^* = BEB^*.$$

The other cases $a_1 = 0$, $a_2 \neq 0$, $b_1 = 0$, $b_2 \neq 0$ are similar and hence omitted. The three cases combined show that the construction of this lemma generates all self-adjoint boundary conditions, see [**620**].

EXAMPLE 14.6.2. Let $n = 4$. Consider the equation

$$My = [(p_2y'')' + p_1y']' + qy = \lambda wy \quad on \ J = (a, b), \quad -\infty \le a < b \le \infty,$$

and its adjoint equation

$$M^+z = [(\overline{p}_2z'')' + \overline{p}_1z']' + \overline{q}z = \overline{\lambda}\overline{w}z \quad on \ J$$

with

$$\frac{1}{p_2}, p_1, q, w \in L^1(J, \mathbb{C}), \quad \lambda \in \mathbb{C}.$$

Lemma. Let $A, B \in M_4(\mathbb{C})$, the set of 4×4 matrices over the complex numbers, with

$$\text{rank}(A : B) = 4$$

and let

$$Y = \begin{bmatrix} y \\ y' \\ p_2 y'' \\ (p_2 y'')' + p_1 y' \end{bmatrix}, \quad Z = \begin{bmatrix} z \\ z' \\ \overline{p}_2 z'' \\ (\overline{p}_2 z'')' + \overline{p}_1 z' \end{bmatrix}.$$

Choose any 4×4 matrices α, β such that the block matrix

$$\begin{bmatrix} A & B \\ \alpha & \beta \end{bmatrix}$$

is nonsingular; then choose 4×4 matrices F, G, H, K such that

$$\begin{bmatrix} F & G \\ H & K \end{bmatrix} \begin{bmatrix} A & B \\ \alpha & \beta \end{bmatrix} = \begin{bmatrix} E_4 & 0 \\ 0 & -E_4 \end{bmatrix}, \quad E_4 = \begin{bmatrix} 0 & 0 & 0 & -1 \\ 0 & 0 & 1 & 0 \\ 0 & -1 & 0 & 0 \\ 1 & 0 & 0 & 0 \end{bmatrix}. \tag{14.6.5}$$

Then the boundary conditions

$$G^* Z(a) + K^* Z(b) = 0$$

are adjoint to the conditions

$$AY(a) + BY(b) = 0.$$

Proof. Let $y \in D(M)$, $z \in D(M^+)$. Then

$$\begin{aligned} \overline{z} My - y\overline{M^+ z} &= \overline{z}\{[(p_2 y'')' + p_1 y']' + qy\} - y\overline{\{[(\overline{p}_2 z'')' + \overline{p}_1 z']' + \overline{q} z\}} \\ &= \{\overline{z}[(p_2 y'')' + p_1 y'] - y[(p_2 \overline{z}'')' + p_1 \overline{z}'] - (p_2 y'')\overline{z}' + p_2 \overline{z}'' y'\}'. \end{aligned}$$

However

$$\begin{aligned} &\overline{z}[(p_2 y'')' + p_1 y'] - y[(p_2 \overline{z}'')' + p_1 \overline{z}'] - (p_2 y'')\overline{z}' + p_2 \overline{z}'' y' \\ &= (\overline{z} \ \ \overline{z}' \ \ p_2 \overline{z}'' \ \ (p_2 \overline{z}'')' + p_1 \overline{z}') \begin{bmatrix} 0 & 0 & 0 & 1 \\ 0 & 0 & -1 & 0 \\ 0 & 1 & 0 & 0 \\ -1 & 0 & 0 & 0 \end{bmatrix} \begin{pmatrix} y \\ y' \\ p_2 y'' \\ (p_2 y'')' + p_1 y' \end{pmatrix} \\ &= - Z^* E_4 Y. \end{aligned}$$

Hence

$$\begin{aligned} &\int_a^b \{\overline{z} My - y\overline{M^+ z}\} \\ &\quad = \{\overline{z}[(p_2 y'')' + p_1 y'] - y[(p_2 \overline{z}'')' + p_1 \overline{z}'] - (p_2 y'')\overline{z}' + p_2 \overline{z}'' y'\}(b) \\ &\quad\quad - \{\overline{z}[(p_2 y'')' + p_1 y'] - y[(p_2 \overline{z}'')' + p_1 \overline{z}'] - (p_2 y'')\overline{z}' + p_2 \overline{z}'' y'\}(a) \\ &\quad = Z^*(a) E_4 Y(a) - Z^*(b) E_4 Y(b) \\ &\quad = [Z^*(a), Z^*(b)] \begin{bmatrix} E_4 & 0 \\ 0 & -E_4 \end{bmatrix} \begin{bmatrix} Y(a) \\ Y(b) \end{bmatrix} \\ &\quad = [Z^*(a), Z^*(b)] \begin{bmatrix} F & G \\ H & K \end{bmatrix} \begin{bmatrix} A & B \\ \alpha & \beta \end{bmatrix} \begin{bmatrix} Y(a) \\ Y(b) \end{bmatrix} \\ &\quad = [Z^*(a) F + Z^*(b) H][AY(a) + BY(b)] \\ &\quad\quad + [Z^*(a) G + Z^*(b) K][\alpha Y(a) + \beta Y(b)]. \end{aligned}$$

And $Z^*(a)G + Z^*(b)K = 0$ if and only if $G^* Z(a) + K^* Z(b) = 0$.

Illustration. By (14.6.5) we have

$$FA + G\alpha = E_4, \qquad FB + G\beta = 0, \\ HA + K\alpha = 0, \qquad HB + K\beta = -E_4. \tag{14.6.6}$$

Case 1. Set $A = A$, $B = -I$, $\alpha = -E_4$, $\beta = 0$, then by the equations (14.6.6), we have

$$G = -I, \quad F = 0, \quad K = -E_4AE_4, \quad H = E_4.$$

Hence

$$AE_4C^* = AE_4G = AE_4(-I) = -AE_4,$$
$$BE_4D^* = BE_4K = -IE_4(-E_4AE_4) = -AE_4.$$

So

$$AE_4C^* = BE_4D^*.$$

In the following, we let

$$\frac{1}{p_2}, p_1, q, w \in L^1(J, \mathbb{R}).$$

Case 2. Self-adjoint boundary conditions.

Set

$$A = \begin{pmatrix} I_2 & \gamma \\ 0 & I_2 \end{pmatrix}, \quad \gamma = \begin{pmatrix} \gamma_{11} & \gamma_{12} \\ \gamma_{21} & \gamma_{22} \end{pmatrix},$$

where γ_{ij} satisfy $\gamma_{11} = -\overline{\gamma}_{22}$ and γ_{12}, γ_{21} are real numbers. Note that $\gamma = E_2\gamma^*E_2$. And set $B = I_4$, $\alpha = 0_4$, $\beta = -E_4$. By (14.6.6) we have

$$F = E_4A^{-1}, \quad G = -E_4A^{-1}E_4, \quad H = 0, \quad K = I.$$

In terms of $\gamma = E_2\gamma^*E_2$, we can easily obtain that $AE_4A^* = E_4$.

Note that

$$G = -E_4\, A^{-1}E_4 = -E_4\, A^{-1}AE_4\, A^* = A^*,$$
$$C = G^* = A, \qquad D = K^* = B.$$

So

$$AE_4C^* = AE_4A^* = E_4,$$
$$BE_4D^* = BE_4B^* = IE_4I = E_4.$$

Therefore

$$AE_4A^* = BE_4B^*.$$

Case 3. Separated self-adjoint boundary conditions.

Set

$$A = \begin{pmatrix} I & A_1 \\ 0 & 0 \end{pmatrix}, \quad B = \begin{pmatrix} 0 & 0 \\ B_1 & I \end{pmatrix},$$

where I is the 2×2 unit matrix, 0 is the 2×2 zero matrix and

$$A_1 = \begin{pmatrix} a_1 & a_2 \\ a_3 & a_4 \end{pmatrix}, \quad B_1 = \begin{pmatrix} b_1 & b_2 \\ b_3 & b_4 \end{pmatrix}$$

satisfy

$$A_1 = E_2A_1^*E_2, \quad B_1 = E_2B_1^*E_2,$$

i.e. a_2, a_3, b_2, b_3 are real numbers and $a_1 = -\overline{a}_4$, $b_1 = -\overline{b}_4$. And set

$$\alpha = \begin{pmatrix} 0 & E_2 \\ E_2 & E_2A_1 \end{pmatrix}, \quad \beta = \begin{pmatrix} E_2B_1 & E_2 \\ -E_2 & 0 \end{pmatrix}.$$

Then by (14.6.6), we have

$$C = G^* = A, \quad D = K^* = B.$$

So

$$AE_4C^* = AE_4A^* = 0,$$
$$BE_4D^* = BE_4B^* = 0.$$

Hence

$$AE_4A^* = BE_4B^*.$$

7. The Green's Function of the Legendre Equation

As an illustration of some of the above results we construct the singular Legendre Green's function in this section. This seems to be new in [**404**] even though the Legendre equation

$$-(py')' = \lambda y, \quad p(t) = 1 - t^2, \quad on\ J = (-1, 1) \tag{14.7.1}$$

is one of the simplest singular differential equations and there is a voluminous literature associated with it in Pure and Applied Mathematics. Its potential function q is zero, its weight function w is the constant 1, and its leading coefficient p is a simple quadratic. It is singular at both endpoits -1 and $+1$. The singularities are due to the fact that $1/p$ is not Lebesgue integrable in left and right neighborhoods of these points. In spite of its simple appearance this equation (14.7.1) and its associated self-adjoint operators exhibit a surprisingly wide variety of interesting phenomena.

The above construction of singular Green's functions is a five step procedure:

(1) Formulate the singular second order scalar equation (14.7.1) as a first order singular system.
(2) 'Regularize' this singular system by constructing regular systems which are equivalent to it.
(3) Construct the Green's matrix for boundary value problems of the regular system.
(4) Construct the singular Green's matrix for the equivalent singular system from the regular one.
(5) Extract the upper right corner element from the singular Green's matrix. This is the Green's function for singular scalar boundary value problems for equation (14.7.1).

For $\lambda = 0$ two linearly independent solutions of (14.7.1) are given by

$$u(t) = 1, \quad v(t) = \frac{-1}{2}\ln(|\frac{1-t}{t+1}|). \tag{14.7.2}$$

The standard system formulation of (14.7.1) has the form

$$Y' = (P - \lambda W)Y \quad on\ (-1, 1), \tag{14.7.3}$$

where

$$Y = \begin{pmatrix} y \\ py' \end{pmatrix}, \quad P = \begin{pmatrix} 0 & 1/p \\ 0 & 0 \end{pmatrix}, \quad W = \begin{pmatrix} 0 & 0 \\ 1 & 0 \end{pmatrix}.$$

Let u and v be given by (14.7.2) and let

$$U = \begin{pmatrix} u & v \\ pu' & pv' \end{pmatrix} = \begin{pmatrix} 1 & v \\ 0 & 1 \end{pmatrix}.$$

Note that $\det U(t) = 1$, for $t \in J = (-1, 1)$, and set

$$Z = U^{-1}Y.$$

Then

$$\begin{aligned} Z' &= (U^{-1})'Y + U^{-1}Y' \\ &= -U^{-1}U'U^{-1}Y + (U^{-1})(P - \lambda W)Y \\ &= -U^{-1}U'Z + (U^{-1})(P - \lambda W)UZ \\ &= -U^{-1}(PU)Z + U^{-1}(PU)Z - \lambda(U^{-1}WU)Z \\ &= -\lambda(U^{-1}WU)Z. \end{aligned}$$

Letting $G = U^{-1}WU$ we may conclude that

$$Z' = -\lambda GZ, \tag{14.7.4}$$

where

$$G = U^{-1}W\,U = \begin{pmatrix} -v & -v^2 \\ 1 & v \end{pmatrix}. \tag{14.7.5}$$

DEFINITION 14.7.1. We call (14.7.4) a 'regularized' Legendre system.

The next theorem justifies this definition and gives the relationship between this 'regularized' system and equation (14.7.1).

THEOREM 14.7.1. *Let $\lambda \in \mathbb{C}$ and let G be given by* (14.7.5).

(1) *Every component of G is in $L^1(-1, 1)$ and therefore* (14.7.4) *is a regular system.*
(2) *For any $c_1, c_2 \in \mathbb{C}$ the initial value problem*

$$Z' = -\lambda GZ, \quad Z(-1) = \begin{pmatrix} c_1 \\ c_2 \end{pmatrix}$$

has a unique solution Z defined on the closed interval $[-1, 1]$.
(3) *If $Y = \begin{pmatrix} y(t,\lambda) \\ (py')(t,\lambda) \end{pmatrix}$ is a solution of* (14.7.3) *and $Z = U^{-1}Y = \begin{pmatrix} z_1(t,\lambda) \\ z_2(t,\lambda) \end{pmatrix}$, then Z is a solution of* (14.7.4) *and for all $t \in (-1, 1)$ we have*

$$\begin{aligned} y(t,\lambda) &= uz_1(t,\lambda) + v(t)z_2(t,\lambda) = z_1(t,\lambda) + v(t)z_2(t,\lambda), \\ (py')(t,\lambda) &= (pu')\,z_1(t,\lambda) + (pv')(t)\,z_2(t,\lambda) = z_2(t,\lambda). \end{aligned} \tag{14.7.6}$$

(4) *For every solution $y(t,\lambda)$ of the singular scalar Legendre equation* (14.7.1) *the quasi-derivative $(py')(t,\lambda)$ is continuous on the compact interval $[-1, 1]$. More specifically we have*

$$\lim_{t\to -1^+} (py')(t,\lambda) = z_2(-1,\lambda), \qquad \lim_{t\to 1^-} (py')(t,\lambda) = z_2(1,\lambda).$$

Thus the quasi-derivative is a continuous function on the closed interval $[-1, 1]$ for every $\lambda \in \mathbb{C}$.
(5) *Let $y(t,\lambda)$ be given by* (14.7.6). *If $z_2(1,\lambda) \neq 0$ then $y(t,\lambda)$ is unbounded at 1; If $z_2(-1,\lambda) \neq 0$ then $y(t,\lambda)$ is unbounded at -1.*

(6) *Fix* $t \in [-1,1]$, *let* $c_1, c_2 \in \mathbb{C}$. *If* $Z = \begin{pmatrix} z_1(t,\lambda) \\ z_2(t,\lambda) \end{pmatrix}$ *is the solution of* (14.7.4) *determined by the initial conditions* $z_1(-1,\lambda) = c_1,\ z_2(-1,\lambda) = c_2$, *then* $z_i(t,\lambda)$ *is an entire function of* λ, $i = 1, 2$. *Similarly for the initial condition* $z_1(1,\lambda) = c_1,\ z_2(1,\lambda) = c_2$.

(7) *For each* $\lambda \in \mathbb{C}$ *there is a nontrivial solution which is bounded in a (two sided) neighborhood of* 1; *and there is a (generally different) nontrivial solution which is bounded in a (two sided) neighborhood of* -1.

(8) *A nontrivial solution* $y(t,\lambda)$ *of the singular scalar Legendre equation* (14.7.1) *is bounded at* 1 *if and only if* $z_2(1,\lambda) = 0$; *A nontrivial solution* $y(t,\lambda)$ *of the singular scalar Legendre equation* (14.7.1) *is bounded at* -1 *if and only if* $z_2(-1,\lambda) = 0$.

PROOF. Part (1) follows from (14.7.5), (2) is a direct consequence of (1) and the theory of regular systems, $Y = UZ$ implies (3)$\Longrightarrow$(4) and (5); (6) follows from (2) and the basic theory of regular systems. For (7) determine solutions $y_1(t,\lambda)$, $y_{-1}(t,\lambda)$ by applying the Frobenius method to obtain power series solutions of (14.7.1) in the form: (see [**182**], page 5 with different notations)

$$y_1(t,\lambda) = 1 + \sum_{n=1}^{\infty} a_n(\lambda)(t-1)^n, \quad |t-1| < 2;$$

$$y_{-1}(t,\lambda) = 1 + \sum_{n=1}^{\infty} b_n(\lambda)(t+1)^n, \quad |t+1| < 2.$$

To prove (8) it follows from (14.7.6) that if $z_2(1,\lambda) \neq 0$, then $y(t,\lambda)$ is not bounded at 1. Suppose $z_2(1,\lambda) = 0$. If the corresponding $y(t,\lambda)$ is not bounded at 1 then there are two linearly unbounded solutions at 1 and hence all nontrivial solutions are unbounded at 1. This contradiction establishes (8) and completes the proof of the theorem. □

REMARK 14.7.1. From Theorem 14.7.1 we see that, *for every* $\lambda \in \mathbb{C}$, the equation (14.7.1) has a solution y_1 which is bounded at 1 and has a solution y_{-1} which is bounded at -1.

It is well known that for $\lambda_n = n(n+1) : n \in \mathbb{N}_0 = \{0, 1, 2, ...\}$ the Legendre polynomials P_n are solutions on $(-1, 1)$ and hence are bounded at -1 and at $+1$.

For later reference we introduce the primary fundamental matrix of the system (14.7.4).

DEFINITION 14.7.2. Fix $\lambda \in \mathbb{C}$. Let $\Phi(\cdot,\cdot,\lambda)$ be the primary fundamental matrix of (14.7.4); i.e. for each $s \in [-1,1]$, $\Phi(\cdot, s, \lambda)$ is the unique matrix solution of the initial value problem:

$$\Phi(s,s,\lambda) = I$$

where I is the 2×2 identity matrix. Since (14.7.4) is regular, $\Phi(t,s,\lambda)$ is defined for all $t, s \in [-1,1]$ and, for each fixed t, s, $\Phi(t,s,\lambda)$ is an entire function of λ.

We now consider two point boundary conditions for (14.7.4); later we will relate these to singular boundary conditions for (14.7.1).

Let $A, B \in M_2(\mathbb{C})$, the set of 2×2 complex matrices, and consider the boundary value problem:

$$Z' = -\lambda G Z, \quad AZ(-1) + B\,Z(1) = 0. \tag{14.7.7}$$

LEMMA 14.7.1. *A complex number* $-\lambda$ *is an eigenvalue of* (14.7.7) *if and only if*

$$\Delta(\lambda) = \det[A + B\,\Phi(1,-1,-\lambda)] = 0.$$

Furthermore, a complex number $-\lambda$ *is an eigenvalue of geometric multiplicity two if and only if*

$$A + B\,\Phi(1,-1,-\lambda) = 0.$$

PROOF. Note that a solution for the initial condition $Z(-1) = C$ is given by

$$Z(t) = \Phi(t,-1,-\lambda)\,C, \quad t \in [-1,1].$$

The boundary value problem (14.7.7) has a nontrivial solution for Z if and only if the algebraic system

$$[A + B\,\Phi(1,-1,-\lambda)]\,Z(-1) = 0 \tag{14.7.8}$$

has a nontrivial solution for $Z(-1)$.

To prove the furthermore part, observe that two linearly independent solutions of the algebraic system (14.7.8) for $Z(-1)$ yield two linearly independent solutions $Z(t)$ of the differential system and conversely. □

Given any $\lambda \in \mathbb{R}$ and any solutions y, z of (14.7.1) the Lagrange form $[y,z](t)$ is defined by

$$[y,z](t) = y(t)(p\overline{z'})(t) - \overline{z}(t)(py')(t).$$

So, in particular, we have

$$[u,v](t) = +1, \quad [v,u](t) = -1, \quad [y,u](t) = -(py')(t), \quad t \in \mathbb{R},$$
$$[y,v](t) = y(t) - v(t)(py')(t), \quad t \in \mathbb{R}, \quad t \neq \pm 1.$$

We will see below that, although v blows up at ± 1, the form $[y,v](t)$ is well defined at -1 and $+1$ since the limits

$$\lim_{t\to -1}[y,v](t), \quad \lim_{t\to +1}[y,v](t)$$

exist and are finite from both sides. This holds for any solution y of equation (14.7.1) for any $\lambda \in \mathbb{R}$. Note that, since v blows up at 1, this means that y must blow up at 1 except, possibly when $(py')(1) = 0$.

We are now ready to construct the Green's function of the singular scalar Legendre problem consisting of the equation

$$My = -(py')' = \lambda y + h \;\; \text{on } J = (-1,1), \quad p(t) = 1 - t^2, \quad -1 < t < 1, \tag{14.7.9}$$

together with two point boundary conditions

$$A\begin{bmatrix} (-py')(-1) \\ (ypv' - v(py'))(-1) \end{bmatrix} + B\begin{bmatrix} (-py')(1) \\ (ypv' - v(py'))(1) \end{bmatrix} = \begin{bmatrix} 0 \\ 0 \end{bmatrix}, \tag{14.7.10}$$

where u, v are given by (14.7.2) and A, B are 2×2 complex matrices. This construction is based on the system regularization discussed above and we will use the notation from above. Consider the regular nonhomogeneous system

$$Z' = -\lambda G Z + F, \quad AZ(-1) + BZ(1) = 0. \tag{14.7.11}$$

where

$$F = \begin{pmatrix} f_1 \\ f_2 \end{pmatrix}, \quad f_j \in L^1(J,\mathbb{C}), \quad j = 1,2.$$

THEOREM 14.7.2. *Let* $-\lambda \in \mathbb{C}$ *and let* $\Delta(-\lambda) = [A + B\,\Phi(1,-1,-\lambda)]$. *Then the following statements are equivalent:*

(1) *For* $F = 0$ *on* $J = (-1,1)$, *the homogeneous problem* (14.7.11) *has only the trivial solution.*
(2) $\Delta(-\lambda)$ *is nonsingular.*
(3) *For every* $F \in L^1(-1,1)$ *the nonhomogeneous problem* (14.7.11) *has a unique solution* Z *and this solution is given by*

$$Z(t,-\lambda) = \int_{-1}^{1} K(t,s,-\lambda)\,F(s)\mathrm{d}s, \quad -1 \le t \le 1, \tag{14.7.12}$$

where

$$K(t,s,-\lambda) = \begin{cases} \Phi(t,-1,-\lambda)\,\Delta^{-1}(-\lambda)(-B)\,\Phi(1,s,-\lambda), & -1 \le t < s \le 1, \\ \Phi(t,-1,-\lambda)\,\Delta^{-1}(-\lambda)(-B)\,\Phi(1,s,-\lambda) + \phi(t,s-\lambda), & -1 \le s < t \le 1, \\ \Phi(t,-1,-\lambda)\,\Delta^{-1}(-\lambda)(-B)\,\Phi(1,s,-\lambda) + \frac{1}{2}\phi(t,s-\lambda), & -1 \le s = t \le 1. \end{cases}$$

PROOF. See Theorem 14.2.1. □

DEFINITION 14.7.3. Let

$$L(t,s,\lambda) = U(t)\,K(t,s,-\lambda)\,U^{-1}(s), \quad -1 \le t,s \le 1.$$

The next theorem shows that L_{12}, the upper right component of L, is the Green's function of the singular scalar Legendre problem (14.7.9), (14.7.10).

THEOREM 14.7.3. *Assume that* $[A + B\,\Phi(1,-1,-\lambda)]$ *is nonsingular. Then for every function* h *satisfying*

$$h,\ vh \in L^1(J,\mathbb{C}),$$

the singular scalar Legendre problem (14.7.9), (14.7.10) *has a unique solution* $y(\cdot,\lambda)$ *given by*

$$y(t,\lambda) = \int_{-1}^{1} L_{12}(t,s)\,h(s)\mathrm{d}s, \quad -1 < t < 1.$$

PROOF. Let

$$F = \begin{pmatrix} f_1 \\ f_2 \end{pmatrix} = U^{-1}H, \quad H = \begin{pmatrix} 0 \\ -h \end{pmatrix}.$$

Then $f_j \in L^1(J,\mathbb{C})$, $j = 1,2$. Since $Y(t,\lambda) = U(t)\,Z(t,-\lambda)$ we get from (14.7.12)

$$\begin{aligned} Y(t,\lambda) &= U(t)\,Z(t,-\lambda) \\ &= U(t)\int_{-1}^{1} K(t,s,-\lambda)\,F(s)\mathrm{d}s \\ &= \int_{-1}^{1} U(t)\,K(t,s,-\lambda)U^{-1}(s)\,H(s)\mathrm{d}s \\ &= \int_{-1}^{1} L(t,s,\lambda)H(s)\mathrm{d}s, \quad -1 < t < 1, \end{aligned}$$

and therefore $y(t,\lambda) = -\int_{-1}^{1} L_{12}(t,s,\lambda)h(s)\mathrm{d}s$, $-1 < t < 1$. □

8. Comments

The construction of the Green's function for singular self-adjoint and non-self-adjoint boundary value problems with maximal deficiency index given here is not the usual one as found, for example in the well known book by Coddington-Levinson [**102**], which uses a selection theorem to select a sequence of regular Green's functions on truncated intervals whose limit as the truncated interval endpoints converge to the singular endpoints is the Green's function of the singular problem. The construction given here is based on the 2012 paper by Wang-Ridenhour-Zettl [**550**]. It is direct, elementary, and explicit in terms of solutions and based on a simple transformation of the dependent variable which leaves the underlying interval unchanged and transforms the singular problem into a regular one. See also the recent paper by the authors [**549**] on the Green's function for two-interval problems.

The construction of the Green's function for regular problems given here is also not the usual one (but is equivalent to the ususal one since the Green's function is unique). It has its roots in a construction of Neuberger [**442**] for second order problems which was extended to higher order problems by Coddington-Zettl [**103**]. The 'simple' transformation from singular problems to regular ones has its roots in the paper by Fulton and Krall [**219**] where it was used for $n = 4$. For the Legendre equation the solutions used in the characterization of the self-adjoint operators can be computed explicitly and then used to describe all singular self-adjoint domains explicitly, not just the special one

$$(py')(-1) = (py')(1)$$

whose eigenfunctions are the Legendre polynomials.

CHAPTER 15

Notation

Russel, Bertrand (1872-1970)

A good notation has a subtly and suggestiveness which at times make it almost seem like a live teacher.

In J. R. Newman (ed.) The world of Mathematics, New York: Simon and Schuster, 1956.

$\mathbb{R}$ The set of real numbers

$\mathbb{C}$ The set of complex numbers

$\mathbb{N}_0 = \{0, 1, 2, 3, ...\}$

$\mathbb{N} = \{1, 2, 3, ...\}$

$\mathbb{N}_2 = \{2, 3, ...\}$

$M_{n,m}(S)$ The n by m matrices with entries from the set S; if $n = m$ we abbreviate this to $M_n(S)$;

Also if $m = 1$ we sometimes write S^n for $M_{n,1}(S)$

(a, b) denotes the open interval with finite or infinite endpoints, $-\infty \leq a < b \leq \infty$

$[a, b]$ denotes the closed interval with finite or infinite endpoints; thus f continuous on $[a, b]$ with $a = -\infty$ means that f has a finite limit at $-\infty$

$[a, b)$ includes a but not b; similarly for $(a, b]$

$L^2(J, w) = \{f : J \to \mathbb{C}, \int_J |f|^2 w < \infty\}$ This is the Hilbert space of square-integrable functions with weight w if $w > 0$ a.e. on J.

$d(\lambda)$ The number of linearly independent solutions lying in $H = L^2(J, w)$ for $\lambda \in \mathbb{C}$

$r(\lambda)$ The number of linearly independent solutions lying in $H = L^2(J, w)$ for $\lambda \in \mathbb{R}$

d^+ $d^+ = d(i)$, $i = \sqrt{-1}$

d^- $d^- = d(-i)$, $i = \sqrt{-1}$

d The deficiency index on (a, b), i.e. $d = d^+ = d^-$

d_a The deficiency index on (a, c), $a < c < b$, $d_a = d_a^+ = d_a^-$

d_b The deficiency index on (c, b), $a < c < b$, $d_b = d_b^+ = d_b^-$

J The interval $J = (a, b)$

J_a The interval $J = (a, c)$, $a < c < b$

J_b The interval $J = (c, b)$, $a < c < b$

$L(J, \mathbb{C})$ is the set of Lebesgue integrable complex valued functions defined almost everywhere on a Lebesgue measurable subset J of $\mathbb{R}$

$L(J, \mathbb{R})$ is the set of real valued Lebesgue integrable functions on a Lebesgue measurable set J

$L_{loc}(J, \mathbb{R})$ is the set of functions y satisfying $y \in L([\alpha, \beta], \mathbb{R})$ for every compact subinterval $[\alpha, \beta]$ of J

$L_{loc}(J, \mathbb{C})$ denotes the functions y satisfying $y \in L([\alpha, \beta], \mathbb{C})$ for every compact subinterval $[\alpha, \beta]$ of J

$AC_{loc}(J)$ denotes the collection of complex valued functions which are absolutely continuous on all compact subintervals of J.

$|P|$ denotes the absolute value of P if P is a real or complex number or function. If P is a real or a complex matrix constant or function then $|P|$ denotes a matrix norm.

Since all matrix norms are topologically equivalent (in a finite dimensional vector space) this matrix norm can be taken as the $1 - norm$:

$$|P| = \sum |p_{ij}|.$$

$||Y||$ denotes a norm in a vector space; this space is either specified or is clear from the context

$\{X_n : n \in \mathbb{N}\}$ denotes the sequence $X_1, X_2, X_3, ...$

$\Phi(t, u)$ or $\Phi(t, u, P)$ This is the "primary fundamental matrix"of the system $Y' = PY$. The primary fundamental matrix Φ is a matrix solution satisfying: for each fixed $u \in J$, $\Phi(u, u) = I$. Here I is the identity matrix.

$T'(x)$ The Frechet derivative in Banach spaces, see Section 1.7

$o(|h|)$ See Section 1.7.

$Z_n(J)$ For $n > 1$ denotes the set of matrix functions Q which generate quasi-differential expressions $M = M_Q$ on the interval J as defined in Chapter 2.

Q^+ is the Lagrange adjoint or L-adjoint matrix of Q,

$$Q^+ = -E^{-1}Q^*E, \; where \; E = E_n = ((-1)^r \delta_{r,n+1-s})_{r,s=1}^n.$$

$M^+ = M_{Q^+}$ is the adjoint or L-adjoint differential expression generated by Q^+.

$C_0^\infty(J)$ The set of infinitely differentiable functions on J.

rank(A) The rank of the matrix A.

$\det(A)$ The determinant of the matrix A.

$\mathrm{Im}(\lambda)$ The imaginary part of the complex number λ.

$\mathcal{N}(A)$ The null space of the matrix A, i.e. the set of all vectors x such that $Ax = 0$.

$\mathcal{R}(A)$ The range of the matrix A.

$\sigma(S)$ The spectrum of an operator S.

$\sigma_e(S)$ The essential spectrum of an operator S.

$\sigma_d(S)$ The discrete spectrum of an operator S.

$\sigma_p(S)$ The point spectrum of an operator S.

$\sigma_c(S)$ The continuous spectrum of an operator S.

$\emptyset$ The empty set.

$[\cdot, \cdot]$ The Lagrange bracket.

$(A : B)$ A matrix whose first columns are those of A and the next columns are those of B.

CHAPTER 16

Topics Not Covered and Open Problems

So far as the mere imparting of information is concerned, no university has had any justification for existence since the popularization of printing in the fifteenth century.

Alfred North Whitehead, The Aims of Education.

In this chapter we comment on some topics not covered and give a few references. These references are not comprehensive and may not be up to date. Also some open problems are stated at the end.

1. Topics Not Covered

(1) J-self-adjoint operators. There is an extensive literature on this topic. See Galindo [**221**], Dijksma [**109**], Knowles [**336**], Race [**482**], Shang [**509**], Wang-Sun [**565**], [**563**], [**551**], Cascaval and Gesztesy [**95**], Fu-Wang [**216**].

(2) Absolutely continuous spectrum. We have not discussed this spectrum. We have only considered the simplest division of the spectrum into its discrete, essential, and continuous, parts. See the seminal paper of Gilbert and Pearson [**236**] for criteria involving subordinate solutions - an extension of the notion of principal solutions - for the absolutely continuous spectrum; see also Hinton and Shaw [**299**], [**300**], [**301**], Gesztesy, Gurarie, Holden, Klaus, Sadun, Simon, and Vogel [**226**] and the references therein. Also see Weidmann [**573**], [**574**] for some operator theory background and see [**7**], [**570**].

(3) There is an extensive literature on left definite problems, when the weight function w is allowed to change sign. See the monograph by Mingarelli [**422**] and its references. Such problems can be studied in the setting of Krein and Pontryagen spaces rather than Hilbert space. Although the abstract operator theory in these spaces is well developed, explicit applications to ordinary differential operators are fragmentary, especially in the higher order case.

    For oscillatory properties of the eigenfunctions when the weight function changes sign or is identically zero on subintervals, see Everitt, Kwong and Zettl [**179**]. Other papers for problems with an indefinite weight function [**9**], [**10**], [**29**].

(4) See Volkmer [**543**] and the papers by Binding and Brown [**61**], [**63**], [**64**].

(5) In Chapters 12 and 13 we discuss two interval symmetric operators with the self-adjoint operators as a special case. The extension of the two interval self-adjoint theory to any finite number of intervals is routine. The extension of the self-adjoint theory to an infinite number of intervals is not

routine since convergence problems arise, among other things. See Everitt, Shubin, Stolz and Zettl [**197**] for an introduction to problems on infinitely many intervals. Also see Gesztesy and Kirsch [**229**], [**230**], [**225**]. If an equation has an interior singularity then, in general, the solutions and their quasi-derivatives do not exist at this singularity. We study this problem on two separate intervals each of which has this singular point on the boundary. Take the direct sum of two self-adjoint operators from the two intervals and you have a two-interval self-adjoint operator which is not particularly interesting because it doesn't "connect" the two intervals together. More interesting operators are obtained by connecting solutions through the singular point, even if they blow up there, in such a way as to get a new "two interval" self-adjoint operator. Actually this construction is also of interest when the interior point is regular, obtaining what is known as "point interactions" [**594**].

For a symplectic algebra approach to the single interval and multi-interval theory for ordinary and partial self-adjoint differential equations see the papers and monographs by Everitt and Markus [**184**], [**185**].

(6) Discreteness criteria. See [**380**], [**437**], [**240**], [**25**], [**138**], [**292**], [**373**], [**376**], [**375**], [**377**], [**372**], [**379**], [**371**], [**529**].
(7) Inverse spectral theory. See the landmark paper of Gelfand and Levitan [**224**], the elegant exposition of Pöschel and Trubowitz [**471**] and the references therein. See the list of open problems below. See also the seminal paper of Borg [**73**], and [**223**], [**231**], [**399**], [**588**], [**587**], [**214**], [**564**], [**567**], [**248**], [**246**], [**66**].
(8) Eigenparameter dependent boundary conditions. See [**218**], [**54**], [**66**], [**60**], [**109**], [**110**], [**217**], [**290**], [**67**], [**65**], [**62**], [**547**], [**623**], [**435**], [**587**], [**564**], [**248**], [**246**], [**14**].
(9) Expansion Theorems. For problems which are self-adjoint in a Hilbert space the spectral theorem can be applied. Most of the standard books discuss expansion theorems: [**574**], [**573**], [**536**], [**535**], [**464**], [**440**], [**400**], [**102**], [**23**], [**111**], [**5**], [**69**], [**127**]. Some expansion theorems for non-self-adjoint (right definite or left-definite) problems can be found in Mennicken and Möller [**421**], Locker [**406**], [**407**], Eberhard and Freiling [**127**], Eberhard, Freiling and Zettl [**128**].
(10) Half-Range Expansions. Perhaps these problems should be called 'half-domain expansions' since they involve expansions in terms of eigenfunctions restricted to a subinterval of the domain interval [**318**].
(11) Integral Inequalities for derivatives. A Sturm-Liouville equation involves a function y, its derivative y' or quasi-derivative (py') and y'' or $(py\prime)\prime$. For the study of such equations relationships, particularly relationships involving integrals are important. For instance inequalities of the form

$$\left(\int_J |y'|^2 w\right)^2 \leq K \int_J |y|^2 w \int_J |y''|^2 w$$

play an important role in the study of discreteness conditions for the spectrum. The names of Landau, Hardy-Littlewood, Kolmogorov, Schonberg-Cavaretta, Hadamar, Gabushin, Kallman-Rota, are often associated with these kinds of inequalities and their extensions. See the monograph of Kwong and Zettl [**385**] and its references for some information on these kinds of inequalities. Also see [**11**], [**85**], [**100**], [**137**], [**148**], [**147**], [**145**], [**136**], [**135**], [**134**], [**174**], [**198**], [**143**], [**203**], [**140**], [**209**], [**208**], [**243**], [**378**], [**374**], [**383**], [**382**], [**384**], [**380**].

(12) Non-Linear Sturm Liouville problems. This is not a well defined area but there is a considerable literature available but this writer is not well informed on this literature. [**68**], [**78**], [**583**], [**584**], [**595**].

(13) Strong limit-point conditions. These were introduced by Everitt and have been studied by several authors. [**157**], [**163**], [**169**], [**369**].

(14) Separation of SL equations. This topic was also introduced by Everitt and has an active literature. [**170**], [**139**].

(15) Oscillation theory for nonlinear SL equations.This is also a currently active research area. An example is the Emden-Fowler equation which has received a lot of attention in the literature. [**583**], [**584**].

(16) Eigenvalues below the essential spectrum. [**82**], [**83**], [**272**], [**263**].

(17) Gaps in the essential spectrum. [**29**], [**232**], [**277**], [**273**], [**276**], [**501**], [**523**], [**522**].

(18) Essential spectrum.[**264**], [**269**], [**284**], [**297**], [**460**], [**461**], [**603**], [**600**], [**625**], [**480**], [**481**].

(19) Bounds for the starting point of the essential spectrum. [**153**], [**138**].

(20) Perturbation theory for the spectrum. [**85**], [**79**], [**88**], [**87**], [**207**], [**225**], [**512**], [**626**].

(21) Systems and higher order linear ordinary differential equations. [**1**], [**2**], [**3**], [**108**], [**112**], [**113**], [**114**], [**126**], [**131**], [**159**], [**183**], [**195**], [**144**], [**140**], [**202**], [**204**], [**200**], [**201**], [**210**], [**211**], [**219**], [**234**], [**245**], [**278**], [**302**], [**303**], [**314**], [**330**], [**416**], [**420**], [**428**], [**426**], [**485**], [**483**], [**484**], [**497**], [**507**], [**568**], [**595**], [**609**], [**617**], [**594**], [**596**], [**611**], [**606**], [**605**], [**604**], [**608**], [**599**], [**607**], [**610**], [**601**], [**613**], [**602**], [**616**], [**598**], [**597**], [**614**], [**615**], [**612**], [**127**].

(22) Transformation theory. Transforming one equation into another by a change of dependent or independent variable or some other way, e.g. by 'regularization', [**184**], [**28**], [**4**], [**158**], [**171**], [**193**], [**194**], [**320**], [**444**], [**443**].

(23) Scattering theory. [**17**], [**41**], [**40**], [**45**], [**70**], [**108**], [**327**], [**445**], [**446**].

(24) The m-function. [**27**], [**22**], [**21**], [**58**], [**57**], [**98**], [**116**], [**153**], [**166**], [**164**], [**172**], [**173**], [**274**], [**259**], [**260**], [**258**], [**261**], [**262**], [**266**], [**265**], [**267**], [**295**], [**296**], [**316**], [**317**].

(25) Orthogonal Polynomials. [**26**], [**20**], [**177**], [**180**], [**181**].

(26) Numerical Approximations . [**31**], [**32**], [**33**], [**34**], [**35**], [**36**], [**39**], [**44**], [**53**], [**80**], [**38**], [**37**], [**478**], [**82**], [**84**], [**83**], [**97**], [**476**], [**220**], [**244**], [**245**], [**413**], [**414**], [**411**], [**415**], [**469**], [**472**], [**474**], [**473**], [**475**], [**477**], [**600**], [**8**].

(27) Stability theory. [**75**], [**74**], [**86**].

(28) Asymptotic form of solutions. [**271**].

(29) Other topics. [**196**], [**94**], [**99**], [**176**], [**358**], [**152**], [**151**], [**156**], [**162**], [**150**], [**165**], [**149**], [**178**], [**190**], [**228**], [**233**], [**226**], [**222**], [**230**], [**235**], [**249**], [**247**], [**251**], [**270**], [**275**], [**279**], [**282**], [**281**], [**286**], [**294**], [**304**], [**308**], [**309**], [**310**], [**312**], [**321**], [**319**], [**313**], [**328**], [**331**], [**332**], [**340**], [**356**], [**342**], [**343**], [**345**], [**344**], [**346**], [**250**], [**349**], [**90**], [**89**], [**348**],

[347], **[339]**, **[354]**, **[353]**, **[350]**, **[352]**, **[351]**, **[357]**, **[362]**, **[363]**, **[364]**, **[359]**, **[360]**, **[366]**, **[392]**, **[391]**, **[388]**, **[389]**, **[387]**, **[393]**, **[390]**, **[395]**, **[397]**, **[396]**, **[394]**, **[402]**, **[527]**, **[405]**, **[403]**, **[408]**, **[412]**, **[418]**, **[419]**, **[422]**, **[424]**, **[430]**, **[431]**, **[434]**, **[432]**, **[429]**, **[438]**, **[441]**, **[442]**, **[447]**, **[449]**, **[452]**, **[451]**, **[450]**, **[457]**, **[459]**, **[463]**, **[468]**, **[479]**, **[496]**, **[500]**, **[504]**, **[508]**, **[511]**, **[518]**, **[517]**, **[519]**, **[521]**, **[525]**, **[526]**, **[537]**, **[534]**, **[538]**, **[539]**, **[542]**, **[541]**, **[540]**, **[545]**, **[572]**, **[571]**, **[569]**, **[578]**, **[618]**, **[619]**, **[506]**, **[6]**, **[5]**, **[16]**, **[18]**, **[19]**, **[30]**, **[46]**, **[55]**, **[69]**, **[71]**, **[72]**, **[76]**, **[96]**, **[104]**, **[105]**, **[106]**, **[121]**, **[122]**, **[123]**, **[129]**, **[130]**, **[43]**, **[206]**, **[227]**, **[241]**, **[240]**, **[242]**, **[423]**, **[283]**, **[285]**, **[287]**, **[288]**, **[298]**, **[306]**, **[307]**, **[311]**, **[322]**, **[315]**, **[329]**, **[323]**, **[355]**, **[361]**, **[365]**, **[385]**, **[386]**, **[400]**, **[406]**, **[407]**, **[409]**, **[410]**, **[421]**, **[433]**, **[437]**, **[439]**, **[440]**, **[464]**, **[462]**, **[470]**, **[471]**, **[493]**, **[494]**, **[495]**, **[490]**, **[491]**, **[499]**, **[503]**, **[505]**, **[520]**, **[533]**, **[535]**, **[536]**, **[543]**, **[548]**, **[566]**, **[573]**, **[574]**, **[575]**, **[576]**, **[579]**, **[577]**, **[581]**, **[586]**, **[115]**, **[107]**, **[101]**, **[92]**, **[15]**, **[13]**, **[12]**, **[50]**, **[52]**, **[215]**, **[188]**, **[186]**, **[189]**, **[187]**,**[142]**,**[128]**, **[592]**, **[593]**, **[124]**, **[81]**, **[254]**, **[255]**,**[401]**, **[91]**, **[93]**, **[510]**, **[622]**, **[624]**.

2. Open Problems

Unfortunately what is little recognized is that the most worthwhile scientific books are those in which the author clearly indicates what he does not know; for an author most hurts his readers by concealing difficulties.

`Galois, Everiste, in N. Rose (ed), Mathematical Maxims and Minims, Raleigh N.C.: Rome Press, 1988.`

I. The Deficiency Index.

Let $Q=(q_{ij})\in Z_n(J,\mathbb{R})$, $n=2k$, $k>1$, $Q=Q^+$, $M=M_Q$, $J=(a,b)$, $-\infty\le a<b\le\infty$, and let w be a weight function on J. Consider the equation

$$My=\lambda wy \quad on\ J.$$

(1) Given s, $k\le s\le 2k$, find Q,w such that $d(Q,w)=s$.
(2) Given Q does there exist a w such that $d(Q,w)=s$?
(3) Given $s,t\in\{0,1,2,\ldots,n-1\}$ find $Q=(q_{ij})\in Z_n(J,\mathbb{C})$ and w such that

$$s=d^+(Q,w),\quad t=d^-(Q,w).$$

II. Given any symmetric extension of the minimal operator $S_{\min}$ find its symmetric and self-adjoint extensions. In particular the Friedrichs extension. See the Niessen-Zettl papers [453], [454] for special cases.

III. Properties of Eigenvalues and Eigenfunctions of Self-Adjoint Operators when n >2. For n $=2$, see the book [620] for continuous dependence on coefficients, continuous and discontinuous dependence on the boundary conditions, differentiable dependence on coefficients and boundary conditions, monotone properties of eigenvalues, inequalities among eigenvalues, the number of zeros of eigenfunctions, etc.

Bibliography

[1] A. A. Abramov, N. B. Konyukhova, and K. Balla, *Stable initial manifolds and singular boundary value problems for systems of ordinary differential equations* (Russian), Computational mathematics (Warsaw, 1980), Banach Center Publ., vol. 13, PWN, Warsaw, 1984, pp. 319–351. MR798107

[2] C. D. Ahlbrandt, *Disconjugacy criteria for self-adjoint differential systems*, J. Differential Equations **6** (1969), 271–295, DOI 10.1016/0022-0396(69)90018-7. MR0244541

[3] C. D. Ahlbrandt, *Equivalent boundary value problems for self-adjoint differential systems*, J. Differential Equations **9** (1971), 420–435, DOI 10.1016/0022-0396(71)90015-5. MR0284636

[4] C. D. Ahlbrandt, D. B. Hinton, and R. T. Lewis, *The effect of variable change on oscillation and disconjugacy criteria with applications to spectral theory and asymptotic theory*, J. Math. Anal. Appl. **81** (1981), no. 1, 234–277, DOI 10.1016/0022-247X(81)90060-3. MR618771

[5] N. I. Akhiezer and I. M. Glazman, *Theory of linear operators in Hilbert space. Vol. II*, Translated from the Russian by Merlynd Nestell, Frederick Ungar Publishing Co., New York, 1963. MR0264421

[6] N. I. Akhiezer and I. M. Glazman, *Theory of linear operators in Hilbert space, volumes I and II*, Pitman and Scottish Academic Press, London and Edinburgh, 1980.

[7] I. Al-Naggar and D. B. Pearson, *A new asymptotic condition for absolutely continuous spectrum of the Sturm-Liouville operator on the half-line*, Helv. Phys. Acta **67** (1994), no. 2, 144–166. MR1286387

[8] G. Alefeld and J. Herzberger, *Introduction to interval computations*, Computer Science and Applied Mathematics, Academic Press, Inc. [Harcourt Brace Jovanovich, Publishers], New York, 1983. Translated from the German by Jon Rokne. MR733988

[9] W. Allegretto and A. B. Mingarelli, *Boundary problems of the second order with an indefinite weight-function*, J. Reine Angew. Math. **398** (1989), 1–24, DOI 10.1515/crll.1989.398.1. MR998469

[10] W. Allegretto and A. B. Mingarelli, *On the nonexistence of positive solutions for a schrödinger equation with an indefinite weight function*, C. R. Math. Rep. Acad. Sci. Canada **8** (1986), 69-73.

[11] T. G. Anderson and D. B. Hinton, *Relative boundedness and compactness theory for second-order differential operators*, J. Inequal. Appl. **1** (1997), no. 4, 375–400, DOI 10.1155/S1025583497000271. MR1732634

[12] J.-j. Ao, J. Sun, and A. Zettl, *Equivalence of fourth order boundary value problems and matrix eigenvalue problems*, Results Math. **63** (2013), no. 1-2, 581–595, DOI 10.1007/s00025-011-0219-5. MR3009707

[13] J.-j. Ao, J. Sun, and A. Zettl, *Matrix representations of fourth order boundary value problems with finite spectrum*, Linear Algebra Appl. **436** (2012), no. 7, 2359–2365, DOI 10.1016/j.laa.2011.10.001. MR2889996

[14] J.-j. Ao, J. Sun, and M.-z. Zhang, *The finite spectrum of Sturm-Liouville problems with transmission conditions and eigenparameter-dependent boundary conditions*, Results Math. **63** (2013), no. 3-4, 1057–1070, DOI 10.1007/s00025-012-0252-z. MR3057354

[15] J.-j. Ao, J. Sun, and M.-z. Zhang, *The finite spectrum of Sturm-Liouville problems with transmission conditions*, Appl. Math. Comput. **218** (2011), no. 4, 1166–1173, DOI 10.1016/j.amc.2011.05.033. MR2831624

[16] F. M. Arscott, *Periodic differential equations*, Pergamon, 1964.

[17] A. A. Arsen′ev, *Resonances in the scattering problem for the Sturm-Liouville operator* (Russian, with Russian summary), Mat. Sb. **191** (2000), no. 3, 3–12, DOI

10.1070/SM2000v191n03ABEH000459; English transl., Sb. Math. **191** (2000), no. 3-4, 319–328. MR1773250

[18] M. Ashbaugh, R. Brown, and D. Hinton, *Interpolation inequalities and nonoscillatory differential equations*, General inequalities, 6 (Oberwolfach, 1990), Internat. Ser. Numer. Math., vol. 103, Birkhäuser, Basel, 1992, pp. 243–255, DOI 10.1007/978-3-0348-7565-3_21. MR1213011

[19] R. A. Askey, T. H. Koornwinder, and W. Schempp (eds.), *Special functions: group theoretical aspects and applications*, Mathematics and its Applications, D. Reidel Publishing Co., Dordrecht, 1984. MR774053

[20] F. V. Atkinson, *Estimation of an eigen-value occurring in a stability problem*, Math. Z. **68** (1957), 82–99, DOI 10.1007/BF01160333. MR92044

[21] F. V. Atkinson, *On the asymptotic behaviour of the Titchmarsh-Weyl m-coefficient and the spectral function for scalar second-order differential expressions*, Ordinary and partial differential equations (Dundee, 1982), Lecture Notes in Math., vol. 964, Springer, Berlin, 1982, pp. 1–27, DOI 10.1007/BFb0064985. MR693099

[22] F. V. Atkinson, *On the location of the Weyl circles*, Proc. Roy. Soc. Edinburgh Sect. A **88** (1981), no. 3-4, 345–356, DOI 10.1017/S0308210500020163. MR616784

[23] F. V. Atkinson, *Discrete and continuous boundary value problems*, Academic Press, New York–London, 1964.

[24] F. V. Atkinson, M. S. P. Eastham, and J. B. McLeod, *The limit-point, limit-cirlce nature of rapidly oscillating potentials*, Proc. Roy. Soc. Edinburgh Sect. A **76** (1977), no. 3, 183-196.

[25] F. V. Atkinson and W. N. Everitt, *Bounds for the point spectrum for a Sturm-Liouville equation*, Proc. Roy. Soc. Edinburgh Sect. A **80** (1978), no. 1-2, 57–66, DOI 10.1017/S0308210500010131. MR529569

[26] F. V. Atkinson and W. N. Everitt, *Orthongonal polynomials which satisfy second-order differential equations*, Prodeedings of the Christoffel Symposium, (1979), 11-19. Birkhauser-Verlag; Basel, 1981.

[27] F. V. Atkinson, W. N. Everitt, and K. S. Ong, *On the m-coefficient of Weyl for a differential equation with an indefinite weight function*, Proc. London Math. Soc. (3) **29** (1974), 368–384, DOI 10.1112/plms/s3-29.2.368. MR0404746

[28] F. V. Atkinson, W. N. Everitt, and A. Zettl, *Regularization of a Sturm-Liouville problem with an interior singularity using quasiderivatives*, Differential Integral Equations **1** (1988), no. 2, 213–221. MR922562

[29] F. V. Atkinson and D. Jabon, *Indefinite Sturm-Liouville problems*, Argonne Reports ANL-87-27, v. I, edited by Kaper, Kwong and Zettl, (1987), 31-45.

[30] T. Ya. Azizov and I. S. Iokhvidov, *Linear operators in spaces with an indefinite metric*, Pure and Applied Mathematics (New York), John Wiley & Sons, Ltd., Chichester, 1989. Translated from the Russian by E. R. Dawson; A Wiley-Interscience Publication. MR1033489

[31] P. Bailey, *On the approximation of eigenvalues of singular Sturm-Liouville problems by those of suitably chosen regular problems*, Spectral theory and computational methods of Sturm-Liouville problems (Knoxville, TN, 1996), Lecture Notes in Pure and Appl. Math., vol. 191, Dekker, New York, 1997, pp. 171–182. MR1460551

[32] P. Bailey, *SLEIGN: an eigenvalue-eigenfunction code for Sturm-Liouville problems*, Report SAND77-2044 (Sandia National Laboratory, Albuquerque, New Mexico, USA: 1978).

[33] P. B. Bailey, *A slightly modified Prüfer transformation useful for calculating Sturm-Liouville eigenvalues*, J. Comput. Phys. **29** (1978), no. 2, 306–310, DOI 10.1016/0021-9991(78)90163-8. MR511106

[34] P. B. Bailey, *Sturm-Liouville eigenvalues via a phase function*, SIAM J. Appl. Math. **14** (1966), 242–249, DOI 10.1137/0114023. MR0208825

[35] P. B. Bailey, W. N. Everitt, and A. Zettl, *Computing eigenvalues of singular Sturm-Liouville problems*, Results Math. **20** (1991), no. 1-2, 391–423, DOI 10.1007/BF03323182. MR1122349

[36] P. B. Bailey, W. N. Everitt, and A. Zettl, *The SLEIGN2 Sturm-Liouville code*, ACM Trans. Math. Software **21** (2001), 143-192.

[37] P. B. Bailey, B. S. Garbow, H. G. Kaper, and A. Zettl, *Algorithm 700: a FORTRAN software package for Sturm-Liouville problems*, ACM Trans. Math. Software **17** (1991), no. 4, 500–501, DOI 10.1145/210232.210239. MR1140037

[38] P. B. Bailey, B. S. Garbow, H. G. Kaper, and A. Zettl, *Eigenvalue and eigenfunction computations for Sturm-Liouville problems*, ACM Trans. Math. Software **17** (1991), no. 4, 491–499, DOI 10.1145/210232.210238. MR1140036

[39] P. B. Bailey, M. K. Gordon, and L. F. Shampine, *Solving Sturm-Liouville eigenvalues problem*, Report SAND76-0560 (Sandia National Laboratory, Albuquerque, New Mexico, USA; 1976).

[40] V. Bargmann, *On the connection between phase shifts and scattering potential*, Rev. Modern Physics **21** (1949), 488–493, DOI 10.1103/revmodphys.21.488. MR0032069

[41] V. Bargmann, *Remarks on the determination of a central field of force from the elastic scattering phase shifts*, Phys. Rev. (2) **75** (1949), 301–303. MR32070

[42] J. H. Barrett, *Oscillation theory of ordinary linear differential equations*, Advances in Math. **3** (1969), 415–509, DOI 10.1016/0001-8708(69)90008-5. MR0257462

[43] H. Bateman, *Higher transcendental functions: I, II and III*, McGraw-Hill, New York, 1953.

[44] J. V. Baxley, *Eigenvalues of singular differential operators by finite difference methods. I, II*, J. Math. Anal. Appl. **38** (1972), 244–254; ibid. 38 (1972), 257–275, DOI 10.1016/0022-247X(72)90132-1. MR303354

[45] R. Beals, *On an equation of mixed type from electron scattering theory*, J. Math. Anal. Appl. **58** (1977), no. 1, 32–45, DOI 10.1016/0022-247X(77)90225-6. MR492921

[46] P. R. Beesack, *Gronwall inequalities*, Carleton University, Ottawa, Ont., 1975. Carleton Mathematical Lecture Notes, No. 11. MR0486735

[47] H. Behncke, *Spectral analysis of fourth order differential operators. II*, Math. Nachr. **279** (2006), no. 1-2, 73–85, DOI 10.1002/mana.200310346. MR2193608

[48] H. Behncke, *Spectral theory of higher order differential operators*, Proc. London Math. Soc. (3) **92** (2006), no. 1, 139–160, DOI 10.1017/S0024611505015480. MR2192387

[49] H. Behncke and H. Focke, *Deficiency indices of singular Schrödinger operators*, Math. Z. **158** (1978), no. 1, 87–98, DOI 10.1007/BF01214569. MR477498

[50] H. Behncke and D. B. Hinton, *Deficiency indices and spectral theory of third order differential operators on the half line*, Math. Nachr. **278** (2005), no. 12-13, 1430–1457, DOI 10.1002/mana.200310314. MR2169692

[51] H. Behncke and D. B. Hinton, *Eigenfunctions, deficiency indices and spectra of odd-order differential operators*, Proc. Lond. Math. Soc. (3) **97** (2008), no. 2, 425–449, DOI 10.1112/plms/pdn002. MR2439668

[52] H. Behncke and D. B. Hinton, *Transformation theory of symmetric differential expressions*, Adv. Differential Equations **11** (2006), no. 6, 601–626. MR2238021

[53] H. Behnke and F. Goerisch, *Inclusions for eigenvalues of selfadjoint problems*, Topics in validated computations (Oldenburg, 1993), Stud. Comput. Math., vol. 5, North-Holland, Amsterdam, 1994, pp. 277–322, DOI 10.1016/0021-8502(94)90369-7. MR1318957

[54] B. P. Belinskiy and J. P. Dauer, *On a regular Sturm-Liouville problem on a finite interval with the eigenvalue parameter appearing linearly in the boundary conditions*, Spectral theory and computational methods of Sturm-Liouville problems (Knoxville, TN, 1996), Lecture Notes in Pure and Appl. Math., vol. 191, Dekker, New York, 1997, pp. 183–196. MR1460552

[55] W. W. Bell, *Special functions for scientists and engineers*, D. Van Nostrand Co., Ltd., London- Princeton, N.J.-Toronto, Ont., 1968. MR0302944

[56] C. Bennewitz, *A generalisation of Niessen's limit-circle criterion*, Proc. Roy. Soc. Edinburgh Sect. A **78** (1977/78), no. 1-2, 81–90, DOI 10.1017/S0308210500009823. MR0492649

[57] C. Bennewitz, *A note on the Titchmarsh-Weyl m-funciton*, Proceedings, 1987 Symposium, Argonne National Laboratory, Argonne, IL., Report ANL-87-26, vol. 2, 105-111.

[58] C. Bennewitz and W. N. Everitt, *Some remarks on the Tichmarsh-Weyl m-coefficient*, In Tribute to Ake Pleijel, Mathematics Department, University of Uppsala, Sweden, (1980), 49-108.

[59] A. Bielecki, *Une remarque sur la méthode de Banach-Cacciopoli-Tikhonov dans la théorie des équations différentielles ordinaires* (French), Bull. Acad. Polon. Sci. Cl. III. **4** (1956), 261–264. MR0082073

[60] P. Binding, *A hierarchy of Sturm-Liouville problems*, Math. Methods Appl. Sci. **26** (2003), no. 4, 349–357, DOI 10.1002/mma.358. MR1953300

[61] P. Binding and P. J. Browne, *Applications of two parameter spectral theory to symmetric generalised eigenvalue problems*, Appl. Anal. **29** (1988), no. 1-2, 107–142, DOI 10.1080/00036818808839776. MR960581

[62] P. A. Binding and P. J. Browne, *Left definite Sturm-Liouville problems with eigenparameter dependent boundary conditions*, Differential Integral Equations **12** (1999), no. 2, 167–182. MR1672742

[63] P. A. Binding and P. J. Browne, *Multiparameter Sturm theory*, Proc. Roy. Soc. Edinburgh Sect. A **99** (1984), 173–184.

[64] P. A. Binding and P. J. Browne, *Spectral properties of two parameter eigenvalue problems II*, Proc. Roy. Soc. Edinburgh Sect. A **106** (1987), 39–51.

[65] P. A. Binding, P. J. Browne, and B. A. Watson, *Equivalence of inverse Sturm-Liouville problems with boundary conditions rationally dependent on the eigenparameter*, J. Math. Anal. Appl. **291** (2004), no. 1, 246–261, DOI 10.1016/j.jmaa.2003.11.025. MR2034071

[66] P. A. Binding, P. J. Browne, and B. A. Watson, *Inverse spectral problems for Sturm-Liouville equations with eigenparameter dependent boundary conditons*, J. London Math. Soc., **62** (2000), 161–182.

[67] P. A. Binding, P. J. Browne, and B. A. Watson, *Sturm-Liouville problems with boundary conditions rationally dependent on the eigenparameter. II*, J. Comput. Appl. Math. **148** (2002), no. 1, 147–168, DOI 10.1016/S0377-0427(02)00579-4. On the occasion of the 65th birthday of Professor Michael Eastham. MR1946193

[68] P. A. Binding and Y. X. Huang, *Existence and nonexistence of positive eigenfunctions for the p-Laplacian*, Proc. Amer. Math. Soc. **123** (1995), no. 6, 1833–1838, DOI 10.2307/2160998. MR1260160

[69] G. Birkhoff and G.-C. Rota, *Ordinary differential equations*, Introductions to Higher Mathematics, Ginn and Company, Boston, Mass.-New York-Toronto, 1962. MR0138810

[70] F. Bloch, *Über die quantenmechanik der elektronen in kristallgittern*, Zeit. für Physik, **52** (1929), 555–600.

[71] M. Bôcher, *Lecons sur les méthodes de Sturm dans la théorie des équations différentielles linéaires, et leurs développements modernes*, Gauthier-Villars, Paris, 1917.

[72] J. Bognár, *Indefinite inner product spaces*, Springer-Verlag, New York-Heidelberg, 1974. Ergebnisse der Mathematik und ihrer Grenzgebiete, Band 78. MR0467261

[73] G. Borg, *Eine Umkehrung der Sturm-Liouvilleschen Eigenwertaufgabe. Bestimmung der Differentialgleichung durch die Eigenwerte* (German), Acta Math. **78** (1946), 1–96, DOI 10.1007/BF02421600. MR0015185

[74] G. Borg, *On a Liapounoff criterion of stability*, Amer. J. Math. **71** (1949), 67–70, DOI 10.2307/2372093. MR28500

[75] G. Borg, *Über die Stabilität gewisser Klassen von linearen Differentialgleichungen* (German), Ark. Mat. Astr. Fys. **31A** (1944), no. 1, 31. MR0016803

[76] O. Boruvka, *Linear differential transformations of the second order*, The English Universities Press, Ltd., London, 1971. Translated from the German by F. M. Arscott. MR0463539

[77] J. P. Boyd, *Sturm-Liouville eigenproblems with an interior pole*, J. Math. Phys. **22** (1981), no. 8, 1575–1590, DOI 10.1063/1.525100. MR628532

[78] J. S. Bradley, *Comparison theorems for the square integrability of solutions of $(r(t)y')' + q(t)y = f(t, y)$*, Glasgow Math. J. **13** (1972), 75–79, DOI 10.1017/S0017089500001415. MR0313579

[79] I. Brinck, *Self-adjointness and spectra of Sturm-Liouville operators*, Math. Scand. **7** (1959), 219–239, DOI 10.7146/math.scand.a-10575. MR0112999

[80] B. M. Brown, V. G. Kirby, W. D. Evans, and M. Plum, *Safe numerical bounds for the Titchmarsh-Weyl $m(\lambda)$-function*, Math. Proc. Cambridge Philos. Soc. **113** (1993), no. 3, 583–599, DOI 10.1017/S0305004100076222. MR1207522

[81] B. M. Brown, D. K. R. McCormack, and A. Zettl, *On a computer assisted proof of the existence of eigenvalues below the essential spectrum of the Sturm-Liouville problem*, J. Comput. Appl. Math. **125** (2000), no. 1-2, 385–393, DOI 10.1016/S0377-0427(00)00481-7. Numerical analysis 2000, Vol. VI, Ordinary differential equations and integral equations. MR1803204

[82] B. M. Brown, D. K. R. McCormick, and M. Marletta, *On computing enclosures for the eigenvalues of Sturm-Liouville problems*, Math. Nach. **213** (1999), 17–33.

[83] B. M. Brown, D. K. R. McCormack, and A. Zettl, *On a computer assisted proof of the existence of eigenvalues below the essential spectrum of the Sturm-Liouville problem*, J. Comput. Appl. Math. **125** (2000), no. 1-2, 385–393, DOI 10.1016/S0377-0427(00)00481-7.

Numerical analysis 2000, Vol. VI, Ordinary differential equations and integral equations. MR1803204

[84] B. M. Brown, D. K. R. McCormick, and A. Zettl, *On the existence of an eigenvalue below the essential spectrum*, Proc. Roy. Soc. London Ser. A (1999), 2229–2234.

[85] B. M. Brown and M. S. P. Eastham, *The Hurwitz theorem for Bessel functions and antibound states in spectral theory*, Proc. Roy. Soc. Lond. Proc. Ser. A: Math. Phys. Eng. Sci. **459** (2003), no. 2038, 2431–2448, DOI 10.1098/rspa.2003.1129. MR2011349

[86] R. Brown, S. Clark, and D. Hinton, *Some function space inequalities and their application to oscillation and stability problems in differential equations*, Analysis and applications (Ujjain, 1999), Narosa, New Delhi, 2002, pp. 21–41. MR1970585

[87] M. Burnat, *Die Spektraldarstellung einiger Differentialoperatoren mit periodischen Koeffiezienten im Raume der fastperiodischen Funktionen* (German), Studia Math. **25** (1964/1965), 33–64, DOI 10.4064/sm-25-1-33-64. MR0181791

[88] M. Burnat, *Stability of eigenfunctions and the spectrum for Hill's equation* (Russian, with English summary), Bull. Acad. Polon. Sci. Sér. Sci. Math. Astronom. Phys. **9** (1961), 795–798. MR0132262

[89] X. Cao, Q. Kong, H. Wu, and A. Zettl, *Geometric aspects of Sturm-Liouville problems. III. Level surfaces of the nth eigenvalue*, J. Comput. Appl. Math. **208** (2007), no. 1, 176–193, DOI 10.1016/j.cam.2006.10.040. MR2347744

[90] X. Cao, Q. Kong, H. Wu, and A. Zettl, *Sturm-Liouville problems whose leading coefficient function changes sign*, Canad. J. Math. **55** (2003), no. 4, 724–749, DOI 10.4153/CJM-2003-031-0. MR1994071

[91] Z. J. Cao, *Ordinary differential operators*, Shanghai Science Tech. Press, 1987 (in Chinese).

[92] Z. J. Cao and J. L. Liu, *The deficiency index theory of singular symmetric differential operators* (Chinese), Adv. in Math. (Beijing) **12** (1983), no. 3, 161–178. MR744709

[93] Z. J. Cao and J. Sun, *Collection of differential operators*, Inner Mongolia Univ. Press, Hohhot, 1992.

[94] R. Carmona, *One-dimensional Schrödinger operators with random or deterministic potentials: new spectral types*, J. Funct. Anal. **51** (1983), no. 2, 229–258, DOI 10.1016/0022-1236(83)90027-7. MR701057

[95] R. Cascaval and F. Gesztesy, *$\mathcal{I}$-self-adjointness of a class of Dirac-type operators*, J. Math. Anal. Appl. **294** (2004), no. 1, 113–121, DOI 10.1016/j.jmaa.2004.02.002. MR2059793

[96] K. Chadan and P. C. Sabatier, *Inverse problems in quantum scattering theory*, 2nd ed., Texts and Monographs in Physics, Springer-Verlag, New York, 1989. With a foreword by R. G. Newton. MR985100

[97] B. Chanane, *Computing eigenvalues of regular Sturm-Liouville problems*, Appl. Math. Lett. **12** (1999), no. 7, 119–125, DOI 10.1016/S0893-9659(99)00111-1. MR1750070

[98] J. Chaudhuri and W. N. Everitt, *The spectrum of a fourth-order differential operator*, Proc. Roy. Soc. Edinburgh Sect. A **68** (1970), 185–210. MR0636388

[99] J. Chaudhuri and W. N. Everitt, *On the square of a formally self-adjoint differential expression*, J. London Math. Soc. (2) 1 (1969), 661–673, DOI 10.1112/jlms/s2-1.1.661. MR0248562

[100] R. S. Chisholm and W. N. Everitt, *On bounded integral operators in the space of integrable-square functions*, Proc. Roy. Soc. Edinburgh Sect. A **69** (1970/71), 199–204. MR0295153

[101] R. S. Chisholm, W. N. Everitt, and L. L. Littlejohn, *An integral operator inequality with applications*, J. Inequal. Appl. **3** (1999), no. 3, 245–266, DOI 10.1155/S1025583499000168. MR1732931

[102] E. A. Coddington and N. Levinson, *Theory of ordinary differential equations*, McGraw-Hill Book Company, Inc., New York-Toronto-London, 1955. MR0069338

[103] E. A. Coddington and A. Zettl, *Hermitian and anti-hermitian properties of Green's matrices*, Pacific J. Math. **18** (1966), 451–454. MR197832

[104] J. B. Conway, *Functions of one complex variable*, Springer-Verlag, New York-Heidelberg, 1973. Graduate Texts in Mathematics, 11. MR0447532

[105] W. A. Coppel, *Stability and asymptotic behavior of differential equations*, D. C. Heath and Co., Boston, Mass., 1965. MR0190463

[106] E. T. Copson, *Theory of functions of a complex variable*, Oxford University Press, Oxford, 1946.

[107] B. Ćurgus and H. Langer, *A Kreĭn space approach to symmetric ordinary differential operators with an indefinite weight function*, J. Differential Equations **79** (1989), no. 1, 31–61, DOI 10.1016/0022-0396(89)90112-5. MR997608

[108] P. Deift and E. Trubowitz, *Inverse scattering on the line*, Comm. Pure Appl. Math. **32** (1979), no. 2, 121–251, DOI 10.1002/cpa.3160320202. MR512420

[109] A. Dijksma, *Eigenfunction expansions for a class of J-selfadjoint ordinary differential operators with boundary conditions containing the eigenvalue parameter*, Proc. Roy. Soc. Edinburgh Sect. A **86** (1980), no. 1-2, 1–27, DOI 10.1017/S0308210500011951. MR580241

[110] A. Dijksma, H. Langer, and H. de Snoo, *Symmetric Sturm-Liouville operators with eigenvalue depending boundary conditions*, Oscillations, bifurcation and chaos (Toronto, Ont., 1986), CMS Conf. Proc., vol. 8, Amer. Math. Soc., Providence, RI, 1987, pp. 87–116. MR909902

[111] N. Dunford and J. T. Schwartz, *Linear operators. Part II: Spectral theory. Self adjoint operators in Hilbert space*, With the assistance of William G. Bade and Robert G. Bartle, Interscience Publishers John Wiley & Sons New York-London, 1963. MR0188745

[112] H. I. Dwyer, *Eigenvalues of matrix Sturm-Liouville problems with separated or coupled boundary conditions*, ProQuest LLC, Ann Arbor, MI, 1993. Thesis (Ph.D.)–Northern Illinois University. MR2689292

[113] H. I. Dwyer and A. Zettl, *Computing eigenvalues of regular Sturm-Liouville problems*, Electron. J. Differential Equations (1994), No. 06, approx. 10 pp. MR1292109

[114] H. I. Dwyer and A. Zettl, *Eigenvalue computations for regular matrix Sturm-Liouville problems*, Electronic J. Differential Equations 1995 (1995), 1–13.

[115] M. S. P. Eastham, *Antibound states and exponentially decaying Sturm-Liouville potentials*, J. London Math. Soc. (2) **65** (2002), no. 3, 624–638, DOI 10.1112/S0024610702003216. MR1895737

[116] M. S. P. Eastham, *A connection formula for Sturm-Liouville spectral functions*, Proc. Roy. Soc. Edinburgh Sect. A **130** (2000), no. 4, 789–791, DOI 10.1017/S030821050000041X. MR1776676

[117] M. S. P. Eastham, *The limit-3 case of self-adjoint differential expressions of fourth order with oscillating coefficients*, J. Lond. Math. Soc. **2** (1974), 427–437.

[118] M. S. P. Eastham, *Limit-circle differential expressions of the second-order with an oscillating coefficient*, Quart. J. Math. Oxford **24** (1973), no. 2, 257–263.

[119] M. S. P. Eastham, *On a limit-point method of Hartman*, Bull. London Math. Soc. **4** (1972), 340–344, DOI 10.1112/blms/4.3.340. MR0316801

[120] M. S. P. Eastham, *Semi-bounded second-order differential operators*, Proc. Roy. Soc. Edinburgh Sect. A **72** (1973), no. 2, 9–16.

[121] M. S. P. Eastham, *Theory of ordinary differential equations*, Van Nostrand Reinhold, London–New York, 1970.

[122] M. S. P. Eastham, *The spectral theory of periodic differential equations*, Scottish Academic Press, Edinburgh/London, 1973.

[123] M. S. P. Eastham and H. Kalf, *Schrödinger-type operators with continuous spectra*, Research Notes in Mathematics, vol. 65, Pitman (Advanced Publishing Program), Boston, Mass.-London, 1982. MR667015

[124] M. S. P. Eastham, Q. Kong, H. Wu, and A. Zettl, *Inequalities among eigenvalues of Sturm-Liouville problems*, J. Inequal. Appl. **3** (1999), no. 1, 25–43, DOI 10.1155/S1025583499000028. MR1731667

[125] M. S. P. Eastham and M. L. Thompson, *On the limit-point, limit-circle classification of second-order ordinary differential equations*, Quart. J. Math. Oxford Ser. (2) **24** (1973), 531–535, DOI 10.1093/qmath/24.1.531. MR0417481

[126] M. S. P. Eastham and A. Zettl, *Second order differential expressions whose squares are limit-3*, Proc. Roy. Soc. Edinburgh Sect. A **16** (1977), 223–238.

[127] W. Eberhard and G. Freiling, *Stone-reguläre Eigenwertprobleme* (German), Math. Z. **160** (1978), no. 2, 139–161, DOI 10.1007/BF01214265. MR497942

[128] W. Eberhard, G. Freiling, and A. Zettl, *Sturm-Liouville problems with singular nonselfadjoint boundary conditions*, Math. Nachr. **278** (2005), no. 12-13, 1509–1523, DOI 10.1002/mana.200310318. MR2169696

[129] D. E. Edmunds and W. D. Evans, *Spectral theory and differential operators*, Oxford Mathematical Monographs, The Clarendon Press, Oxford University Press, New York, 1987. Oxford Science Publications. MR929030

[130] U. Elias, *Oscillation theory of two-term differential equations*, Mathematics and its Applications, vol. 396, Kluwer Academic Publishers Group, Dordrecht, 1997. MR1445292

[131] L. H. Erbe, Q. Kong, and S. G. Ruan, *Kamenev type theorems for second-order matrix differential systems*, Proc. Amer. Math. Soc. **117** (1993), no. 4, 957–962, DOI 10.2307/2159522. MR1154244

[132] W. D. Evans, *On limit-point and Dirichlet-type results for second-order differential expressions*, Ordinary and partial differential equations (Proc. Fourth Conf., Univ. Dundee, Dundee, 1976), Springer, Berlin, 1976, pp. 78–92. Lecture Notes in Math., Vol. 564. MR0593161

[133] W. D. Evans, *On the limit-point, limit-circle classification of a second-order differential equation with a complex coefficient*, J. London Math. Soc. (2) **4** (1971), 245–256, DOI 10.1112/jlms/s2-4.2.245. MR0291538

[134] W. D. Evans and W. N. Everitt, *HELP inequalities for limit-circle and regular problems*, Proc. Roy. Soc. London Ser. A **432** (1991), no. 1886, 367–390, DOI 10.1098/rspa.1991.0022. MR1116538

[135] W. D. Evans and W. N. Everitt, *On an inequality of Hardy-Littlewood type: I*, Proc. R. Soc. Edinburgh Sect. A **101** (1985), 131–140.

[136] W. D. Evans and W. N. Everitt, *A return to the Hardy-Littlewood integral inequality*, Proc. Roy. Soc. London Ser. A **380** (1982), 447–486.

[137] W. D. Evans, W. N. Everitt, W. K. Hayman, and D. S. Jones, *Five integral inequalities; an inheritance from Hardy and Littlewood*, J. Inequal. Appl. **2** (1998), no. 1, 1–36, DOI 10.1155/S1025583498000010. MR1671721

[138] W. D. Evans, M. K. Kwong, and A. Zettl, *Lower bounds for the spectrum of ordinary differential operators*, J. Differential Equations **48** (1983), no. 1, 123–155, DOI 10.1016/0022-0396(83)90062-1. MR692847

[139] W. D. Evans and A. Zettl, *Dirichlet and separation results for Schrödinger type operators*, Proc. Roy. Soc. Edinburgh Sect. A **80** (1978), 151–162.

[140] W. D. Evans and A. Zettl, *Interval limit-point criteria for differential expressions and their powers*, J. London Math. Soc. (2) **15** (1977), no. 1, 119–133, DOI 10.1112/jlms/s2-15.1.119. MR0442348

[141] W. D. Evans and A. Zettl, *Levinson's limit-point criterion and powers*, J. Math. Anal. Appl. **62** (1978), no. 3, 629–639, DOI 10.1016/0022-247X(78)90155-5. MR488696

[142] W. D. Evans and A. Zettl, *Levinson's limit-point criterion and powers*, J. Math. Anal. Appl. **62** (1978), no. 3, 629–639, DOI 10.1016/0022-247X(78)90155-5. MR488696

[143] W. D. Evans and A. Zettl, *Norm inequalities involving derivatives*, Proc. Roy. Soc. Edinburgh Sect. A **82** (1978/79), no. 1-2, 51–70, DOI 10.1017/S0308210500011033. MR524672

[144] W. D. Evans and A. Zettl, *On the deficiency indices of powers of real 2nth-order symmetric differential expressions*, J. London Math. Soc. (2) **13** (1976), no. 3, 543–556, DOI 10.1112/jlms/s2-13.3.543. MR0412513

[145] W. N. Everitt, *A general integral inequality associated with certain ordinary differential operators*, Quaestiones Math. **2** (1977/78), no. 4, 479–494. MR0486761

[146] W. N. Everitt, *Integrable-square solutions of ordinary differential equations. III*, Quart. J. Math. Oxford Ser. (2) **14** (1963), 170–180, DOI 10.1093/qmath/14.1.170. MR0151660

[147] W. N. Everitt, *Integral inequalities and spectral theory*, Spectral theory and differential equations (Proc. Sympos., Dundee, 1974; dedicated to Konrad Jörgens), Springer, Berlin, 1975, pp. 148–166. Lecture Notes in Math., Vol. 448. MR0393391

[148] W. N. Everitt, *Integral inequalities and the Liouville transformation*, in: Ordinary and Partial Differential Equations, Lecture Notes in Mathematics, vol 415, 338–352, Springer, Berlin, Heidelberg, 1974.

[149] W. N. Everitt, *Legendre polynomials and singular differential operators*, Ordinary and partial differential equations (Proc. Fifth Conf., Univ. Dundee, Dundee, 1978), Lecture Notes in Math., vol. 827, Springer, Berlin, 1980, pp. 83–106. MR610812

[150] W. N. Everitt, *A note on the Dirichlet condition for second-order differential expressions*, Canadian J. Math. **28** (1976), no. 2, 312–320, DOI 10.4153/CJM-1976-033-3. MR430391

[151] W. N. Everitt, *A note on the self-adjoint domains of second-order differential equations*, Quart. J. Math. Oxford Ser. (2) **14** (1963), 41–45, DOI 10.1093/qmath/14.1.41. MR0143986

[152] W. N. Everitt, *On a generalization of Bessel functions and a resulting class of Fourier kernels*, Quart. J. Math. Oxford Ser. (2) **10** (1959), 270–279, DOI 10.1093/qmath/10.1.270. MR0117507

[153] W. N. Everitt, *On a property of the m-coefficient of a second-order linear differential equation*, J. Lond. Math. Soc. **4** (1972), no. 2, 443–457.

[154] W. N. Everitt, *On the limit-circle classification of second-order differential expressions*, Quart. J. Math. Oxford Ser. (2) **23** (1972), 193–196, DOI 10.1093/qmath/23.2.193. MR0299861

[155] W. N. Everitt, *On the limit-point classification of second-order differential operators*, J. London Math. Soc. **41** (1966), 531–534, DOI 10.1112/jlms/s1-41.1.531. MR0200519

[156] W. N. Everitt, *On the spectrum of a second order linear differential equation with a p-integrable coefficient*, Applicable Anal. **2** (1972), 143–160, DOI 10.1080/00036817208839034. Collection of articles dedicated to Wolfgang Haack on the occasion of his 70th birthday. MR0397072

[157] W. N. Everitt, *On the strong limit-point condition of second-order differential expressions*, Proceedings of the International Conference on Differential Equations, Los Angeles, (1974), 287–307. Academic Press, Inc., New York.

[158] W. N. Everitt,*On the transformation theory of ordinary second-order linear symmetric differential equations*, Czechoslovak Mathematical Journal **32** (1982), 275–306.

[159] W. N. Everitt, *Self-adjoint boundary value problems on finite intervals*, J. London Math. Soc. **37** (1962), 372–384, DOI 10.1112/jlms/s1-37.1.372. MR0138819

[160] W. N. Everitt, *Singular differential equations. I. The even order case*, Math. Ann. **156** (1964), 9–24, DOI 10.1007/BF01359977. MR166433

[161] W. N. Everitt, *Singular differential equations II: some self-adjoint even cases*, Quart. J. Math. **18** (1967), 13–32.

[162] W. N. Everitt, *Some remarks on a differential expression with an indefinite weight function*, North-Holland Mathematics Studies, **13** (1974), 13–28.

[163] W. N. Everitt, *Some remarks on the strong limit-point condition of second-order linear differential expressions* (English, with Russian and Czech summaries), Časopis Pěst. Mat. **111** (1986), no. 2, 137–145. MR847313

[164] W. N. Everitt, *Some remarks on the Titchmarsh-Weyl m-coefficient and associated differential operators*, Differential equations, dynamical systems, and control science, Lecture Notes in Pure and Appl. Math., vol. 152, Dekker, New York, 1994, pp. 33–53. MR1243192

[165] W. N. Everitt, *Spectral theory of the Wirtinger inequality*, Ordinary and partial differential equations (Proc. Fourth Conf., Univ. Dundee, Dundee, 1976), Springer, Berlin, 1976, pp. 93–105. Lecture Notes in Math., Vol. 564. MR0583175

[166] W. N. Everitt, *The Titchmarsh-Weyl m-coefficient for second-order linear ordinary differential equations: a short survey of properties and applications*, Proceedings of the 5th Spanish National Congress on Differential Equations and Applications, (1985), 249–265. Pulblished by the University of La Laguna, Tenerife.

[167] W. N. Everitt and M. Giertz, *On some properties of the powers of a formally self-adjoint differential expression*, Proc. London Math. Soc. (3) **24** (1972), 149–170, DOI 10.1112/plms/s3-24.1.149. MR0289841

[168] W. N. Everitt and M. Giertz, *On the deficiency indices of powers of formally symmetric differential expressions*, Spectral theory and differential equations (Proc. Sympos., Dundee, 1974; dedicated to Konrad Jörgens), Springer, Berlin, 1975, pp. 167–181. Lecture Notes in Math., Vol. 448. With addenda by T. T. Reed, R. M. Kaufman and Anton Zettl. MR0450661

[169] W. N. Everitt, M. Giertz, and J. B. McLeod, *On the strong and weak limit-point classfication of second-order differential expressions*, Proc. London Math. Soc. (3) **29** (1974), 142–158, DOI 10.1112/plms/s3-29.1.142. MR0361255

[170] W. N. Everitt, M. Giertz, and J. Weidmann, *Some remarks on a separation and limit-point criterion of second-order, ordinary differential expressions*, Math. Ann. **200** (1973), 335–346, DOI 10.1007/BF01428264. MR326047

[171] W. N. Everitt, J. Gunson, and A. Zettl, *Some comments on Sturm-Liouville eigenvalue problems with interior singularities*, Z. Angew. Math. Phys. **38** (1987), no. 6, 813–838, DOI 10.1007/BF00945820. MR928586

[172] W. N. Everitt and S. G. Halvorsen, *On the asymptotic form of the Titchmarsh-Weyl m-coefficient*, Applicable Anal. **8** (1978/79), no. 2, 153–169, DOI 10.1080/00036817808839223. MR523952

[173] W. N. Everitt, D. B. Hinton, and J. K. Shaw, *The asymptotic form of the Titchmarsh-Weyl coefficient for Dirac systems*, J. London Math. Soc. (2) **27** (1983), no. 3, 465–476, DOI 10.1112/jlms/s2-27.3.465. MR697139

[174] W. N. Everitt and D. S. Jones, *On an integral inequality*, Proc. Roy. Soc. London Ser. A **357** (1977), no. 1690, 271–288, DOI 10.1098/rspa.1977.0167. MR0450475

[175] W. N. Everitt, I. W. Knowles, and T. T. Read, *Limit-point and limit-circle criteria for Sturm-Liouville equations with intermittently negative principal coefficients*, Proc. Roy. Soc. Edinburgh Sect. A **103** (1986), no. 3-4, 215–228, DOI 10.1017/S0308210500018874. MR866835

[176] W. N. Everitt, A. M. Krall, L. L. Littlejohn, and V. P. Onyango-Otieno, *Differential operators and the Laguerre type polynomials*, SIAM J. Math. Anal. **23** (1992), no. 3, 722–736, DOI 10.1137/0523037. MR1158830

[177] W. N. Everitt, K. H. Kwon, L. L. Littlejohn, and R. Wellman, *On the spectral analysis of the Laguerre polynomials* $\{L_n^{-k}(x)\}$ *for positive integers* k, Spectral theory and computational methods of Sturm-Liouville problems (Knoxville, TN, 1996), Lecture Notes in Pure and Appl. Math., vol. 191, Dekker, New York, 1997, pp. 251–283. MR1460555

[178] W. N. Everitt, M. K. Kwong, and A. Zettl, *Differential operators and quadratic inequalities with a degenerate weight*, J. Math. Anal. Appl. **98** (1984), no. 2, 378–399, DOI 10.1016/0022-247X(84)90256-7. MR730514

[179] W. N. Everitt, M. K. Kwong, and A. Zettl, *Oscillation on eigenfunctions of weighted regular Sturm-Liouville problems*, J. London Math. Soc. **27** (1983), 106–120.

[180] W. N. Everitt and L. L. Littlejohn, *Differential operators and the Legendre type polynomials*, Differential Integral Equations **1** (1988), no. 1, 97–116. MR920492

[181] W. N. Everitt and L. L. Littlejohn, *Orthogonal polynomials and spectral theory: A survey*, Orthogonal Polynomials and their Applications, C. Berzinski, L. Gori and A. Ronveaux (Editors), Proceedings of the third International Symposium held in Erice, Italy 1990 (1991), 21–55.

[182] W. N. Everitt, L. L. Littlejohn, and V. Marić, *On properties of the Legendre differential expression*, Results Math. **42** (2002), no. 1-2, 42–68, DOI 10.1007/BF03323553. MR1934224

[183] W. N. Everitt and C. Markett, *On a generalization of Bessel functions satisfying higher-order differential equations*, J. Comput. Appl. Math. **54** (1994), no. 3, 325–349, DOI 10.1016/0377-0427(94)90255-0. MR1321078

[184] W. N. Everitt and L. Markus, *Boundary value problems and symplectic algebra for ordinary differential and quasi-differential operators*, Mathematical Surveys and Monographs, vol. 61, American Mathematical Society, Providence, RI, 1999. MR1647856

[185] W. N. Everitt and L. Markus, *Complex symplectic geometry with applications to ordinary differential operators*, Trans. Amer. Math. Soc. **351** (1999), no. 12, 4905–4945, DOI 10.1090/S0002-9947-99-02418-6. MR1637066

[186] W. N. Everitt and L. Markus, *The Glazman-Krein-Naimark theorem for ordinary differential operators*, New results in operator theory and its applications, Oper. Theory Adv. Appl., vol. 98, Birkhäuser, Basel, 1997, pp. 118–130. MR1478469

[187] W. N. Everitt, M. Marletta, and A. Zettl, *Inequalities and eigenvalues of Sturm-Liouville problems near a singular boundary*, J. Inequal. Appl. **6** (2001), no. 4, 405–413, DOI 10.1155/S1025583401000248. MR1888433

[188] W. N. Everitt, M. Möller, and A. Zettl, *Discontinuous dependence of the nth Sturm-Liouville eigenvalue*, General inequalities, 7 (Oberwolfach, 1995), Internat. Ser. Numer. Math., vol. 123, Birkhäuser, Basel, 1997, pp. 145–150, DOI 10.1007/978-3-0348-8942-1_12. MR1457275

[189] W. N. Everitt, M. Möller, and A. Zettl, *Sturm-Liouville problems and discontinuous eigenvalues*, Proc. Roy. Soc. Edinburgh Sect. A **129** (1999), 707–716.

[190] W. N. Everitt and G. Nasri-Roudsari, *Sturm-Liouville problems with coupled boundary conditions and Lagrange interpolation series*, J. Comput. Anal. Appl. **1** (1999), no. 4, 319–347, DOI 10.1023/A:1022628422429. MR1758281

[191] W. N. Everitt and F. Neuman, *A concept of adjointness and symmetry of differential expressions based on the generalised Lagrange identity and Green's formula*, Ordinary differential equations and operators (Dundee, 1982), Lecture Notes in Math., vol. 1032, Springer, Berlin, 1983, pp. 161–169, DOI 10.1007/BFb0076797. MR742639

[192] W. N. Everitt and D. Race, *On necessary and sufficient conditions for the existence of Carathéodory solutions of ordinary differential equations*, Quaestiones Math. **2** (1977/78), no. 4, 507–512. MR0477222

[193] W. N. Everitt and D. Race, *The regular representation of singular second-order differential expressions using quasi-derivatives*, Proc. London Math. Soc. (3) **65** (1992), no. 2, 383–404, DOI 10.1112/plms/s3-65.2.383. MR1168193

[194] W. N. Everitt and D. Race, *The regularization of singular second-order differential expressions using Frentzen-type quasi-derivatives*, Quart. J. Math. Oxford Ser. (2) **44** (1993), no. 175, 301–313, DOI 10.1093/qmath/44.3.301. MR1240473

[195] W. N. Everitt and D. Race, *Some remarks on linear ordinary quasidifferential expressions*, Proc. London Math. Soc. (3) **54** (1987), no. 2, 300–320, DOI 10.1112/plms/s3-54.2.300. MR872809

[196] W. N. Everitt, G. Schöttler, and P. L. Butzer, *Sturm-Liouville boundary value problems and Lagrange interpolation series* (English, with English and Italian summaries), Rend. Mat. Appl. (7) **14** (1994), no. 1, 87–126. MR1284905

[197] W. N. Everitt, C. Shubin, G. Stolz, and A. Zettl, *Sturm-Liouville problems with an infinite number of interior singularities*, Spectral theory and computational methods of Sturm-Liouville problems (Knoxville, TN, 1996), Lecture Notes in Pure and Appl. Math., vol. 191, Dekker, New York, 1997, pp. 211–249. MR1460554

[198] W. N. Everitt and S. D. Wray, *On quadratic integral inequalities associated with second-order symmetric differential expressions*, Ordinary differential equations and operators (Dundee, 1982), Lecture Notes in Math., vol. 1032, Springer, Berlin, 1983, pp. 170–223, DOI 10.1007/BFb0076798. MR742640

[199] W. N. Everitt and A. Zettl, *Differential operators generated by a countable number of quasi-differential expressions on the real line*, Proc. London Math. Soc. (3) **64** (1992), no. 3, 524–544, DOI 10.1112/plms/s3-64.3.524. MR1152996

[200] W. N. Everitt and A. Zettl, *Generalized symmetric ordinary differential expressions. I. The general theory*, Nieuw Arch. Wisk. (3) **27** (1979), no. 3, 363–397. MR553264

[201] W. N. Everitt amd A. Zettl, *The Kallman-Rota inequality; a survey*, Gian-Carlo Rota on Analysis and Probability; Selected papers and Commentaries. Chapter 5; Pages 227–250. Birkhauser, Boston, 2003.

[202] W. N. Everitt amd A. Zettl, *The number of integrable-square solutions of products of differential expressions*, Proc. Roy. Soc. Edinburgh Sect. A **76** (1977), 215–226.

[203] W. N. Everitt amd A. Zettl, *On a class of integral inequalities*, J. London Math. Soc. **17** (1978), 291–303.

[204] W. N. Everitt and A. Zettl, *Products of differential expressions without smoothness assumptions*, Quaestiones Math. **3** (1978/79), no. 1, 67–82. MR508600

[205] W. N. Everitt and A. Zettl, *Sturm-Liouville differential operators in direct sum spaces*, Rocky Mountain J. Math. **16** (1986), no. 3, 497–516, DOI 10.1216/RMJ-1986-16-3-497. MR862277

[206] G. Fichera, *Numerical and quantitative analysis*, Surveys and Reference Works in Mathematics, vol. 3, Pitman (Advanced Publishing Program), Boston, Mass.-London, 1978. Translated from the Italian by Sandro Graffi. MR519677

[207] N. E. Firsova, *The Levinson formula for a perturbed Hill operator* (Russian, with English summary), Teoret. Mat. Fiz. **62** (1985), no. 2, 196–209. MR783052

[208] Z. M. Franco, H. G. Kaper, M. K. Kwong, and A. Zettl, *Best constants in norm inequalities for derivatives on a half-line*, Proc. Roy. Soc. Edinburgh Sect. A **100** (1985), no. 1-2, 67–84, DOI 10.1017/S0308210500013640. MR801845

[209] Z. M. Franco, H. G. Kaper, M. K. Kwong, and A. Zettl, *Bounds for the best constants in Landau's inequality on the line*, Proc. Roy. Soc. Edinburgh Sect. A **95** (1983), 257–262.

[210] H. Frentzen, *Equivalence, adjoints and symmetry of quasidifferential expressions with matrix-valued coefficients and polynomials in them*, Proc. Roy. Soc. Edinburgh Sect. A **92** (1982), no. 1-2, 123–146, DOI 10.1017/S0308210500019995. MR667131

[211] H. Frentzen, D. Race, and A. Zettl, *On the commutativity of certain quasi-differential expressions. II*, Proc. Roy. Soc. Edinburgh Sect. A **123** (1993), no. 1, 27–43, DOI 10.1017/S0308210500021223. MR1204850

[212] K. Friedrichs, *Spektraltheorie halbbeschränkter Operatoren und Anwendung auf die Spektralzerlegung von Differentialoperatoren* (German), Math. Ann. **109** (1934), no. 1, 465–487, DOI 10.1007/BF01449150. MR1512905

[213] K. Friedrichs, *Über die ausgezeichnete Randbedingung in der Spektraltheorie der halbbeschränkten gewöhnlichen Differentialoperatoren zweiter Ordnung* (German), Math. Ann. **112** (1936), no. 1, 1–23, DOI 10.1007/BF01565401. MR1513033

[214] S. Fu, Z. Wang, and G. Wei, *Three spectra inverse Sturm-Liouville problems with overlapping eigenvalues*, Electron. J. Qual. Theory Differ. Equ., posted on 2017, Paper No. 31, 7, DOI 10.14232/ejqtde.2017.1.31. MR3650202

[215] S. Z. Fu, *On the selfadjoint extensions of symmetric ordinary differential operators in direct sum spaces*, J. Differential Equations **100** (1992), no. 2, 269–291, DOI 10.1016/0022-0396(92)90115-4. MR1194811

[216] S. Fu and Z. Wang, *The relationships among multiplicities of a J-self-adjoint differential operator's eigenvalue*, Pac. J. Appl. Math. **4** (2012), no. 4, 293–303 (2013). MR3099065

[217] C. T. Fulton, *Singular eigenvalue problems with eigenvalue parameter contained in the boundary conditions*, Proc. Roy. Soc. Edinburgh Sect. A **87** (1980/81), no. 1-2, 1–34, DOI 10.1017/S0308210500012312. MR600446

[218] C. T. Fulton, *Two-point boundary value problems with eigenvalue parameter contained in the boundary conditions*, Proc. Roy. Soc. Edinburgh Sect. A **77** (1977), no. 3-4, 293–308, DOI 10.1017/S030821050002521X. MR0593172

[219] C. T. Fulton and A. M. Krall, *Selfadjoint 4th order boundary value problem in the limit-4 case*, Ordinary differential equations and operators (Dundee, 1982), Lecture Notes in Math., vol. 1032, Springer, Berlin, 1983, pp. 240–256, DOI 10.1007/BFb0076800. MR742642

[220] C. T. Fulton and S. A. Pruess, *Eigenvalue and eigenfunction asymptotics for regular Sturm-Liouville problems*, J. Math. Anal. Appl. **188** (1994), no. 1, 297–340, DOI 10.1006/jmaa.1994.1429. MR1301734

[221] A. Galindo, *On the existence of J-selfadjoint extensions of J-symmetric operators with adjoint*, Comm. Pure Appl. Math. **15** (1962), 423–425, DOI 10.1002/cpa.3160150405. MR0149305

[222] C. S. Gardner, J. M. Greene, M. D. Kruskal, and R. M. Miura, *Korteweg-deVries equation and generalization. VI. Methods for exact solution*, Comm. Pure Appl. Math. **27** (1974), 97–133, DOI 10.1002/cpa.3160270108. MR0336122

[223] Z. M. Gasimov, *Inverse singular periodic problem of Sturm-Liouville*, Trans. Acad. Sci. Azerb. Ser. Phys.-Tech. Math. Sci. **19** (1999), no. 5, Math. Mech., 27–31 (2000). Translated by V. K. Panarina. MR1784880

[224] I. M. Gel′fand and B. M. Levitan, *On the determination of a differential equation from its spectral function* (Russian), Izvestiya Akad. Nauk SSSR. Ser. Mat. **15** (1951), 309–360. MR0045281

[225] F. Gesztesy, *On the one-dimensional Coulomb Hamiltonian*, J. Phys. A **13** (1980), no. 3, 867–875. MR560542

[226] F. Gesztesy, D. Gurarie, H. Holden, M. Klaus, L. Sadun, B. Simon, and P. Vogl, *Trapping and cascading of eigenvalues in the large coupling limit*, Comm. Math. Phys. **118** (1988), no. 4, 597–634. MR962490

[227] F. Gesztesy and H. Holden, *Soliton equations and their algebro-geometric solutions. Vol. I: (1 + 1)-dimensional continuous models*, Cambridge Studies in Advanced Mathematics, vol. 79, Cambridge University Press, Cambridge, 2003. MR1992536

[228] F. Gesztesy, W. Karwowski, and Z. Zhao, *Limits of soliton solutions*, Duke Math. J. **68** (1992), no. 1, 101–150, DOI 10.1215/S0012-7094-92-06805-0. MR1185820

[229] F. Gesztesy and W. Kirsch, *One-dimensional Schrödinger operators with interactions singular on a discrete set*, J. Reine Angew. Math. **362** (1985), 28–50. MR809964

[230] F. Gesztesy, C. Macdeo, and L. Streit, *An exactly solvable periodic Schrödinger operator*, J. Phys. A **18** (1985), no. 9, L503–L507. MR796384

[231] F. Gesztesy and B. Simon, *A new approach to inverse spectral theory. II. General real potentials and the connection to the spectral measure*, Ann. of Math. (2) **152** (2000), no. 2, 593–643, DOI 10.2307/2661393. MR1804532

[232] F. Gesztesy, B. Simon, and G. Teschl, *Zeros of the Wronskian and renormalized oscillation theory*, Amer. J. Math. **118** (1996), no. 3, 571–594. MR1393260

[233] F. Gesztesy and R. Weikard, *Elliptic algebro-geometric solutions of the KdV and AKNS hierarchies—an analytic approach*, Bull. Amer. Math. Soc. (N.S.) **35** (1998), no. 4, 271–317, DOI 10.1090/S0273-0979-98-00765-4. MR1638298

[234] M. Giertz, M. K. Kwong, and A. Zettl, *Commuting linear differential expressions*, Proc. Roy. Soc. Edinburgh Sect. A **87** (1980/81), no. 3-4, 331–347, DOI 10.1017/S0308210500015250. MR606340

[235] D. J. Gilbert and B. J. Harris, *Connection formulae for spectral functions associated with singular Sturm-Liouville equations*, Proc. Roy. Soc. Edinburgh Sect. A **130** (2000), no. 1, 25–34, DOI 10.1017/S0308210500000020. MR1742578

[236] D. J. Gilbert and D. B. Pearson, *On subordinacy and analysis of the spectrum of one-dimensional Schrödinger operators*, J. Math. Anal. Appl. **128** (1987), no. 1, 30–56, DOI 10.1016/0022-247X(87)90212-5. MR915965

[237] R. C. Gilbert, *Asymptotic formulas for solutions of a singular linear ordinary differential equation*, Proc. Roy. Soc. Edinburgh Sect. A **81** (1978), no. 1-2, 57–70, DOI 10.1017/S0308210500010441. MR529377

[238] R. C. Gilbert, *A class of symmetric ordinary differential operators whose deficiency numbers differ by an integer*, Proc. Roy. Soc. Edinburgh Sect. A **82** (1979), 117–134.

[239] I. M. Glazman, *On the theory of singular differential operators*, Uspek. Math. Nauk. **40** (1950), 102–135; Trans. Amer. Math. Soc. **4** (1962), 331–372.

[240] I. M. Glazman, *Direct methods for the qualitative spectral analysis of singular differential operators*, Eng. transl., Israel Program for Scientific Translations, Jerusalem, 1965.

[241] I. C. Gohberg and M. G. Kreĭn, *Theory and applications of Volterra operators in Hilbert space*, Translated from the Russian by A. Feinstein. Translations of Mathematical Monographs, Vol. 24, American Mathematical Society, Providence, R.I., 1970. MR0264447

[242] S. Goldberg, *Unbounded linear operators: Theory and applications*, McGraw-Hill Book Co., New York-Toronto, Ont.-London, 1966. MR0200692

[243] J. A. Goldstein, M. K. Kwong, and A. Zettl, *Weighted Landau inequalities*, J. Math. Anal. Appl. **95** (1983), no. 1, 20–28, DOI 10.1016/0022-247X(83)90133-6. MR710417

[244] L. Greenberg, *A Prüfer method for calculating eigenvalues of self-adjoint systems of ordinary differential equations*, Parts 1 and 2, University of Maryland Technical Report TR91-24, 1991.

[245] L. Greenberg and M. Marletta, *Algorithm 775: the code SLEUTH for solving fourth-order Sturm-Liouville problems*, ACM Trans. Math. Software **23** (1997), no. 4, 453–493, DOI 10.1145/279232.279231. MR1671714

[246] T. Gulsen, E. Yilmaz, and H. Koyunbakan, *An inverse nodal problem for differential pencils with complex spectral parameter dependent boundary conditions*, New Trends in Mathematical Sciences, **5** (2017), no. 1, 137–144.

[247] J. Gunson, *Perturbation theory for a Sturm-Liouville problem with an interior singularity*, Proc. Roy. Soc. London Ser. A **414** (1987), no. 1846, 255–269. MR919724

[248] Y. Guo and G. Wei, *Inverse nodal problem for Dirac equations with boundary conditions polynomially dependent on the spectral parameter*, Results Math. **67** (2015), no. 1-2, 95–110, DOI 10.1007/s00025-014-0396-0. MR3304033

[249] I. M. Guseĭnov, A. A. Nabiev, and R. T. Pashaev, *Transformation operators and asymptotic formulas for the eigenvalues of a polynomial pencil of Sturm-Liouville operators* (Russian, with Russian summary), Sibirsk. Mat. Zh. **41** (2000), no. 3, 554–566, ii, DOI 10.1007/BF02674102; English transl., Siberian Math. J. **41** (2000), no. 3, 453–464. MR1778672

[250] K. Haertzen, Q. Kong, H. Wu, and A. Zettl, *Geometric aspects of Sturm-Liouville problems. II. Space of boundary conditions for left-definiteness*, Trans. Amer. Math. Soc. **356** (2004), no. 1, 135–157, DOI 10.1090/S0002-9947-03-03028-9. MR2020027

[251] S. G. Halvorsen, *A function-theoretic property of solutions of the equation* $x''+(\lambda w-q)x=0$, Quart. J. Math. Oxford Ser. (2) **38** (1987), no. 149, 73–76, DOI 10.1093/qmath/38.1.73. MR876264

[252] S. Halvorsen, *On the quadratic integrability of solutions of* $d^2x/dt^2+f(t)x=0$, Math. Scand. **14** (1964), 111–119, DOI 10.7146/math.scand.a-10711. MR0168846

[253] X. Hao, J. Sun, A. Wang, and A. Zettl, *Characterization of domains of self-adjoint ordinary differential operators II*, Results Math. **61** (2012), no. 3-4, 255–281, DOI 10.1007/s00025-011-0096-y. MR2925120

[254] X. Hao, J. Sun, and A. Zettl, *Canonical forms of self-adjoint boundary conditions for differential operators of order four*, J. Math. Anal. Appl. **387** (2012), no. 2, 1176–1187, DOI 10.1016/j.jmaa.2011.10.025. MR2853205

[255] X. Hao, J. Sun, and A. Zettl, *Fourth order canonical forms of singular self-adjoint boundary conditions*, Linear Algebra Appl. **437** (2012), no. 3, 899–916, DOI 10.1016/j.laa.2012.03.022. MR2921744

[256] X. Hao, J. Sun, and A. Zettl, *Real-parameter square-integrable solutions and the spectrum of differential operators*, J. Math. Anal. Appl. **376** (2011), no. 2, 696–712, DOI 10.1016/j.jmaa.2010.11.052. MR2747790

[257] X. Hao, J. Sun, and A. Zettl, *The spectrum of differential operators and square-integrable solutions*, J. Funct. Anal. **262** (2012), no. 4, 1630–1644, DOI 10.1016/j.jfa.2011.11.015. MR2873853

[258] B. J. Harris, *The asymptotic form of the spectral functions associated with a class of Sturm-Liouville equations*, Proc. Roy. Soc. Edinburgh Sect. A **100** (1985), no. 3-4, 343–360, DOI 10.1017/S030821050001386X. MR807711

[259] B. J. Harris, *The asymptotic form of the Titchmarsh-Weyl m-function*, J. London Math. Soc. (2) **30** (1984), no. 1, 110–118, DOI 10.1112/jlms/s2-30.1.110. MR760880

[260] B. J. Harris, *The asymptotic form of the Titchmarsh-Weyl m-function associated with a Dirac system*, J. London Math. Soc. (2) **31** (1985), no. 2, 321–330, DOI 10.1112/jlms/s2-31.2.321. MR809953

[261] B. J. Harris, *The asymptotic form of the Titchmarsh-Weyl m-function associated with a second-order differential equation with locally integrable coefficient*, Proc. Roy. Soc. Edinburgh Sect. A **102** (1986), 243–251.

[262] B. J. Harris, *The asymptotic form of the Titchmarsh-Weyl m-function for second-order linear differential equations with analytic coefficients*, J. Differential Equations **65** (1986), no. 2, 219–234, DOI 10.1016/0022-0396(86)90034-3. MR861517

[263] B. J. Harris, *Bounds for the point spectra of Sturm-Liouville equations*, J. London Math. Soc. (2) **25** (1982), no. 1, 145–161, DOI 10.1112/jlms/s2-25.1.145. MR645872

[264] B. J. Harris, *Criteria for Sturm-Liouville equations to have continuous spectra*, J. Math. Anal. Appl. **93** (1983), no. 1, 235–249, DOI 10.1016/0022-247X(83)90228-7. MR699711

[265] B. J. Harris, *An exact method for the calculation of certain Tischmarsh-Weyl m-functions*, Proc. Roy. Soc. Edinburgh Sect. A **106** (1987), 137–142.

[266] B. J. Harris, *The form of the spectral functions associated with a class of Sturm-Liouville equations with integrable coefficient*, Proc. Roy. Soc. Edinburgh Sect. A **105** (1987), 215–227, DOI 10.1017/S0308210500022058. MR890057

[267] B. J. Harris, *An inverse problem involving the Titchmarsh-Weyl m-function*, Proc. Roy. Soc. Edinburgh Sect. A **110** (1988), no. 3-4, 305–309, DOI 10.1017/S0308210500022290. MR974746

[268] B. J. Harris, *Limit-circle criteria for second-order differential expressions*, Quart. J. Math. Oxford Ser. (2) **35** (1984), no. 140, 415–427, DOI 10.1093/qmath/35.4.415. MR767772

[269] B. J. Harris, *Lower bounds for the spectrum of second-order linear differential equation with a coefficient whose negative part is p-integrable*, Proc. Roy. Soc Edinburgh Sect. A **97** (1984), 105–107.

[270] B. J. Harris, *A note on a paper of Atkinson concerning the asyptotics of an eigenvalue problem with interior singularity*, Proc. Roy. Soc. Edinburgh Sect. A **110** (1988), 63–71.

[271] B. J. Harris, *On the asymptotic properties of linear differential equations*, Mathematika **34** (1987), no. 2, 187–198, DOI 10.1112/S0025579300013449. MR933498

[272] B. J. Harris, *On the essential spectrum of self-adjoint operators*, Proc. Roy. Soc. Edinburgh Sect. A **86** (1980), 261–274.

[273] B. J. Harris, *On the spectra and stability of periodic differential equations*, Proc. London Math. Soc. (3) **41** (1980), no. 1, 161–192, DOI 10.1112/plms/s3-41.1.161. MR579720

[274] B. J. Harris, *On the Titchmarsh-Weyl m-function*, Proc. Roy. Soc. Edinburgh Sect. A **95** (1983), no. 3-4, 223–237, DOI 10.1017/S0308210500012932. MR726873

[275] B. J. Harris, *A series solution for certain Ricatti equations with applications to Sturm-Liouville problems*, J. Math. Anal. Appl. **137** (1989), no. 2, 462–470, DOI 10.1016/0022-247X(89)90256-4. MR984970

[276] B. J. Harris, *Some spectral gap results*, Ordinary and partial differential equations (Proc. Sixth Conf., Univ. Dundee, Dundee, 1980), Lecture Notes in Math., vol. 846, Springer, Berlin, 1981, pp. 148–157. MR610642

[277] B. J. Harris, *A systematic method of estimating gaps in the essential spectrum of self-adjoint operators*, J. London Math. Soc. (2) **18** (1978), no. 1, 115–132, DOI 10.1112/jlms/s2-18.1.115. MR0492491

[278] P. Hartman, *Comparison theorems for selfadjoint second order systems and uniqueness of eigenvalues of scalar boundary value problems*, Contributions to analysis and geometry (Baltimore, Md., 1980), Johns Hopkins Univ. Press, Baltimore, Md., 1981, pp. 1–22. MR648451

[279] P. Hartman, *Differential equations with non-oscillatory eigenfunctions*, Duke Math. J. **15** (1948), 697–709. MR27927

[280] P. Hartman, *The number of L_2-solutions of $x''(t) + q(t)x = 0$*, Amer. J. Math. **73** (1951), 635–646.

[281] P. Hartman, *On an ordinary differential equation involving a convex function*, Trans. Amer. Math. Soc. **146** (1969), 179–202, DOI 10.2307/1995167. MR276539

[282] P. Hartman, *On linear second order differential equations with small coefficients*, Amer. J. Math. **73** (1951), 955–962, DOI 10.2307/2372126. MR45896

[283] P. Hartman, *Ordinary differential equations*, Wiley and Sons, Inc., New York, 1964.

[284] P. Hartman and A. Wintner, *A separation theorem for continuous spectra*, Amer. J. Math. **71** (1949), 650–662, DOI 10.2307/2372356. MR31159

[285] O. Haupt, *Untersuchungen uber oszillationtheoreme*, Teubner, Leipzig, 1911.

[286] S. W. Hawking and R. Penrose, *The singularities of gravitational collapse and cosmology*, Proc. Roy. Soc. London Ser. A **314** (1970), 529–548, DOI 10.1098/rspa.1970.0021. MR0264959

[287] D. Hilbert, *Grundzüge einer allgemeinen Theorie der linearen Integralgleichungen* (German), Chelsea Publishing Company, New York, N.Y., 1953. MR0056184

[288] E. Hille, *Lectures on ordinary differential equations*, Addison-Wesley Publ. Co., Reading, Mass.-London-Don Mills, Ont., 1969. MR0249698

[289] D. B. Hinton, *Disconjugate properties of a system of differential equations*, J. Differential Equations **2** (1966), 420–437, DOI 10.1016/0022-0396(66)90052-0. MR0208046

[290] D. B. Hinton, *An expansion theorem for an eigenvalue problem with eigenvalue parameter in the boundary condition*, Quart. J. Math. Oxford Ser. (2) **30** (1979), no. 117, 33–42, DOI 10.1093/qmath/30.1.33. MR528889

[291] D. B. Hinton, *Limit point-limit circle criteria for $(py')' + qy = \lambda ky$*, Ordinary and partial differential equations (Proc. Conf., Univ. Dundee, Dundee, 1974), Springer, Berlin, 1974, pp. 173–183. Lecture Notes in Math., Vol. 415. MR0425236

[292] D. Hinton, *On the location of the least point of the essential spectrum*, J. Comput. Appl. Math. **148** (2002), no. 1, 77–89, DOI 10.1016/S0377-0427(02)00574-5. On the occasion of the 65th birthday of Professor Michael Eastham. MR1946188

[293] D. Hinton, *Solutions of $(ry^{(n)})^{(n)} + qy = 0$ of class $\mathcal{L}_p\,[0, \infty)$*, Proc. Amer. Math. Soc. **32** (1972), 134–138, DOI 10.2307/2038319. MR288348

[294] D. Hinton, *Sturm's 1836 oscillation results: evolution of the theory*, Sturm-Liouville theory, Birkhäuser, Basel, 2005, pp. 1–27, DOI 10.1007/3-7643-7359-8_1. MR2145075

[295] D. B. Hinton, M. Klaus, and J. K. Shaw, *Levinson's theorem and Titchmarsh-Weyl $m(\lambda)$ theory for Dirac systems*, Proc. Roy. Soc. Edinburgh Sect. A **109** (1988), no. 1-2, 173–186, DOI 10.1017/S0308210500026743. MR952335

[296] D. B. Hinton, M. Klaus, and J. K. Shaw, *On the Titchmarsh-Weyl function for the half-line perturbed periodic Hill's equation*, Quart. J. Math. Oxford Ser. (2) **41** (1990), no. 162, 189–224, DOI 10.1093/qmath/41.2.189. MR1053662

[297] D. B. Hinton and R. T. Lewis, *Singular differential operators with spectra discrete and bounded below*, Proc. Roy. Soc. Edinburgh Sect. A **84** (1979), no. 1-2, 117–134, DOI 10.1017/S0308210500016991. MR549875

[298] D. Hinton and P. W. Schaefer (eds.), *Spectral theory and computational methods of Sturm-Liouville problems*, Lecture Notes in Pure and Applied Mathematics, vol. 191, Marcel Dekker, Inc., New York, 1997. MR1460546

[299] D. B. Hinton and J. K. Shaw, *Absolutely continuous spectra of perturbed periodic Hamiltonian systems*, Rocky Mountain J. Math. **17** (1987), no. 4, 727–748, DOI 10.1216/RMJ-1987-17-4-727. MR923743

[300] D. B. Hinton and J. K. Shaw, *Absolutely continuous spectra of second order differential operators with short and long range potentials*, SIAM J. Math. Anal. **17** (1986), no. 1, 182–196, DOI 10.1137/0517017. MR819222

[301] D. B. Hinton and J. K. Shaw, *On the absolutely continuous spectrum of the perturbed Hill's equation*, Proc. London Math. Soc. (3) **50** (1985), no. 1, 175–192, DOI 10.1112/plms/s3-50.1.175. MR765373

[302] D. B. Hinton and J. K. Shaw, *On the spectrum of a singular Hamiltonian system*, Quaestiones Math. **5** (1982/83), no. 1, 29–81. MR644777

[303] D. B. Hinton and J. K. Shaw, *Spectrum of a Hamiltonian system with spectral parameter in a boundary condition*, Oscillations, bifurcation and chaos (Toronto, Ont., 1986), CMS Conf. Proc., vol. 8, Amer. Math. Soc., Providence, RI, 1987, pp. 171–186. MR909908

[304] H. Hochstadt, *A special Hill's equation with discontinuous coefficients*, Amer. Math. Monthly **70** (1963), 18–26, DOI 10.2307/2312778. MR145145

[305] R. A. Horn and C. R. Johnson, *Matrix analysis*, Cambridge University Press, Cambridge, 1985. MR832183

[306] E. L. Ince, *Ordinary Differential Equations*, Dover Publications, New York, 1944. MR0010757

[307] I. S. Iohvidov, M. G. Kreĭn, and H. Langer, *Introduction to the spectral theory of operators in spaces with an indefinite metric*, Mathematical Research, vol. 9, Akademie-Verlag, Berlin, 1982. MR691137

[308] R. S. Ismagilov, *On the self-adjointness of the Sturm-Liouville operator* (Russian), Uspehi Mat. Nauk **18** (1963), no. 5 (113), 161–166. MR0155037

[309] C. Jacobi, *Zur theorie der variationsrechnung und die differentialgleichungen*, J. Reine Angew. Math., **17** (1837), 68–82.

[310] K. Jörgens, *Spectral theory of second order ordinary differential operators*, Lectures delivered at Aarhus Universitet 1962/63, Matematisk Institut Aarhus 1964.

[311] K. Jörgens and F. Rellich, *Eigenwerttheorie gewöhnlicher Differentialgleichungen* (German), Springer-Verlag, Berlin-New York, 1976. Überarbeitete und ergänzte Fassung der Vorlesungsausarbeitung "Eigenwerttheorie partieller Differentialgleichungen, Teil 1" von Franz Rellich (Wintersemester 1952/53); Bearbeitet von J. Weidmann; Hochschultext. MR0499411

[312] I. S. Kac, *Power-asymptotic estimates for sprectral functions of generalized boundry value problems of second-order*, Sov. Math. Dokl. 13 (2) (1972), 453-457.

[313] H. Kalf, *Remarks on some Dirichlet type results for semibounded Sturm-Liouville operators*, Math. Ann. **210** (1974), 197–205, DOI 10.1007/BF01350583. MR355177

[314] E. Kamke, *Über die definiten selbstadjungierten Eigenwertaufgaben bei gewöhnlichen linearen Differentialgleichungen. II und III* (German), Math. Z. **46** (1940), 251–286, DOI 10.1007/BF01181441. MR2432

[315] E. Kamke, *Differentialgleichungen: Lösungsmethoden und Lösungen. I: Gewöhnliche Differentialgleichungen; Neunte Auflage; Mit einem Vorwort von Detlef Kamke* (German), B. G. Teubner, Stuttgart, 1977. MR0466672

[316] H. G. Kaper and M. K. Kwong, *Asymptotics of the Titchmarsh-Weyl m-coefficient for integrable potentials*, Proc. Roy. Soc. Edinburgh Sect. A **103** (1986), no. 3-4, 347–358, DOI 10.1017/S0308210500018990. MR866847

[317] H. G. Kaper and M. K. Kwong, *Asymptotics of the Titchmarsh-Weyl m-coefficient for integrable potentials. II*, Differential equations and mathematical physics (Birmingham, Ala., 1986), Lecture Notes in Math., vol. 1285, Springer, Berlin, 1987, pp. 222–229, DOI 10.1007/BFb0080601. MR921273

[318] H. G. Kaper, M. K. Kwong, C. G. Lekkerkerker, and A. Zettl, *Full- and partial-range eigenfunction expansions for Sturm-Liouville problems with indefinite weights*, Proc. Roy. Soc. Edinburgh Sect. A **98** (1984), no. 1-2, 69–88, DOI 10.1017/S0308210500025567. MR765489

[319] H. G. Kaper, M. K. Kwong, and A. Zettl, *Characterizations of the Friedrichs extensions of singular Sturm-Liouville expressions*, SIAM J. Math. Anal. **17** (1986), no. 4, 772–777, DOI 10.1137/0517056. MR846388

[320] H. G. Kaper, M. K. Kwong, and A. Zettl, *Regularizing transformations for certain singular Sturm-Liouville boundary value problems*, SIAM J. Math. Anal. **15** (1984), no. 5, 957–963, DOI 10.1137/0515072. MR755855

[321] H. G. Kaper, M. K. Kwong, and A. Zettl, *Singular Sturm-Liouville problems with nonnegative and indefinite weights*, Monatsh. Math. **97** (1984), no. 3, 177–189, DOI 10.1007/BF01299145. MR753441

[322] H. G. Kaper, M. K. Kwong, and A. Zettl, *Proceedings of the focused research program on spectral theory and boundary value problems*, Argonne Reports ANL-87-26, vol. 1-4. (Edited by Kaper, Kwong and Zettl), 1988 and 1989.

[323] T. Kato, *Perturbation theory for linear operators, 2nd ed.*, Springer, Heidelberg, 1980.

[324] R. M. Kauffman, *Polynomials and the limit point condition*, Trans. Amer. Math. Soc. **201** (1975), 347–366, DOI 10.2307/1997342. MR358438

[325] R. M. Kauffman, *A rule relating the deficiency indices of L^j to those of L^k*, Proc. Roy. Soc. Edinburgh **74** (1976), 115–118.

[326] R. M. Kauffman, T. T. Read, and A. Zettl, *The deficiency index problem for powers of ordinary differential expressions*, Lecture Notes in Mathematics, Vol. 621, Springer-Verlag, Berlin-New York, 1977. MR0481243

[327] I. Kay and H. E. Moses, *Reflectionless transmission through dielectrics and scattering potentials*, J. Appl. Phys. **27** (1956), 1503–1508.

[328] M. V. Keldyš, *On the characteristic values and characteristic functions of certain classes of non-self-adjoint equations* (Russian), Doklady Akad. Nauk SSSR (N.S.) **77** (1951), 11–14. MR0041353

[329] H. B. Keller, *Numerical methods for two-point boundary-value problems*, Blaisdell Publishing Co. Ginn and Co., Waltham, Mass.-Toronto, Ont.-London, 1968. MR0230476

[330] V. I. Hrabustovskiĭ, *The discrete spectrum of perturbed differential operators of arbitrary order with periodic matrix coefficients* (Russian), Mat. Zametki **21** (1977), no. 6, 829–838. MR0458248

[331] M. Klaus, *On the variation-diminishing property of Schrödinger operators*, Oscillations, bifurcation and chaos (Toronto, Ont., 1986), CMS Conf. Proc., vol. 8, Amer. Math. Soc., Providence, RI, 1987, pp. 199–211. MR909910

[332] M. Klaus and J. K. Shaw, *On the eigenvalues of Zakharov-Shabat systems*, SIAM J. Math. Anal. **34** (2003), no. 4, 759–773, DOI 10.1137/S0036141002403067. MR1969601

[333] I. Knowles, *A limit-point criterion for a second-order linear differential operator*, J. London Math. Soc. (2) **8** (1974), 719–727, DOI 10.1112/jlms/s2-8.4.719. MR0367355

[334] I. Knowles, *Note on a limit-point criterion*, Proc. Amer. Math. Soc. **41** (1973), 117–119, DOI 10.2307/2038825. MR320425

[335] I. Knowles, *On second-order differential operators of limit-circle type*, in: Sleeman B. D., Michael I. M. (eds), Ordinary and Partial Differential Equations. Lecture Notes in Mathematics 415, Springer-Verlag, (1974), 184–187.

[336] I. Knowles, *On the boundary conditions characterizing J-selfadjoint extensions of J-symmetric operators*, J. Differential Equations **40** (1981), no. 2, 193–216, DOI 10.1016/0022-0396(81)90018-8. MR619134

[337] V. I. Kogan and F. S. Rofe-Beketov, *On square-integrable solutions of symmetric systems of differential equations of arbitrary order*, Proc. Roy. Soc. Edinburgh Sect. A **74** (1974/75), 5–40 (1976), DOI 10.1017/s0308210500016516. MR0454141

[338] V. I. Kogan and F. S. Rofe-Beketov, *On the question of the deficiency indices of differential operators with complex coefficients*, Proc. Roy. Soc. Edinburgh Sect. A **72** (1975), no. 4, 281–298. Translated from the Russian by E. R. Dawson (Mat. Fiz. i Funkcional Anal. No. 2 (1971), 45–60). MR0394303

[339] L. Kong, Q. Kong, H. Wu, and A. Zettl, *Regular approximations of singular Sturm-Liouville problems with limit-circle endpoints*, Results Math. **45** (2004), no. 3-4, 274–292, DOI 10.1007/BF03323382. MR2078454

[340] Q. Kong, Q. Lin, H. Wu, and A. Zettl, *A new proof of the inequalities among Sturm-Liouville eigenvalues*, PanAmer. Math. J. **10** (2000), no. 2, 1–11. MR1754507

[341] Q. Kong, H. Volkmer, and A. Zettl, *Matrix representations of Sturm-Liouville problems with finite spectrum*, Results Math. **54** (2009), no. 1-2, 103–116, DOI 10.1007/s00025-009-0371-3. MR2529630

[342] Q. Kong, H. Wu, and A. Zettl, *Dependence of eigenvalues on the problem*, Math. Nachr. **188** (1997), 173–201, DOI 10.1002/mana.19971880111. MR1484674

[343] Q. Kong, H. Wu, and A. Zettl, *Dependence of the nth Sturm-Liouville eigenvalue on the problem*, J. Differential Equations **156** (1999), no. 2, 328–354, DOI 10.1006/jdeq.1998.3613. MR1705395

[344] Q. Kong, H. Wu, and A. Zettl, *Geometric aspects of Sturm-Liouville problems I. structures on spaces of boundary conditions*, Proceedings of the Roy. Society of Edinburgh Sect. A **130** (2000), 561–589.

[345] Q. Kong, H. Wu, and A. Zettl, *Inequalities among eigenvalues of singular Sturm-Liouville problems*, Dynam. Systems Appl. **8** (1999), no. 3-4, 517–531. MR1722977

[346] Q. Kong, H. Wu, and A. Zettl, *Left-definite Sturm-Liouville problems*, J. Differential Equations **177** (2001), no. 1, 1–26, DOI 10.1006/jdeq.2001.3997. MR1867611

[347] Q. Kong, H. Wu, and A. Zettl, *Limits of Sturm-Liouville eigenvalues when the interval shrinks to an end point*, Proc. Roy. Soc. Edinburgh Sect. A **138** (2008), no. 2, 323–338, DOI 10.1017/S0308210506001004. MR2406693

[348] Q. Kong, H. Wu, and A. Zettl, *Multiplicity of Sturm-Liouville eigenvalues*, J. Comput. Appl. Math. **171** (2004), no. 1-2, 291–309, DOI 10.1016/j.cam.2004.01.036. MR2077210

[349] Q. Kong, H. Wu, and A. Zettl, *Sturm-Liouville problems with finite spectrum*, J. Math. Anal. Appl. **263** (2001), no. 2, 748–762, DOI 10.1006/jmaa.2001.7661. MR1866077

[350] Q. Kong and A. Zettl, *Dependence of eigenvalues of Sturm-Liouville problems on the boundary*, J. Differential Equations **126** (1996), no. 2, 389–407, DOI 10.1006/jdeq.1996.0056. MR1383983

[351] Q. Kong and A. Zettl, *The derivative of the matrix exponential function*, Cubo Mat. Educ. **3** (2001), no. 2, 121–124. MR1961583

[352] Q. Kong and A. Zettl, *Eigenvalues of regular Sturm-Liouville problems*, J. Differential Equations **131** (1996), no. 1, 1–19, DOI 10.1006/jdeq.1996.0154. MR1415044

[353] Q. Kong and A. Zettl, *Interval oscillation conditions for difference equations*, SIAM J. Math. Anal. **26** (1995), no. 4, 1047–1060, DOI 10.1137/S0036141093251286. MR1338373

[354] Q. Kong and A. Zettl, *Linear ordinary differential equations*, Inequalities and Applications of WSSIAA, **3** (1994), 381–397.

[355] T. H. Koornwinder, *Jacobi functions and analysis on noncompact semisimple Lie groups*, Special functions: group theoretical aspects and applications, Math. Appl., Reidel, Dordrecht, 1984, pp. 1–85. MR774055

[356] S. Kotani, *Lyapunov exponents and spectra for one-dimensional random Schrödinger operators*, Random matrices and their applications (Brunswick, Maine, 1984), Contemp. Math., vol. 50, Amer. Math. Soc., Providence, RI, 1986, pp. 277–286, DOI 10.1090/conm/050/841099. MR841099

[357] A. M. Krall, *Boundary values for an eigenvalue problem with a singular potential*, J. Differential Equations **45** (1982), no. 1, 128–138, DOI 10.1016/0022-0396(82)90059-6. MR662491

[358] A. M. Krall, W. N. Everitt, L. L. Littlejohn, and V. P. Onyango-Otieno, *The Laguerre type operator in a left definite Hilbert space*, J. Math. Anal. Appl. **192** (1995), no. 2, 460–468, DOI 10.1006/jmaa.1995.1183. MR1332221

[359] A. M. Krall and A. Zettl, *Singular selfadjoint Sturm-Liouville problems*, Differential Integral Equations **1** (1988), no. 4, 423–432. MR945819

[360] A. M. Krall and A. Zettl, *Singular selfadjoint Sturm-Liouville problems. II. Interior singular points*, SIAM J. Math. Anal. **19** (1988), no. 5, 1135–1141, DOI 10.1137/0519078. MR957673

[361] W. Kratz, *Quadratic functionals in variational analysis and control theory*, Mathematical Topics, vol. 6, Akademie Verlag, Berlin, 1995. MR1334092

[362] M. Krein, *The theory of self-adjoint extensions of semi-bounded Hermitian transformations and its applications. I* (Russian, with English summary), Rec. Math. [Mat. Sbornik] N.S. **20(62)** (1947), 431–495. MR0024574

[363] M. G. Krein, *The theory of self-adjoint extensions of semi-bounded Hermitian transformations and its applications II*, Mat. Sb. **21** (1947), 365–404. (In Russian).

[364] K. Kreith, *PDE generalizations of the Sturm comparison theorem*, Mem. Amer. Math. Soc. **48** (1984), no. 298, 31–46. Essays in the history of mathematics (San Francisco, Calif., 1981). MR733263

[365] K. Kreith, *Oscillation theory*, Lecture Notes in Mathematics 324, Springer-Verlag, Berlin, 1973.

[366] N. P. Kupcov, *Conditions for non-selfadjointness of a second-order linear differential operator* (Russian), Dokl. Akad. Nauk SSSR **138** (1961), 767–770. MR0132869

[367] H. Kurss, *A limit-point criterion for nonoscillatory Sturm-Liouville differential operators*, Proc. Amer. Math. Soc. **18** (1967), 445–449, DOI 10.2307/2035475. MR213640

[368] M. K. Kwong, *L^p-perturbations of second order linear differential equations*, Math. Ann. **215** (1975), 23–34, DOI 10.1007/BF01351788. MR377206

[369] M. K. Kwong, *Note on the strong limit point condition of second order differential expressions*, Quart. J. Math. Oxford Ser. (2) **28** (1977), no. 110, 201–208, DOI 10.1093/qmath/28.2.201. MR0450658

[370] M. K. Kwong, *On boundedness of solutions of second order differential equations in the limit circle case*, Proc. Amer. Math. Soc. **52** (1975), 242–246, DOI 10.2307/2040138. MR387710

[371] M. K. Kwong and A. Zettl, *An alternate proof of Kato's inequality*, Evolution equations, Lecture Notes in Pure and Appl. Math., vol. 234, Dekker, New York, 2003, pp. 275–279. MR2073751

[372] M. K. Kwong and A. Zettl, *Best constants for discrete Kolmogorov inequalities*, Houston J. Math. **15** (1989), no. 1, 99–119. MR1002084

[373] M. K. Kwong and A. Zettl, *Discreteness conditions for the spectrum of ordinary differential operators*, J. Differential Equations **40** (1981), no. 1, 53–70, DOI 10.1016/0022-0396(81)90010-3. MR614218

[374] M. K. Kwong and A. Zettl, *An extension of the Hardy-Littlewood inequality*, Proc. Amer. Math. Soc. **77** (1979), no. 1, 117–118, DOI 10.2307/2042727. MR539642

[375] M. K. Kwong and A. Zettl, *Extremals in Landau's inequality for the difference operator*, Proc. Roy. Soc. Edinburgh Sect. A **107** (1987), no. 3-4, 299–311, DOI 10.1017/S0308210500031176. MR924523

[376] M. K. Kwong and A. Zettl, *Landau's inequality*, Proc. Roy. Soc. Edinburgh Sect. A **97** (1984), 161–163.

[377] M. K. Kwong and A. Zettl, *Landau's inequality for the difference operator*, Proc. Amer. Math. Soc. **104** (1988), no. 1, 201–206, DOI 10.2307/2047486. MR958067

[378] M. K. Kwong and A. Zettl, *Norm inequalities for dissipative operators on inner product spaces*, Houston J. Math. **5** (1979), no. 4, 543–557. MR567911

[379] M. K. Kwong and A. Zettl, *Norm inequalities for the powers of a matrix*, Amer. Math. Monthly **98** (1991), no. 6, 533–538, DOI 10.2307/2324875. MR1109578

[380] M. K. Kwong and A. Zettl, *Norm inequalities of product form is weighted L^p spaces*, Proc. Roy. Soc. Edinburgh Sect. A **89** (1981), 293–307.

[381] M. K. Kwong and A. Zettl, *On the limit point classification of second order differential equations*, Math. Z. **132** (1973), 297–304, DOI 10.1007/BF01179735. MR322258

[382] M. K. Kwong and A. Zettl, *Ramifications of Landau's inequality*, Proc. Roy. Soc. Edinburgh Sect. A **86** (1980), 175–212.

[383] M. K. Kwong and A. Zettl, *Remarks on best constants for norm inequalities among powers of an operator*, J. Approx. Theory **26** (1979), no. 3, 249–258, DOI 10.1016/0021-9045(79)90062-5. MR551676

[384] M. K. Kwong and A. Zettl, *Weighted norm inequalities of sum form involving derivatives*, Proc. Roy. Soc. Edinburgh Sect. A **88** (1981), 121–134.

[385] M. K. Kwong and A. Zettl, *Norm inequalities for derivatives and differences*, Lecture Notes in Mathematics, vol. 1536, Springer-Verlag, Berlin, 1992. MR1223546

[386] V. Lakshmikantham and S. Leela, *Differential and integral inequalities: Theory and applications. Vol. I: Ordinary differential equations*, Academic Press, New York-London, 1969. Mathematics in Science and Engineering, Vol. 55-I. MR0379933

[387] P. Lancaster, A. Shkalikov, and Q. Ye, *Strongly definitizable linear pencils in Hilbert space*, Integral Equations Operator Theory **17** (1993), no. 3, 338–360, DOI 10.1007/BF01200290. MR1237958

[388] H. Langer and A. Schneider, *On spectral properties of regular quasidefinite pencils* $F - \lambda G$, Results Math. **19** (1991), no. 1-2, 89–109, DOI 10.1007/BF03322419. MR1091959

[389] W. Lay and S. Yu. Slavyanov, *Heun's equation with nearby singularities*, R. Soc. Lond. Proc. Ser. A Math. Phys. Eng. Sci. **455** (1999), no. 1992, 4347–4361, DOI 10.1098/rspa.1999.0504. MR1809364

[390] Arien-Marie Legendre, *Suite des recherches sur la figure des planétes*, Histoire de l'Académie Royale des Sciences avec les Mémoires de Mathématique et de Physique, Paris 1789 (1793), 372–454.

[391] W. Leighton, *Comparison theorems for linear differential equations of second order*, Proc. Amer. Math. Soc. **13** (1962), 603–610, DOI 10.2307/2034834. MR140759

[392] W. Leighton, *On self-adjoint differential equations of second order*, J. London Math. Soc. **27** (1952), 37–47, DOI 10.1112/jlms/s1-27.1.37. MR0046506

[393] W. Leighton, *Principal quadratic functionals*, Trans. Amer. Math. Soc. **67** (1949), 253–274, DOI 10.2307/1990473. MR34535

[394] A. Ju. Levin, *Classification of the cases of non-oscillation for the equation* $\ddot{x}+p(t)\dot{x}+q(t)x = 0$*, where* $q(t)$ *is of constant sign* (Russian), Dokl. Akad. Nauk SSSR **171** (1966), 1037–1040. MR0206374

[395] A. Ju. Levin, *A comparison principle for second-order differential equations*, Soviet Math. Dokl. **1** (1960), 1313–1316. MR0124563

[396] A. Ju. Levin, *Distribution of the zeros of a linear differential equation*, Soviet Math. Dokl., **5** (1964), 818–822.

[397] A. Ju. Levin, *Some properties bearing on the oscillation of linear differential equations*, Soviet Math. Dokl., **4** (1963), 121–124.

[398] N. Levinson, *Criteria for the limit-point case for second order linear differential operators* (English, with Czech summary), Časopis Pěst. Mat. Fys. **74** (1949), 17–20. MR0032066

[399] N. Levinson, *On the uniqueness of the potential in a Schrödinger equation for a given asymptotic phase*, Danske vid. Selsk. Mat.-Fys. Medd. **25** (1949), 1–29.

[400] B. M. Levitan and I. S. Sargsjan, *Introduction to spectral theory: selfadjoint ordinary differential operators*, American Mathematical Society, Providence, R.I., 1975. Translated from the Russian by Amiel Feinstein; Translations of Mathematical Monographs, Vol. 39. MR0369797

[401] W. M. Li, *The higher order differential operators in direct sum spaces*, J. Differential Equations **84** (1990), no. 2, 273–289, DOI 10.1016/0022-0396(90)90079-5. MR1047570

[402] A. Liapunov, *Probléme général de la stabilité du mouvement*, (French translation of a Russian paper dated 1893), Ann. Fac. Sci. Univ. Toulouse, **2** (1907), 27-247; reprinted as Ann. Math. Studies, No. 17, Princton, 1949.

[403] L. L. Littlejohn and A. Zettl, *Left-definite variations of the classical Fourier expansion theorem*, Electron. Trans. Numer. Anal. **27** (2007), 124–139. MR2346153

[404] L. Littlejohn and A. Zettl, *The Legendre equation and its self-adjoint operators*, Electronic J. Differential Equations (69) 2011 (2011), 1-33.

[405] J. E. Littlewood, *On linear differential equations of the second order with a strongly oscillating coefficient of* y, J. London Math. Soc. **41** (1966), 627–638, DOI 10.1112/jlms/s1-41.1.627. MR0200513

[406] J. Locker, *Functional analysis and two-point differential operators*, Pitman Research Notes in Mathematics Series, vol. 144, Longman Scientific & Technical, Harlow; John Wiley & Sons, Inc., New York, 1986. MR865984

[407] J. Locker, *Spectral theory of non-self-adjoint two-point differential operators*, Mathematical Surveys and Monographs, vol. 73, American Mathematical Society, Providence, RI, 2000. MR1721499

[408] J. Lützen, *Sturm and Liouville's work on ordinary linear differential equations. The emergence of Sturm-Liouville theory*, Arch. Hist. Exact Sci. **29** (1984), no. 4, 309–376, DOI 10.1007/BF00348405. MR745152

[409] J. Lützen, *Joseph Liouville 1809–1882: master of pure and applied mathematics*, Studies in the History of Mathematics and Physical Sciences, vol. 15, Springer-Verlag, New York, 1990. MR1066463

[410] V. A. Marchenko, *Sturm-Liouville operators and applications*, Operator Theory: Advances and Applications, vol. 22, Birkhäuser Verlag, Basel, 1986. Translated from the Russian by A. Iacob. MR897106

[411] P. A. Markowich, *Eigenvalue problems on infinite intervals*, Math. Comp. **39** (1982), no. 160, 421–441, DOI 10.2307/2007322. MR669637

[412] L. Markus and R. A. Moore, *Oscillation and disconjugacy for linear differential equations with almost periodic coefficients*, Acta Math. **96** (1956), 99–123, DOI 10.1007/BF02392359. MR0080813

[413] M. Marletta, *Certification of algorithm 700: Numerical tests of the SLEIGN software for Sturm-Liouville problems*, ACM TOMS, 17 (1991), 481-490.

[414] M. Marletta, *Numerical solution of eigenvalue problems for Hamiltonian systems*, Adv. Comput. Math. **2** (1994), no. 2, 155–184, DOI 10.1007/BF02521106. MR1269379

[415] M. Marletta and A. Zettl, *Counting and computing eigenvalues of left-definite Sturm-Liouville problems*, J. Comput. Appl. Math. **148** (2002), no. 1, 65–75, DOI 10.1016/S0377-0427(02)00573-3. On the occasion of the 65th birthday of Professor Michael Eastham. MR1946187

[416] M. Marletta and A. Zettl, *The Friedrichs extension of singular differential operators*, J. Differential Equations **160** (2000), no. 2, 404–421, DOI 10.1006/jdeq.1999.3685. MR1736997

[417] J. B. McLeod, *The number of integrable-square solutions of ordinary differential equations*, Quart. J. Math. Oxford Ser. (2) **17** (1966), 285–290, DOI 10.1093/qmath/17.1.285. MR0199467

[418] J. B. McLeod, *On the spectrum of wildly oscillating functions*, J. London Math. Soc. **39** (1964), 623–634, DOI 10.1112/jlms/s1-39.1.623. MR0180720

[419] J. B. McLeod, *Some examples of wildly oscillating potentials*, J. London Math. Soc. **43** (1968), 647–654, DOI 10.1112/jlms/s1-43.1.647. MR0230964

[420] F. Meng and A. B. Mingarelli, *Oscillation of linear Hamiltonian systems*, Proc. Amer. Math. Soc. **131** (2003), no. 3, 897–904, DOI 10.1090/S0002-9939-02-06614-5. MR1937428

[421] R. Mennicken and M. Möller, *Non-self-adjoint boundary eigenvalue problems*, North-Holland Mathematics Studies, vol. 192, North-Holland Publishing Co., Amsterdam, 2003. MR1995773

[422] A. B. Mingarelli, *A survey of the regular weighted Sturm-Liouville problem—the nondefinite case*, International workshop on applied differential equations (Beijing, 1985), World Sci. Publishing, Singapore, 1986, pp. 109–137. MR901329

[423] A. B. Mingarelli and S. G. Halvorsen, *Nonoscillation domains of differential equations with two parameters*, Lecture Notes in Mathematics, vol. 1338, Springer-Verlag, Berlin, 1988. MR959733

[424] M. Möller, *On the essential spectrum of a class of operators in Hilbert space*, Math. Nachr. **194** (1998), 185–196, DOI 10.1002/mana.19981940112. MR1653098

[425] M. Möller, *On the unboundedness below of the Sturm-Liouville operator*, Proc. Roy. Soc. Edinburgh Sect. A **129** (1999), 1011–1015.

[426] M. Möller and A. Zettl, *Differentiable dependence of eigenvalues of operators in Banach spaces*, J. Operator Theory **36** (1996), no. 2, 335–355. MR1432122

[427] M. Möller and A. Zettl, *Semi-boundedness of ordinary differential operators*, J. Differential Equations **115** (1995), no. 1, 24–49, DOI 10.1006/jdeq.1995.1002. MR1308603

[428] M. Möller and A. Zettl, *Symmetric differential operators and their Friedrichs extension*, J. Differential Equations **115** (1995), no. 1, 50–69, DOI 10.1006/jdeq.1995.1003. MR1308604

[429] M. Möller and A. Zettl, *Weighted norm inequalities for the quasi-derivatives of ordinary differential operators*, Results Math. **24** (1993), no. 1-2, 153–160, DOI 10.1007/BF03322324. MR1229066

[430] R. A. Moore, *The behavior of solutions of a linear differential equation of second order*, Pacific J. Math. **5** (1955), 125–145. MR68690

[431] R. A. Moore, *The least eigenvalue of Hill's equation*, J. Analyse Math. **5** (1956/57), 183–196, DOI 10.1007/BF02937345. MR0086200

[432] M. Morse, *A generalization of the Sturm separation and comparison theorems in n-space*, Math. Ann. **103** (1930), no. 1, 52–69, DOI 10.1007/BF01455690. MR1512617

[433] M. Morse, *Variational analysis: critical extremals and Sturmian extensions*, Interscience Publishers [John Wiley & Sons, Inc.], New York-London-Sydney, 1973. Pure and Applied Mathematics. MR0420368

[434] P. M. Morse, *Diatomic molecules according to the wave mechanics; II: Vibration levels.*, Phys. Rev. 34 (1929), 57-64.

[435] O. Mukhtarov, *Discontinuous boundary value problem with spectral parameter in boundary conditions* (English, with English and Turkish summaries), Turkish J. Math. **18** (1994), no. 2, 183–192. MR1281382

[436] O. Muhtarov and S. Yakubov, *Problems for ordinary differential equations with transmission conditions*, Appl. Anal. **81** (2002), no. 5, 1033–1064, DOI 10.1080/0003681021000029853. MR1948030

[437] E. Müller-Pfeiffer, *Spectral theory of ordinary differential operators*, Ellis Horwood Ltd., Chichester; Halsted Press [John Wiley & Sons, Inc.], New York, 1981. Translated from the German by the author; Translation edited by M. S. P. Eastham; Ellis Horwood Series in Mathematics and its Applications. MR606197

[438] I. M. Nabiev, *Multiplicity and relative position of the eigenvalues of a quadratic pencil of Sturm-Liouville operators* (Russian, with Russian summary), Mat. Zametki **67** (2000), no. 3, 369–381, DOI 10.1007/BF02676667; English transl., Math. Notes **67** (2000), no. 3-4, 309–319. MR1779470

[439] M. A. Naimark, *Lineare differentialoperatoren. Mathematische lehrbüecher und monographien*, Akadmie-Verlag, Berlin, 1960.

[440] M. A. Naimark, *Linear differential operators II*, Ungar Publishing Company, New York, 1968.

[441] H. Narnhofer, *Quantum theory for $1/r^2$-potentials*, Acta Phys. Austriaca **40** (1974), 306–322. MR0368660

[442] J. W. Neuberger, *Concerning boundary value problems*, Pacific J. Math. **10** (1960), 1385–1392. MR124701

[443] F. Neuman, *On a problem of transformations between limit-circle and limit-point differential equations*, Proc. Roy. Soc. Edinburgh Sect. A **72** (1975), no. 3, 187–193. MR0385226

[444] F. Neuman, *On the Liouville transformation*, Rend. Mat. (6) **3** (1970), 133–140. MR0273090

[445] R. G. Newton, *Inverse scattering by a local impurity in a periodic potential in one dimension*, J. Math. Phys. **24** (1983), no. 8, 2152–2162, DOI 10.1063/1.525968. MR713548

[446] R. G. Newton, *Inverse scattering by a local impurity in a periodic potential in one dimension. II*, J. Math. Phys. **26** (1985), no. 2, 311–316, DOI 10.1063/1.526660. MR776499

[447] R. G. Newton, *The Marchenko and Gelfand-Levitan methods in the inverse scattering problem in one and three dimensions*, In Conference on Inverse Scattering: Theory and Application, Society for Industrial and Applied Mathematics, Philadelphia, 1983.

[448] H.-D. Niessen, *A necessary and sufficient limit-circle criterion for left-definite eigenvalue problems*, Ordinary and partial differential equations (Proc. Conf., Univ. Dundee, Dundee, 1974), Springer, Berlin, 1974, pp. 205–210. Lecture Notes in Math., Vol. 415. MR0430393

[449] H.-D. Niessen, *Zum verallgemeinerten zweiten Weylschen Satz* (German), Arch. Math. (Basel) **22** (1971), 648–656, DOI 10.1007/BF01222630. MR0298109

[450] H. D. Niessen and A. Schneider, *Linksdefinite singuläre kanonische Eigenwertprobleme. II*, J. Reine Angew. Math. **289** (1977), 62–84, DOI 10.1515/crll.1977.289.62. MR435498

[451] H. D. Niessen and A. Schneider, *Linksdefinite singuläre kanonische Eigenwertprobleme. I* (German), J. Reine Angew. Math. **281** (1976), 13–52. MR390358

[452] H.-D. Niessen and A. Schneider, *Spectral theory for left-definiate singular systems of differential equations I and II*, North-Holland Mathematical Studies **13** (1974), 29–56.

[453] H.-D. Niessen and A. Zettl, *The Friedrichs extension of regular ordinary differential operators*, Proc. Roy. Soc. Edinburgh Sect. A **114** (1990), no. 3-4, 229–236, DOI 10.1017/S0308210500024409. MR1055546

[454] H.-D. Niessen and A. Zettl, *Singular Sturm-Liouville problems: the Friedrichs extension and comparison of eigenvalues*, Proc. London Math. Soc. (3) **64** (1992), no. 3, 545–578, DOI 10.1112/plms/s3-64.3.545. MR1152997

[455] K. S. Ong, *On the limit-point and limit-circle theory of second-order differential equations*, Proc. Roy. Soc. Edinburgh Sect. A **72** (1975), no. 3, 245–256. MR0393635

[456] S. A. Orlov, *On the deficiency index of linear differential operators* (Russian), Doklady Akad. Nauk SSSR (N.S.) **92** (1953), 483–486. MR0061277

[457] M. Otelbaev, *The summability with weight of the solution of a Sturm-Liouville equation* (Russian), Mat. Zametki **16** (1974), 969–980. MR0369798

[458] W. T. Patula and P. Waltman, *Limit point classification of second order linear differential equations*, J. London Math. Soc. (2) **8** (1974), 209–216, DOI 10.1112/jlms/s2-8.2.209. MR0344578

[459] W. T. Patula and J. S. W. Wong, *An L^p-analogue of the Weyl alternative*, Math. Ann. **197** (1972), 9–28, DOI 10.1007/BF01427949. MR299865

[460] D. B. Pearson, *Value distribution and spectral analysis of differential operators*, J. Phys. A **26** (1993), no. 16, 4067–4080. MR1236597

[461] D. B. Pearson, *Value distribution and spectral theory*, Proc. London Math. Soc. (3) **68** (1994), no. 1, 127–144, DOI 10.1112/plms/s3-68.1.127. MR1243838

[462] D. B. Pearson, *Quantum scattering and spectral theory*, Academic Press, London, 1988.

[463] W. Peng, M. Racovitan, and H. Wu, *Geometric aspects of Sturm-Liouville problems V. Natural loops of boundary conditions for monotonicity of eigenvalues and their applications*, Pac. J. Appl. Math. **4** (2012), no. 4, 253–273 (2013). MR3088795

[464] I. G. Petrovski, *Ordinary differential equations*, Prentice-Hall, Inc., Englewood Cliffs, N.J., 1966. Revised English edition. Translated from the Russian and edited by Richard A. Silverman. MR0193298

[465] Å. Pleijel, *Complementary remarks about the limit point and limit circle theory.*, Ark. Mat. **8** (1969), 45–47, DOI 10.1007/BF02589534. MR0259224

[466] A. Pleijel, *Generalized Weyl circles*, Lecture Notes in Mathematics 415 (1974), 211-226 (Springer Verlag; Heidelberg, 1974).

[467] Å. Pleijel, *Some remarks about the limit point and limit circle theory*, Ark. Mat. **7** (1969), 543–550, DOI 10.1007/BF02590893. MR0240378

[468] A. Pleijel, *A survey of spectral theory for pairs of ordinary differential operators*, Lecture Notes in Mathematics 448 (1975), 256-272 (Springer, Birlin, Heidelberg, 1975).

[469] M. Plum, *Eigenvalue inclusions for second-order ordinary differential operators by a numerical homotopy method*, Z. Angew. Math. Phys. **41** (1990), no. 2, 205–226, DOI 10.1007/BF00945108. MR1045812

[470] E. G. C. Poole, *Introduction to the theory of linear differential equations*, Oxford University Press, Oxford, 1936.

[471] J. Pöschel and E. Trubowitz, *Inverse spectral theory*, Pure and Applied Mathematics, vol. 130, Academic Press, Inc., Boston, MA, 1987. MR894477

[472] S. Pruess, *Estimating the eigenvalues of Sturm-Liouville problems by approximating the differential equation*, SIAM J. Numer. Anal. **10** (1973), 55–68, DOI 10.1137/0710008. MR0327048

[473] S. Pruess, *High order approximations to Sturm-Liouville eigenvalues*, Numer. Math. **24** (1975), no. 3, 241–247, DOI 10.1007/BF01436595. MR0378431

[474] S. A. Pruess, *Solving linear boundary value problems by approximating the coefficients*, Math. Comp. **27** (1973), 551–561, DOI 10.2307/2005659. MR0371100

[475] S. Pruess, C. T. Fulton, and Y. Xie, *An asymptotic numerical method for a class of singular Sturm-Liouville problems*, SIAM J. Numer. Anal. **32** (1995), no. 5, 1658–1676, DOI 10.1137/0732074. MR1352206

[476] S. Pruess and C. T. Fulton, *Mathematical software for Sturm-Liouville problems*, ACM. Trans. Math. Software 19 (1993), 360-376.

[477] J. D. Pryce, *A test package for Sturm-Liouville solvers*, ACM Trans. Math. Software **25** (1999), no. 1, 21–57, DOI 10.1145/305658.287651. MR1697462

[478] J. D. Pryce, *Numerical solution of Sturm-Liouville problems*, Monographs on Numerical Analysis, The Clarendon Press, Oxford University Press, New York, 1993. Oxford Science Publications. MR1283388

[479] C. R. Putnam, *On the spectra of certain boundary value problems*, Amer. J. Math. **71** (1949), 109–111, DOI 10.2307/2372098. MR28494

[480] J. Qi and S. Chen, *Essential spectra of singular matrix differential operators of mixed order in the limit circle case*, Math. Nachr. **284** (2011), no. 2-3, 342–354, DOI 10.1002/mana.200810062. MR2790893

[481] J. Qi and S. Chen, *On an open problem of Weidmann: essential spectra and square-integrable solutions*, Proc. Roy. Soc. Edinburgh Sect. A **141** (2011), no. 2, 417–430, DOI 10.1017/S0308210509001681. MR2786688

[482] D. Race, *The theory of J-selfadjoint extensions of J-symmetric operators*, J. Differential Equations **57** (1985), no. 2, 258–274, DOI 10.1016/0022-0396(85)90080-4. MR788280

[483] D. Race and A. Zettl, *Characterisation of the factors of quasi-differential expressions*, Proc. Roy. Soc. Edinburgh Sect. A **120** (1992), no. 3-4, 297–312, DOI 10.1017/S0308210500032157. MR1159187

[484] D. Race and A. Zettl, *Nullspaces, representations and factorizations of quasi-differential expressions*, J. Differential Integral Equations **6** (1993), no. 4, 949–960. MR1222312

[485] D. Race and A. Zettl, *On the commutativity of certain quasidifferential expressions. I*, J. London Math. Soc. (2) **42** (1990), no. 3, 489–504, DOI 10.1112/jlms/s2-42.3.489. MR1087223

[486] T. T. Read, *A limit-point criterion for* $-(py')' + qy$, Ordinary and partial differential equations (Proc. Fourth Conf., Univ. Dundee, Dundee, 1976), Springer, Berlin, 1976, pp. 383–390. Lecture Notes in Math., Vol. 564. MR0593164

[487] T. T. Read, *A limit-point criterion for expressions with intermittently positive coefficients*, J. London Math. Soc. (2) **15** (1977), no. 2, 271–276, DOI 10.1112/jlms/s2-15.2.271. MR0437844

[488] T. T. Read, *A limit-point criterion for expressions with oscillatory coefficients*, Pacific J. Math. **66** (1976), no. 1, 243–255. MR442369

[489] T. T. Read, *On the limit point condition for polynomials in a second order differential expression*, J. London Math. Soc. (2) **10** (1975), 357–366, DOI 10.1112/jlms/s2-10.3.357. MR0372310

[490] M. Reed and B. Simon, *Methods of modern mathematical physics. I. Functional analysis*, Academic Press, New York-London, 1972. MR0493419

[491] M. Reed and B. Simon, *Methods of modern mathematical physics, vol. IV: Analysis of Operators*, Academic Press, New York, 1978.

[492] W. T. Reid, *Oscillation criteria for self-adjoint differential systems*, Trans. Amer. Math. Soc. **101** (1961), 91–106, DOI 10.2307/1993413. MR133518

[493] W. T. Reid, *Ordinary differential equations*, John Wiley & Sons, Inc., New York-London-Sydney, 1971. MR0273082

[494] W. T. Reid, *Riccati differential equations*, Academic Press, New York-London, 1972. Mathematics in Science and Engineering, Vol. 86. MR0357936

[495] W. T. Reid, *Sturmian theory for ordinary differential equations*, Applied Mathematical Sciences, vol. 31, Springer-Verlag, New York-Berlin, 1980. With a preface by John Burns. MR606199

[496] F. Rellich, *Die zulässigen Randbedingungen bei den singulären Eigenwertproblemen der mathematischen Physik. (Gewöhnliche Differentialgleichungen zweiter Ordnung.)* (German), Math. Z. **49** (1944), 702–723, DOI 10.1007/BF01174227. MR13183

[497] F. Rellich, *Halbbeschraenkte gewoehnliche diffferentialoperatoren zweiter ordnung*, Math. Ann. **122** (1950/51), 343–368.

[498] C. Remling, *Essential spectrum and* L_2*-solutions of one-dimensional Schrödinger operators*, Proc. Amer. Math. Soc. **124** (1996), no. 7, 2097–2100, DOI 10.1090/S0002-9939-96-03463-6. MR1342044

[499] B. Riemann and H. Weber, *Die partiellen differential gleichungen der mathematischen physik, 5th ed., vol. 2*, Friedr. Vieweg, Braunschweig, 1912.

[500] F. S. Rofe-Beketov, *Non-semibounded differential operators* (Russian), Teor. Funkciĭ Funkcional. Anal. i Priložen. Vyp. **2** (1966), 178–184. MR0199750

[501] F. S. Rofe-Beketov, *A test for the finiteness of the number of discrete levels introduced into the gaps of a continuous spectrum by perturbations of a periodic potential*, Sov. Math. Dokl. **5** (1964), 689–692.

[502] F. S. Rofe-Beketov and A. M. Kholkin, *Spectral analysis of differential operators: Interplay between spectral and oscillatory properties*, World Scientific Monograph Series in Mathematics, vol. 7, World Scientific Publishing Co. Pte. Ltd., Hackensack, NJ, 2005. Translated from the Russian by Ognjen Milatovic and revised by the authors; With a foreword by Vladimir A. Marchenko. MR2175241

[503] A. Ronveaux, *Heun's differential equations*, Oxford University Press, Oxford, 1995.

[504] R. Rosenberger, *Charakterisierung der Friedrichsfortsetzung von halbbeschräenkten Sturm-Louiville opatoren*, Dissertation, Technische Hochschule Darmstadt, 1984.

[505] H. L. Royden, *Real analysis*, 3rd ed., Macmillan Publishing Company, New York, 1988. MR1013117

[506] R. Høegh-Krohn, S. Albeverio, F. Gesztesy, and H. Holden, *Solvable models in quantum mechanics*, Texts and Monographs in Physics, Springer-Verlag, New York, 1988. MR926273

[507] B. Schultze, *Green's matrix and the formula of Titchmarsh-Kodaira for singular left-definite canonical eigenvalue problems*, Proc. Roy. Soc. Edinburgh Sect. A **83** (1979), no. 1-2, 147–183, DOI 10.1017/S030821050001146X. MR538594

[508] D. B. Sears and E. C. Titchmarsh, *Some eigenfunction formulae*, Quart. J. Math., Oxford Ser. (2) **1** (1950), 165–175, DOI 10.1093/qmath/1.1.165. MR0037436

[509] Z. J. Shang, *On J-selfadjoint extensions of J-symmetric ordinary differential operators*, J. Differential Equations **73** (1988), no. 1, 153–177, DOI 10.1016/0022-0396(88)90123-4. MR938220

[510] Z. J. Shang and R. Y. Zhu, *The domains of the selfadjoint extensions of ordinary symmetric differential operators over* $(-\infty, \infty)$ (Chinese, with English summary), Neimenggu Daxue Xuebao **17** (1986), no. 1, 17–28. MR854184

[511] J. K. Shaw, A. P. Baronavski, and H. D. Ladouceur, *Applications of the Walker method*, Spectral theory and computational methods of Sturm-Liouville problems (Knoxville, TN, 1996), Lecture Notes in Pure and Appl. Math., vol. 191, Dekker, New York, 1997, pp. 377–395. MR1460561

[512] D. Shenk and M. A. Shubin, *Asymptotic expansion of state density and the spectral function of the Hill operator* (Russian), Mat. Sb. (N.S.) **128(170)** (1985), no. 4, 474–491, 575. MR820398

[513] D. Shin, *Existence theorems for quasi-differential equations of order n*, Dokl. Akad. Nauk SSSR **18** (1938), 515–518.

[514] D. Shin, *On quasi-differential operators in Hilbert space*, Dokl. Akad. SSSR **18** (1938), 523–526.

[515] D. Shin, *On the solutions of a linear quasi-differential equation of order n*, Mat. Sb. **7** (1940), 479–532.

[516] D. Shin, *Quasi-differential operators in Hilbert space* (Russian, with English summary), Rec. Math. [Mat. Sbornik] N. S. **13(55)** (1943), 39–70. MR0011534

[517] B. Simon, *Resonances in one dimension and Fredholm determinants*, J. Funct. Anal. **178** (2000), no. 2, 396–420, DOI 10.1006/jfan.2000.3669. MR1802901

[518] B. Simon, *Schrödinger semigroups*, Bull. Amer. Math. Soc. (N.S.) **7** (1982), no. 3, 447–526, DOI 10.1090/S0273-0979-1982-15041-8. MR670130

[519] B. Simon and T. Wolff, *Singular continuous spectrum under rank one perturbations and localization for random Hamiltonians*, Comm. Pure Appl. Math. **39** (1986), no. 1, 75–90, DOI 10.1002/cpa.3160390105. MR820340

[520] S. Yu. Slavyanov and W. Lay, *Special functions: A unified theory based on singularities*, Oxford Mathematical Monographs, Oxford Science Publications, Oxford University Press, Oxford, 2000. With a foreword by Alfred Seeger. MR1858237

[521] H. M. Srivastava, V. K. Tuan, and S. B. Yakubovich, *The Cherry transform and its relationship with a singular Sturm-Liouville problem*, Quart. J. Math. **51** (2000), no. 3, 371–383, DOI 10.1093/qjmath/51.3.371. MR1782100

[522] G. Stolz and J. Weidmann, *Approximation of isolated eigenvalues of general singular ordinary differential operators*, Results Math. **28** (1995), no. 3-4, 345–358, DOI 10.1007/BF03322261. MR1356897

[523] G. Stolz and J. Weidmann, *Approximation of isolated eigenvalues of ordinary differential operators*, J. Reine Angew. Math. **445** (1993), 31–44. MR1244968

[524] M. H. Stone, *Linear transformations in Hilbert space*, American Mathematical Society Colloquium Publications, vol. 15, American Mathematical Society, Providence, RI, 1990. Reprint of the 1932 original. MR1451877

[525] C. Sturm, *Sur les equations differentielles lineaires du second ordre*, J. Math. Pures Appl. **1** (1836), 106–186.

[526] C. Sturm, *Sur une classe d'equations à derivée partielle*, J. Math. Pures Appl., **1** (1836), 373–444.

[527] C. Sturm and J. Liouville, *Estrait d'un mémoire sur le développement des fonctions en série dont les différents termes sont assujettis à satisfaire à une même équation différentielle linéaire, contenant un paramètre variable*, Jour. Math. Pures et Appl. de Liouville II (1837), 220-223.

[528] J. Sun, *On the selfadjoint extensions of symmetric ordinary differential operators with middle deficiency indices*, Acta Math. Sinica (N.S.) **2** (1986), no. 2, 152–167, DOI 10.1007/BF02564877. MR877379

[529] J. Sun, D. E. Edmunds, and A. Wang, *The number of Dirichlet solutions and discreteness of spectrum of differential operators with middle deficiency indices*, J. Anal. Appl. **3** (2005), no. 1, 55–66. MR2110500

[530] J. Sun and A. Wang, *Sturm-Liouville operators with interface conditions*, The Progress of Research for Math, Mech, Phys. and High Tech. Science Press, Beijing **12** (2008), 513-516.

[531] J. Sun, A. Wang, and A. Zettl, *Continuous spectrum and square-integrable solutions of differential operators with intermediate deficiency index*, J. Funct. Anal. **255** (2008), no. 11, 3229–3248, DOI 10.1016/j.jfa.2008.08.007. MR2464576

[532] J. Sun, A. Wang, and A. Zettl, *Two-interval Sturm-Liouville operators in direct sum spaces with inner product multiples*, Results Math. **50** (2007), no. 1-2, 155–168, DOI 10.1007/s00025-006-0241-1. MR2313137

[533] C. A. Swanson, *Comparison and oscillation theory of linear differential equations*, Academic Press, New York-London, 1968. Mathematics in Science and Engineering, Vol. 48. MR0463570

[534] F. J. Tipler, *General relativity and conjugate ordinary differential equations*, J. Differential Equations **30** (1978), no. 2, 165–174, DOI 10.1016/0022-0396(78)90012-8. MR513268

[535] E. C. Titchmarsh, *Eigenfunction expansions associated with second-order differential equations. Part I*, Second Edition, Clarendon Press, Oxford, 1962. MR0176151

[536] E. C. Titchmarsh, *Eigenfunction expansions: Part 1: Associated with second-order differential equations (second edition)*, Oxford University Press, 1962.

[537] A. L. Treskunov, *Sharp two-sided estimates for eigenvalues of some class of Sturm-Liouville problems: Nonlinear equations and mathematical analysis*, J. Math. Sci. (New York) **101** (2000), no. 2, 3025–3048, DOI 10.1007/BF02672185. MR1784692

[538] G. M. Tuynman, *The derivation of the exponential map of matrices*, Amer. Math. Monthly **102** (1995), no. 9, 818–820, DOI 10.2307/2974511. MR1357728

[539] M. Venkatesulu and P. K. Baruah, *A classical approach to eigenvalue problems associated with a pair of mixed regular Sturm-Liouville equations. I*, J. Appl. Math. Stochastic Anal. **13** (2000), no. 3, 303–312, DOI 10.1155/S1048953300000277. MR1782688

[540] H. Volkmer, *The coexistence problem for Heun's differential equation*, Math. Nachr. **279** (2006), no. 16, 1823–1834, DOI 10.1002/mana.200410458. MR2274837

[541] H. Volkmer, *Convergence radii for eigenvalues of two-parameter Sturm-Liouville problems*, Analysis (Munich) **20** (2000), no. 3, 225–236, DOI 10.1524/anly.2000.20.3.225. MR1778255

[542] H. Volkmer, *Sturm-Liouville problems with indefinite weights and Everitt's inequality*, Proc. Roy. Soc. Edinburgh Sect. A **126** (1996), 1097–1112.

[543] H. Volkmer, *Multiparameter eigenvalue problems and expansion theorems*, Lecture Notes in Mathematics, vol. 1356, Springer-Verlag, Berlin, 1988. MR973644

[544] P. W. Walker, *A vector-matrix formulation for formally symmetric ordinary differential equations with applications to solutions of integrable square*, J. London Math. Soc. (2) **9** (1974/75), 151–159, DOI 10.1112/jlms/s2-9.1.151. MR0369792

[545] S. Wallach, *The spectra of periodic potentials*, Amer. J. Math. **70** (1948), 842–848, DOI 10.2307/2372215. MR27924

[546] J. Walter, *Bemerkungen zu dem Grenzpunktfallkriterium von N. Levinson* (German), Math. Z. **105** (1968), 345–350, DOI 10.1007/BF01110296. MR229892

[547] J. Walter, *Regular eigenvalue problems with eigenvalue parameter in the boundary condition*, Math. Z. **133** (1973), 301–312, DOI 10.1007/BF01177870. MR335935

[548] W. Walter, *Differential and integral inequalities*, Translated from the German by Lisa Rosenblatt and Lawrence Shampine. Ergebnisse der Mathematik und ihrer Grenzgebiete, Band 55, Springer-Verlag, New York-Berlin, 1970. MR0271508

[549] A. Wang and A. Zettl, *Green's function for two-interval Sturm-Liouville problems*, Electron. J. Differential Equations (2013), No. 76, 13. MR3040653

[550] A. Wang, J. Ridenhour, and A. Zettl, *Construction of regular and singular Green's functions*, Proc. Roy. Soc. Edinburgh Sect. A **142** (2012), no. 1, 171–198, DOI 10.1017/S0308210510001630. MR2887648

[551] A. Wang and J. Sun, *J-self-adjoint extensions of J-symmetric operators with interior singular points*, J. Nanjing Univ. Sci. Tech. 31 (2007), 673-678.

[552] A. Wang, J. Sun, X. Hao, and S. Yao, *Completeness of eigenfunctions of Sturm-Liouville problems with transmission conditions*, Methods Appl. Anal. **16** (2009), no. 3, 299–312, DOI 10.4310/MAA.2009.v16.n3.a2. MR2650798

[553] A. Wang, J. Sun, and A. Zettl, *Characterization of domains of self-adjoint ordinary differential operators*, J. Differential Equations **246** (2009), no. 4, 1600–1622, DOI 10.1016/j.jde.2008.11.001. MR2488698

[554] A. Wang, J. Sun, and A. Zettl, *The classification of self-adjoint boundary conditions of differential operators with two singular endpoints*, J. Math. Anal. Appl. **378** (2011), no. 2, 493–506, DOI 10.1016/j.jmaa.2011.01.070. MR2773260

[555] A. Wang, J. Sun, and A. Zettl, *The classification of self-adjoint boundary conditions: separated, coupled, and mixed*, J. Funct. Anal. **255** (2008), no. 6, 1554–1573, DOI 10.1016/j.jfa.2008.05.003. MR2565718

[556] A. Wang, J. Sun, and A. Zettl, *Two-interval Sturm-Liouville operators in modified Hilbert spaces*, J. Math. Anal. Appl. **328** (2007), no. 1, 390–399, DOI 10.1016/j.jmaa.2006.05.058. MR2285557

[557] A. Wang and A. Zettl, *Characterization of domains of symmetric and self-adjoint ordinary differential operators*, Electron. J. Differential Equations (2018), Paper No. 15, 18. MR3762802

[558] A. Wang and A.Zettl, *Eigenvalues of Sturm-Liouville problems with discontinuous boundary conditions*, Electronic Journal of Differential Equations, **127** (2017), 1–27.

[559] A. Wang and A. Zettl, *A GKN theorem for symmetric operators*, J. Math. Anal. Appl. **452** (2017), no. 2, 780–791, DOI 10.1016/j.jmaa.2017.03.004. MR3632674

[560] A. Wang and A. Zettl, *Proof of the deficiency index conjecture*, J. Math. Anal. Appl. **468** (2018), no. 2, 695–703, DOI 10.1016/j.jmaa.2018.08.041. MR3852548

[561] A. Wang and A. Zettl, *Self-adjoint Sturm-Liouville problems with discontinuous boundary conditions*, Methods Appl. Anal. **22** (2015), no. 1, 37–66, DOI 10.4310/MAA.2015.v22.n1.a2. MR3338826

[562] A. Wang and A. Zettl, *A symmetric GKN-type theorem and symmetric differential operators*, J. Differential Equations **265** (2018), no. 10, 5156–5176, DOI 10.1016/j.jde.2018.06.028. MR3848247

[563] W.-y. Wang and J. Sun, *Complex J-symplectic geometry characterization for J-symmetric extensions of J-symmetric differential operators* (English, with English and Chinese summaries), Adv. Math. (China) **32** (2003), no. 4, 481–484. MR1997425

[564] Y. P. Wang, C. F. Yang, and Z. Y. Huang, *Half inverse problem for Sturm-Liouville operators with boundary conditions dependent on the spectral parameter*, Turkish J. Math. **37** (2013), no. 3, 445–454. MR3055543

[565] Z. Wang and J. Sun, *Qualitative analysis of the spectra of J-selfadjoint differential operators* (Chinese, with English and Chinese summaries), Adv. Math. (China) **30** (2001), no. 5, 405–413. MR1878785

[566] G. N. Watson, *A Treatise on the Theory of Bessel Functions*, Cambridge University Press, Cambridge, England; The Macmillan Company, New York, 1944. MR0010746

[567] Z. Wei and G. Wei, *The uniqueness of inverse problem for the Dirac operators with partial information*, Chin. Ann. Math. Ser. B **36** (2015), no. 2, 253–266, DOI 10.1007/s11401-015-0885-9. MR3305707

[568] J. Weidmann, *Oszillationsmethoden für Systeme gewöhnlicher Differentialgleichungen* (German), Math. Z. **119** (1971), 349–373, DOI 10.1007/BF01109887. MR285758

[569] J. Weidmann, *Spectral theory of Sturm-Liouville operators approximation by regular problems*, Sturm-Liouville Theory: Past and Present (2005), 75–98.

[570] J. Weidmann, *Uniform nonsubordinacy and the absolutely continuous spectrum*, Analysis **16** (1996), 98–99.

[571] J. Weidmann, *Verteilung der Eigenwerte für eine Klasse von Integraloperatoren in $L_2(a, b)$* (German), J. Reine Angew. Math. **276** (1975), 213–220, DOI 10.1515/crll.1975.276.213. MR377608

[572] J. Weidmann, *Zur Spektraltheorie von Sturm-Liouville-Operatoren* (German), Math. Z. **98** (1967), 268–302, DOI 10.1007/BF01112407. MR213915

[573] J. Weidmann, *Linear operators in Hilbert spaces*, Graduate Texts in Mathematics, vol. 68, Springer-Verlag, New York-Berlin, 1980. Translated from the German by Joseph Szücs. MR566954

[574] J. Weidmann, *Spectral theory of ordinary differential operators*, Lecture Notes in Mathematics, vol. 1258, Springer-Verlag, Berlin, 1987. MR923320

[575] J. Weidmann, *Lineare operatoren in Hilbertäumen, Tiel I grundlagen*, Teubner, Stuttgart–Leipiz–Wiesbaden, 2000.

[576] J. Weidmann,*Lineare operatoren in Hilbertäumen, Tiel II anwendungen*, Teubner, Stuttgart–Leipiz–Wiesbaden, 2003.

[577] H. F. Weinberger, *Variational methods for eigenvalue approximation*, Society for Industrial and Applied Mathematics, Philadelphia, Pa., 1974. Based on a series of lectures presented at the NSF-CBMS Regional Conference on Approximation of Eigenvalues of Differential Operators, Vanderbilt University, Nashville, Tenn., June 26–30, 1972; Conference Board of the Mathematical Sciences Regional Conference Series in Applied Mathematics, No. 15. MR0400004

[578] A. Weinstein, *On the Sturm-Liouville theory and the eigenvalues of intermediate problems*, Numer. Math. **5** (1963), 238–245, DOI 10.1007/BF01385895. MR0154421

[579] R. Weinstock, *Calculus of variations: With applications to physics and engineering*, Dover Publications, Inc., New York, 1974. Reprint of the 1952 edition. MR0443487

[580] H. Weyl, *Über gewöhnliche Differentialgleichungen mit Singularitäten und die zugehörigen Entwicklungen willkürlicher Funktionen* (German), Math. Ann. **68** (1910), no. 2, 220–269, DOI 10.1007/BF01474161. MR1511560

[581] E. T. Whittaker and G. N. Watson, *Modern analysis*, Cambridge University Press, Cambridge, 1950.

[582] W. Windau, *Über lineare Differentialgleichungen vierter Ordnung mit Singularitäten und die dazugehörigen Darstellungen willkürlicher Funktionen* (German), Math. Ann. **83** (1921), no. 3-4, 256–279, DOI 10.1007/BF01458384. MR1512012

[583] J. S. W. Wong, *On second order nonlinear oscillation*, Funkcial. Ekvac. **11** (1968), 207–234 (1969). MR0245915

[584] J. S. W. Wong, *Oscillation theorems for second order nonlinear differential equations*, Bull. Inst. Math. Acad. Sinica **3** (1975), no. 2, 283–309. MR0390372

[585] J. S. W. Wong and A. Zettl, *On the limit point classification of second order differential equations*, Math. Z. **132** (1973), 297–304, DOI 10.1007/BF01179735. MR322258

[586] V. A. Yakubovich and V. M. Starzhinskii, *Linear differential equations with periodic coefficients. 1, 2*, Halsted Press [John Wiley & Sons] New York–Toronto; Israel Program for Scientific Translations, Jerusalem-London, 1975. Translated from Russian by D. Louvish. MR0364740

[587] C.-F. Yang and X.-P. Yang, *Inverse nodal problems for the Sturm-Liouville equation with polynomially dependent on the eigenparameter*, Inverse Probl. Sci. Eng. **19** (2011), no. 7, 951–961, DOI 10.1080/17415977.2011.565874. MR2836942

[588] C.-F. Yang and A. Zettl, *Half inverse problems for quadratic pencils of Sturm-Liouville operators*, Taiwanese J. Math. **16** (2012), no. 5, 1829–1846, DOI 10.11650/twjm/1500406800. MR2970688

[589] S. Yao, J. Sun, and A. Zettl, *The Sturm-Liouville Friedrichs extension*, Appl. Math. **60** (2015), no. 3, 299–320, DOI 10.1007/s10492-015-0097-3. MR3419964

[590] S. Yao, J. Sun, and A. Zettl, *Self-adjoint domains, symplectic geometry, and limit-circle solutions*, J. Math. Anal. Appl. **397** (2013), no. 2, 644–657, DOI 10.1016/j.jmaa.2012.07.066. MR2979601

[591] S. Yao, J. Sun, and A. Zettl, *Symplectic geometry and dissipative differential operators*, J. Math. Anal. Appl. **414** (2014), no. 1, 434–449, DOI 10.1016/j.jmaa.2014.01.019. MR3165320

[592] Y. Yuan, J. Sun, and A. Zettl, *Eigenvalues of periodic Sturm Liouville problems*, Linear Algebra Appl. **517** (2017), 148–166, DOI 10.1016/j.laa.2016.11.035. MR3592016

[593] Y. Yuan, J. Sun, and A. Zettl, *Inequalities among eigenvalues of Sturm Liouville equations with periodic coefficients*, Electron. J. Differential Equations (2017), Paper No. 264, 13. MR3723537

[594] A. Zettl, *Adjoint and self-adjoint boundary value problems with interface conditions*, SIAM J. Appl. Math. **16** (1968), 851–859, DOI 10.1137/0116069. MR0234049

[595] A. Zettl, *Adjoint linear differential operators*, Proc. Amer. Math. Soc. **16** (1965), 1239–1241, DOI 10.2307/2035906. MR183920

[596] A. Zettl, *Adjointness in nonadjoint boundary value problems*, SIAM J. Appl. Math. **17** (1969), 1268–1279, DOI 10.1137/0117119. MR0259229

[597] A. Zettl, *An algorithm for the construction of all disconjugate operators*, Proc. Roy. Soc. Edinburgh Sect. A **75** 4 (1976), 33–40.

[598] A. Zettl, *An algorithm for the construction of limit circle expressions*, Proc. Roy. Soc. Edinburgh Sect. A **75** 1 (1976), 1–3.

[599] A. Zettl, *A characterization of the factors of ordinary linear differential operators*, Bull. Amer. Math. Soc., **80** (3), (1974), 498–500.

[600] A. Zettl, *Computing continuous spectrum*, Proc. International Symposium on Trends and developments in ordinary differential equations, Yousef Alavi and Po-Fang Hsieh, editors, World Scientific, (1994), 393–406.

[601] A. Zettl, *A constructive characterization of disconjugacy*, Bull. Amer. Math. Soc. **81** (1975), 145–147, DOI 10.1090/S0002-9904-1975-13677-9. MR352612

[602] A. Zettl, *Deficiency indices of polynomials in symmetric differential expressions, II*, Proc. Roy. Soc. Edinburgh Sect. A **73** (1975), 301–306. MR0422756

[603] A. Zettl, *The essential spectrum of nonselfadjoint ordinary differential operators*, University of Missouri at Rolla Press; edited by J. L. Henderson (1985), 152–168. Proceedings of the Twelfth and Thirteenth Midwest Conference.

[604] A. Zettl, *Explicit conditions for the factorization of nth order linear differential operators*, Proc. Amer. Math. Soc. **41** (1973), 137–145, DOI 10.2307/2038829. MR320413

[605] A. Zettl, *Factorization and disconjugacy of third order differential equations*, Proc. Amer. Math. Soc. **31** (1972), 203–208, DOI 10.2307/2038543. MR296421

[606] A. Zettl, *Factorization of differential operators*, Proc. Amer. Math. Soc. **27** (1971), 425–426, DOI 10.2307/2036335. MR273085

[607] A. Zettl, *Formally self-adjoint quasi-differential operators*, Rocky Mountain J. Math. **5** (1975), 453–474, DOI 10.1216/RMJ-1975-5-3-453. MR0379976

[608] A. Zettl, *General theory of the factorization of ordinary linear differential operators*, Trans. Amer. Math. Soc. **197** (1974), 341–353, DOI 10.2307/1996941. MR364724

[609] A. Zettl, *The lack of self-adjointness in three point boundary value problems*, Proc. Amer. Math. Soc. **17** (1966), 368–371, DOI 10.2307/2035169. MR190422

[610] A. Zettl, *The limit-point and limit-circle cases for polynomials in a differential operator*, Proc. Roy. Society of Edinburgh Sect. A **72** (1975), 219-224.

[611] A. Zettl, *A note on square integrable solutions of linear differential equations*, Proc. Amer. Math. Soc., **21** (1969), 671–672.

[612] A. Zettl, *On the Friedrichs extension of singular differential operators*, Commun. Appl. Anal. **2** (1998), no. 1, 31–36. MR1612893

[613] A. Zettl, *Perturbation of the limit circle case*, Quart. J. Math. Oxford, 26(1), (1975), 355–360.

[614] A. Zettl, *Perturbation theory of deficiency indices of differential operators*, J. London Math. Soc., (2), 12 (1976), 405–412.

[615] A. Zettl, *Powers of real symmetric differential expressions without smoothness assumptions*, Quaestiones Math. **1** (1976), no. 1, 83–94. MR0440116

[616] A. Zettl, *Separation for differential operators and the L^p spaces*, Proc. Amer. Math. Soc. **55** (1976), no. 1, 44–46, DOI 10.2307/2041838. MR393646

[617] A. Zettl, *Some identities related to Pólya's property W for linear differential equations*, Proc. Amer. Math. Soc. **18** (1967), 992–994, DOI 10.2307/2035779. MR222370

[618] A. Zettl, *Square integrable solutions of $Ly = f(t, y)$*, Proc. Amer. Math. Soc. **26** (1970), 635–639, DOI 10.2307/2037125. MR267213

[619] A. Zettl, *Sturm-Liouville problems*, Spectral theory and computational methods of Sturm-Liouville problems (Knoxville, TN, 1996), Lecture Notes in Pure and Appl. Math., vol. 191, Dekker, New York, 1997, pp. 1–104. MR1460547

[620] A. Zettl, *Sturm-Liouville theory*, Mathematical Surveys and Monographs, vol. 121, American Mathematical Society, Providence, RI, 2005. MR2170950

[621] A. Zettl and J. Sun, *Survey article: Self-adjoint ordinary differential operators and their spectrum*, Rocky Mountain J. Math. **45** (2015), no. 3, 763–886, DOI 10.1216/RMJ-2015-45-3-763. MR3385967

[622] M. Zhang, J. Sun, and A. Zettl, *Eigenvalues of limit-point Sturm-Liouville problems*, J. Math. Anal. Appl. **419** (2014), no. 1, 627–642, DOI 10.1016/j.jmaa.2014.05.021. MR3217171

[623] Z. Maozhu, J. Sun, and A. Zettl, *The spectrum of singular Sturm-Liouville problems with eigenparameter dependent boundary conditions and its approximation*, Results Math. **63** (2013), no. 3-4, 1311–1330, DOI 10.1007/s00025-012-0270-x. MR3057371

[624] Y. Zhao, J. Sun, and A. Zettl, *Self-adjoint Sturm-Liouville problems with an infinite number of boundary conditions*, Math. Nachr. **289** (2016), no. 8-9, 1148–1169, DOI 10.1002/mana.201400415. MR3512053

[625] V. A. Zheludev, *Eigenvalues of the perturbed Schrödinger operator with a periodic potential*, Topics in Math. Phys. **2** (1968), 87–101.

[626] V. A. Zheludev, *Perturbation of the spectrum of the one-dimensional self-adjoint Schrödinger operator with a periodic potential*, Topics in Math. Phys. **4** (1971), 55–75.

Index

2-interval self-adjoint domains, 166

Abel's formula, 7
Absolutely continuous spectrum, 215
Adjoint identity for fundamental matrices, 21
Adjointness lemma, 190
Assumption (A), 122

Bielecki norm, 4
boundary condition
  boundary matrix, 61
bounds of solutions, 10

characterization of symmetric domains, 62
Closed finite dimensional extensions, 70
Continuous dependence of solutions of initial value problems on the problem, 12
continuous extension of solutions, 10
Continuous extension of solutions to endpoints, 11
Continuous Spectrum, 119

Derivative of the matrix exponential function, 18
Differentiable dependence of solutions of initial value problems on all the parameters of the problem, 14
Discreteness criteria, 216
dissipative operator in Hilbert space, 144
Dissipative subspace, 139
Dmin characterization, 56

Eigenparameter dependent boundary conditions, 216
Eigenvalues in gaps, 217
Embedded eigenvalues, 217
Essential Spectrum, 119
Examples of symmetric boundary conditions, 113
Examples of symmetric separated boundary conditions, 115
Expansion theorems, 216
exponential law, 15

Frechet derivative, 14
fundamental matrix, 8

Gabushin inequalities, 216
GN Matrix, 46
Green's matrix for regular systems, 191
Gronwall inequality, 8

Hadamar inequalities, 216
Half-Range expansions, 216
Hardy-Littlewood inequalities, 216
Hypotheses (EH), 81

infinite regular endpoints, 29
Initial value problem, 4
Inverse initial value problems, 21
Inverse spectral theory, 216

J-Self-Adjoint Operators, 215

Kallman-Rota inequality, 216, 217
Kolmogorov inequalities, 216

Lagrange adjoint matrix
  adjoint differential expression, 36
Lagrange Identity, 34
Lagrange Symmetric Expressions and Operators, 31
Lagrangian elements
  dissipative elements, 141
Lagrangian subspace, 138, 141
Landau inequalities, 216
LC and LP solutions, 86
LC solutions, 81
leading coefficient, 28

maximal accretive operator in Hilbert space, 144
maximal deficiency index, 91
Maximal Dissipative subspace, 139
  Maximal Accretive subspace, 141
Maximal Operator, 35
Minimal Operator, 36
Multi parameter theory, 215

Nonlinear oscillation, 217

Nonlinear problems, 217
nonreal separated boundary conditions, 110
norm inequalities, 216
Numerical approximations, 217

Operators S(U), 61

partial Separation, 50
Patching Lemma, 37
Primary Fundamental Matrix, 8
primary fundamental matrix, 8, 190
Product formula for solutions, 15
Property BD, 131

Quasi-Differential Expressions, 20

Rank invariance, 6, 7
Rank invariance of solutions at endpoints, 11
Regular Endpoint, 28
Regular Friedrichs Extension, 93
Regularization of singular problem, 94
rigorous definition of separated boundary conditions, 105

Schonberg-Cavaretta inequalities, 216
Self-Adjoint domain characterization, 89
Separated self-adjoint boundary conditions, 99
Separation, 217
singular boundary conditions, 70
Singular Patching Theorem, 56
solution of a system, 3
strictly dissipative operator, 144
Strictly Dissipative subspace, 139
Strong limit-point conditions, 217
Sturm-Liouville Friedrichs extension, 94
Successive approximations, 5
symplectic invariants, 146
  Lagrangeanindex
    Excess index, 140
symplectic matrix E, 31
symplectic ortho-complement spaces, 138

Two-Interval Symmetric Domains, 168
Two-Interval Symmetric GKN, 181

variation of parameters formula, 8

Selected Published Titles in This Series

245 **Aiping Wang and Anton Zettl,** Ordinary Differential Operators, 2019
242 **Bhargav Bhatt, Ana Caraiani, Kiran S. Kedlaya, Peter Scholze, and Jared Weinstein,** Perfectoid Spaces, 2019
241 **Dana P. Williams,** A Tool Kit for Groupoid C^*-Algebras, 2019
240 **Antonio Fernández López,** Jordan Structures in Lie Algebras, 2019
239 **Nicola Arcozzi, Richard Rochberg, Eric T. Sawyer, and Brett D. Wick,** The Dirichlet Space and Related Function Spaces, 2019
238 **Michael Tsfasman, Serge Vlăduţ, and Dmitry Nogin,** Algebraic Geometry Codes: Advanced Chapters, 2019
237 **Dusa McDuff, Mohammad Tehrani, Kenji Fukaya, and Dominic Joyce,** Virtual Fundamental Cycles in Symplectic Topology, 2019
236 **Bernard Host and Bryna Kra,** Nilpotent Structures in Ergodic Theory, 2018
235 **Habib Ammari, Brian Fitzpatrick, Hyeonbae Kang, Matias Ruiz, Sanghyeon Yu, and Hai Zhang,** Mathematical and Computational Methods in Photonics and Phononics, 2018
234 **Vladimir I. Bogachev,** Weak Convergence of Measures, 2018
233 **N. V. Krylov,** Sobolev and Viscosity Solutions for Fully Nonlinear Elliptic and Parabolic Equations, 2018
232 **Dmitry Khavinson and Erik Lundberg,** Linear Holomorphic Partial Differential Equations and Classical Potential Theory, 2018
231 **Eberhard Kaniuth and Anthony To-Ming Lau,** Fourier and Fourier-Stieltjes Algebras on Locally Compact Groups, 2018
230 **Stephen D. Smith,** Applying the Classification of Finite Simple Groups, 2018
229 **Alexander Molev,** Sugawara Operators for Classical Lie Algebras, 2018
228 **Zhenbo Qin,** Hilbert Schemes of Points and Infinite Dimensional Lie Algebras, 2018
227 **Roberto Frigerio,** Bounded Cohomology of Discrete Groups, 2017
226 **Marcelo Aguiar and Swapneel Mahajan,** Topics in Hyperplane Arrangements, 2017
225 **Mario Bonk and Daniel Meyer,** Expanding Thurston Maps, 2017
224 **Ruy Exel,** Partial Dynamical Systems, Fell Bundles and Applications, 2017
223 **Guillaume Aubrun and Stanisław J. Szarek,** Alice and Bob Meet Banach, 2017
222 **Alexandru Buium,** Foundations of Arithmetic Differential Geometry, 2017
221 **Dennis Gaitsgory and Nick Rozenblyum,** A Study in Derived Algebraic Geometry, 2017
220 **A. Shen, V. A. Uspensky, and N. Vereshchagin,** Kolmogorov Complexity and Algorithmic Randomness, 2017
219 **Richard Evan Schwartz,** The Projective Heat Map, 2017
218 **Tushar Das, David Simmons, and Mariusz Urbański,** Geometry and Dynamics in Gromov Hyperbolic Metric Spaces, 2017
217 **Benoit Fresse,** Homotopy of Operads and Grothendieck–Teichmüller Groups, 2017
216 **Frederick W. Gehring, Gaven J. Martin, and Bruce P. Palka,** An Introduction to the Theory of Higher-Dimensional Quasiconformal Mappings, 2017
215 **Robert Bieri and Ralph Strebel,** On Groups of PL-homeomorphisms of the Real Line, 2016
214 **Jared Speck,** Shock Formation in Small-Data Solutions to 3D Quasilinear Wave Equations, 2016
213 **Harold G. Diamond and Wen-Bin Zhang (Cheung Man Ping),** Beurling Generalized Numbers, 2016

For a complete list of titles in this series, visit the
AMS Bookstore at **www.ams.org/bookstore/survseries/**.